SAFETY OF COMPUTER CONTROL SYSTEMS 1990 (SAFECOMP'90)
Safety, Security and Reliability Related Computers for the 1990s

*Proceedings of the IFAC/EWICS/SARS Symposium
Gatwick, UK, 30 October–2 November 1990*

Edited by

B. K. DANIELS

*National Computing Centre Ltd,
Manchester, UK*

Published for the

INTERNATIONAL FEDERATION OF AUTOMATIC CONTROL

by

PERGAMON PRESS

Member of Maxwell Macmillan Pergamon Publishing Corporation

OXFORD · NEW YORK · BEIJING · FRANKFURT
SÃO PAULO · SYDNEY · TOKYO · TORONTO

U.K.	Pergamon Press plc, Headington Hill Hall, Oxford OX3 0BW, England
U.S.A.	Pergamon Press, Inc., Maxwell House, Fairview Park, Elmsford, New York 10523, U.S.A.
PEOPLE'S REPUBLIC OF CHINA	Pergamon Press, Room 4037, Qianmen Hotel, Beijing, People's Republic of China
FEDERAL REPUBLIC OF GERMANY	Pergamon Press GmbH, Hammerweg 6, D-6242 Kronberg, Federal Republic of Germany
BRAZIL	Pergamon Editora Ltda, Rua Eça de Queiros, 346, CEP 04011, Paraiso, São Paulo, Brazil
AUSTRALIA	Pergamon Press Australia Pty Ltd., P.O. Box 544, Potts Point, N.S.W. 2011, Australia
JAPAN	Pergamon Press, 5th Floor, Matsuoka Central Building, 1-7-1 Nishishinjuku, Shinjuku-ku, Tokyo 160, Japan
CANADA	Pergamon Press Canada Ltd., Suite No. 271, 253 College Street, Toronto, Ontario, Canada M5T 1R5

This compilation copyright © 1990 IFAC

All Rights Reserved. No part of this publication may be reproduced, stored in a retrieval system or transmitted in any form or by any means: electronic, electrostatic, magnetic tape, mechanical, photocopying, recording or otherwise, without permission in writing from the copyright holders.

First edition 1990

Library of Congress Cataloging in Publication Data
IFAC/EWICS/SARS Symposium on Safety of Computer Control Systems
(1990: Gatwick, England)
Safety of computer control systems, 1990 (SAFECOMP '90): safety,
security, and reliability related computers for the 1990s:
proceedings of the IFAC/EWICS/SARS Symposium, Gatwick, UK,
30 October–2 November, 1990/edited by B. K. Daniels.—1st ed.
p. cm.—(IFAC symposia series: 1990, no. 17)
Includes indexes.
1. Automatic control—Data processing—Reliability—Congresses.
2. Computer security—Congresses.   3. Computers—Reliability—
Congresses.   I. Daniels, B. K.   II. International Federation of
Automatic Control.   III. European Workshop on Industrial Computer
Systems.   IV. Safety and Reliability Society (Great Britain)   V. Title.   VI. Series.
TJ212.2.I327 1990   629.8'9—dc20   90-7988

British Library Cataloguing in Publication Data
Safety of computer control systems 1990 (SAFECOMP '90).
1. Process control. Applications of computer systems
I. Title   II. Daniels, B. K.   III. International Federation of Automatic
Control   IV. Series   670.437
ISBN 0-08-040953-9

These proceedings were reproduced by means of the photo-offset process using the manuscripts supplied by the authors of the different papers. The manuscripts have been typed using different typewriters and typefaces. The lay-out, figures and tables of some papers did not agree completely with the standard requirements: consequently the reproduction does not display complete uniformity. To ensure rapid publication this discrepancy could not be changed: nor could the English be checked completely. Therefore, the readers are asked to excuse any deficiencies of this publication which may be due to the above mentioned reasons.

The Editor

Printed in Great Britain by BPCC Wheaton Ltd, Exeter

International Federation of Automatic Control

SAFETY OF COMPUTER CONTROL SYSTEMS 1990
Safety, Security and Reliability Related Computers
for the 1990s

IFAC Symposia Series, 1990. Number 17

IFAC SYMPOSIA SERIES

Editor-in-Chief
JANOS GERTLER, Department of Electrical Engineering,
George Mason University, Fairfax, Virginia 22030, USA

JOHNSON et al.: Adaptive Systems in Control and Signal Processing   (*1990, No. 1*)
ISIDORI: Nonlinear Control Systems Design   (*1990, No. 2*)
AMOUROUX & EL JAI: Control of Distributed Parameter Systems   (*1990, No. 3*)
CHRISTODOULAKIS: Dynamic Modelling and Control of National Economies   (*1990, No. 4*)
HUSSON: Advanced Information Processing in Automatic Control   (*1990, No. 5*)
NISHIMURA: Automatic Control in Aerospace   (*1990, No. 6*)
RIJNSDORP et al.: Dynamics and Control of Chemical Reactors, Distillation Columns and Batch Processes (DYCORD '89)   (*1990, No. 7*)
UHI AHN: Power Systems and Power Plant Control   (*1990, No. 8*)
REINISCH & THOMA: Large Scale Systems: Theory and Applications   (*1990, No. 9*)
KOPPEL: Automation in Mining, Mineral and Metal Processing   (*1990, No. 10*)
BAOSHENG HU: Analysis, Design and Evaluation of Man–Machine Systems   (*1990, No. 11*)
PERRIN: Control, Computers, Communications in Transportation   (*1990, No. 12*)
PUENTE & NEMES: Information Control Problems in Manufacturing Technology   (*1990, No. 13*)
NISHIKAWA & KAYA: Energy Systems, Management and Economics   (*1990, No. 14*)
DE CARLI: Low Cost Automation: Components, Instruments, Techniques and Applications   (*1990, No. 15*)
KOPACEK & GENSER: Skill Based Automated Production   (*1990, No. 16*)
DANIELS: Safety of Computer Control Systems 1990 (SAFECOMP'90)   (*1990, No. 17*)

COBELLI & MARIANI: Modelling and Control in Biomedical Systems   (*1989, No. 1*)
MACLEOD & HEHER: Software for Computer Control (SOCOCO '88)   (*1989, No. 2*)
RANTA: Analysis, Design and Evaluation of Man-Machine Systems   (*1989, No. 3*)
MLADENOV: Distributed Intelligence Systems: Methods and Applications   (*1989, No. 4*)
LINKENS & ATHERTON: Trends in Control and Measurement Education   (*1989, No. 5*)
KUMMEL: Adaptive Control of Chemical Processes   (*1989, No. 6*)
CHEN ZHEN-YU: Computer Aided Design in Control Systems   (*1989, No. 7*)
CHEN HAN-FU: Identification and System Parameter Estimation   (*1989, No. 8*)
CALVAER: Power Systems, Modelling and Control Applications   (*1989, No. 9*)
REMBOLD: Robot Control (SYROCO '88)   (*1989, No. 10*)
JELLALI: Systems Analysis Applied to Management of Water Resources   (*1989, No. 11*)

Other IFAC Publications

AUTOMATICA
the journal of IFAC, the International Federation of Automatic Control
Editor-in-Chief: G. S. Axelby, 211 Coronet Drive, North Linthicum, Maryland 21090, USA

IFAC WORKSHOP SERIES
Editor-in-Chief: Pieter Eykhoff, University of Technology, NL-5600 MB Eindhoven, The Netherlands

---

Full list of IFAC Publications appears at the end of this volume

NOTICE TO READERS

If your library is not already a standing/continuation order customer or subscriber to this series, may we recommend that you place a standing/continuation or subscription order to receive immediately upon publication all new volumes. Should you find that these volumes no longer serve your needs your order can be cancelled at any time without notice.

Copies of all previously published volumes are available. A fully descriptive catalogue will be gladly sent on request.

ROBERT MAXWELL
Publisher

IFAC/EWICS/SARS SYMPOSIUM ON SAFETY OF COMPUTER CONTROL SYSTEMS (SAFECOMP'90)
Safety, Security and Reliability Related Computers for the 1990s

Sponsored by
International Federation of Automatic Control (IFAC)
European Workshop on Industrial Computer Systems Technical Committee on Safety, Security and Reliability (EWICS TC7)
The Safety and Reliability Society (SARS)

Co-sponsored by
The International Federation for Information Processing (IFIP)
The Institute of Electrical Engineers (IEE)
The Institute of Measurement and Control (Inst MC)
The British Computer Society
The National Computing Centre Ltd (NCC)
The Centre for Software Reliability (CSR)
Safety Net

Organized by
The Safety and Reliability Society (SARS)

International Programme Committee
B. K. Daniels, UK (Chairman)
L. A. Jones, UK (Secretary)
R. Bell, UK
R. Bloomfield, UK
S. Bologna, Italy
L. Boullart, Belgium
G. Dahll, Norway
W. Ehrenberger, FRG
H. Frey, Switzerland
R. Genser, Austria
J. Górski, Poland
S. L. Hansen, Denmark
R. Isermann, FRG
H. de Kroes, The Netherlands
R. Lauber, FRG
J. F. Lindeberg, Norway
J. McDennid, UK
J. M. A. Rata, France
M. Rodd, UK
K. Sankaran, Australia
B. Stemer, Sweden
B. Tamm, USSR
M. Thomas, UK
A. Toola, Finland

National Organizing Committee
J. Catchpole (Chairman)
B. K. Daniels
J. Jones
L. A. Jones
N. Locke

PREFACE

In an expanding worldwide market for safe, secure and reliable computer systems SAFECOMP'90 provides an opportunity for technical experts, operators and legislators to review the experience of the past decade, to consider the guidance now available, and to identify the skills and technologies required for the 1990's and into the next century.

In the decade since the first SAFECOMP in 1979, the SAFECOMP's held in Germany, France, the UK, USA, Italy and Austria have reflected upon and influenced the significant changes in the technology and scale of use of computer systems in applications which are safety security and reliability related. There is increasing acceptance that computers can be used in life critical applications, provided appropriate technology is used and current guidelines are carefully followed by people with the right training and experience.

It has recently been estimated that the market for software in safety critical systems is currently of the order of £500 Million and is growing at an annual rate of approximately 20%. Interest in the UK is very high. I am very grateful to the IPC for supporting the decision to host this years SAFECOMP in the UK.

Much valuable work has been done by the UK Health and Safety Executive on industrial applications of computers and within UK Government civil and military research programmes on formal methods for specification and for static analysis. A new UK technology transfer and research and development initiative in safety critical systems, SafeIT, has been announced in 1990, and is featured in a panel session[1] in SAFECOMP. The views of the lawyer, the insurer and the consumer are also relevant and these too are featured in a panel session[1].

In the first keynote paper[1], Martyn Thomas looks back to the emergence of interest in safety critical systems and how the technology and marketplace has changed since 1979.

In thirty one contributed papers[1], we access experience of specifying, creating, operating, and licensing computers in safety, security and reliability related applications. Authors review the available guidelines, and provide feedback from their use by industry. Methods and tools for specifying, designing, documenting, analyzing, testing and assessing systems dependent on computers for safety, security and reliability are described.

The final keynote paper by John McDermid addresses the importance of skills and technologies required for the development and evaluation of safety critical systems in the 1990's and into the next century in the expanding worldwide market for safety related computers.

As Chairman of the International Programme Committee (IPC) I would like to thank all members of the Committee for their advice and assistance in constructing the Programme. The Programme Committee included many members of the European Workshop on Industrial Computer Systems Technical Committee 7 (Safety Security and Reliability), who have provided the core of each SAFECOMP IPC since 1979, on this occasion the Committee was grateful for the assistance of a number of additional members who are key workers in this field.

My thanks must also go to the session Chairmen, the authors and co-authors of the papers, and the members of the National Organising Committee.

Barry Daniels,
National Computing Centre Ltd.,
Oxford Road,
Manchester,
M1 7ED,
UK

---

1. Copies of the keynote paper by Martyn Thomas, summaries of the panel sessions, and of the following list of papers not included inthe proceedings are available from the Safety and Reliability Society, Clayton House, Manchester, M1 2AQ., UK.

An Outline of the EDI Security Model Application Model, S.Schindler, D Buchta, TELES GmbH, Berlin, FRG.

Developing and Assessing complex safety and secure systems, J.Elliott, P-E International systems Group, UK.

Static code analysis of (MIL-STD-1750A) Assembly Language. T.Jennings, P.Keenan, SD Scicon plc, UK.

CONTENTS

MATHEMATICAL FORMALISMS

A Formal Model for Safety-critical Computing Systems 1
A. SAEED, T. ANDERSON, M. KOUTNY

Formal Approach to Faults in Safety Related Applications 7
J. GORSKI

Provably Correct Safety Critical Software 13
A.P. RAVN, H. RISCHEL, V. STAVRIDOU

CORRECTNESS

Aspects of Proving Compiler Correctness 19
B.v. KARGER

A Concept of a Computer System for the Execution of Safety Critical 25
Licensable Software Programmed in a High Level Language
W.A. HALANG, SOON-KEY JUNG

Modeling and Verifying Systems and Software in Propositional Logic 31
G. STALMARCK, M. SAFLUND

Logical Foundations of a Probabilistic Theory of Software Correctness 37
T. GRAMS

ISSUES OF SECURITY

Methods of Protection against Computer Viruses 43
K. GAJ, K. GORSKI, R. KOSSOWSKI, J. SOBCZYK

The Need for a Standard Method for Computer Security Requirements Analysis 49
R.J. TALBOT

JUST TESTING!

The Testing of Real-time Embedded Software by Dynamic Analysis Techniques 55
D. HEDLEY

The ELEKTRA Testbed: Architecture of a Real-time Test Environment 59
for High Safety and Reliability Requirements
E. SCHOITSCH, E. DITTRICH, S. GRASEGGER, D. KROPFITSCH, A. ERB,
P. FRITZ, H. KOPP

SOFTWARE METRICS

Software Coverage Metrics and Operational Reliability 67
A. VEEVERS

Quality Measurement of Mission Critical Systems 71
J.B. WRIGHT, F. FICHOT, C. GEORGES, M. ROMAIN

RELIABILITY AND DEPENDABILITY

Software Reliability Assessment - The Need for Process Visibility 77
C. DALE

Assessing Software Reliability in a Changing Environment 83
T. STALHANE

Dependability Evaluation of Watchdog Processors 89
P.J. GIL, J.J. SERRANO, R. ORS, V. SANTONJA

ASSESSMENT AND VALIDATION

Practical Experience in the Assessment of Existing Safety Critical 95
Computer Based Systems
B.W. FINNIE, I.H.A. JOHNSTON

Methodological Aspects of Critics during Safety Validation 99
G. LIST

STATIC ANALYSIS AND SYMBOLIC EXECUTION

SYMBAD: A Symbolic Executor of Sequential Ada Programs 105
A. COEN-PORISINI, F. DE PAOLI

Tools and Methodologies for Quality Assurance 113
U. ANDERS, E.-U. MAINKA, G. RABE

A Comparison of Static and Dynamic Conformance Analysis 119
M.A. HENNELL, E. FERGUS

SAFE OPERATION

Computer Based Training for Contingency Decisions 125
K.H. DRAGER, H. SOMA, R. GULLIKSEN

Qualitative Knowledge in a Diagnostic Expert System for Nuclear Power 131
Plant Safety
I. OBREJA

CASE STUDIES IN INDUSTRY PRACTICE

Management of Computer-aided Control System Design from Concept to 135
Flight Test
B.N. TOMLINSON, G.D. PADFIELD, P.R. SMITH

The History and Development of Computer Based Safety Systems for Offshore Oil and Gas Production Platforms from the Sixties to the Present Day
C.J. GORING 145

Controlling Software Production, from a Customer Point of View
F. FICHEUX, Y. MAYADOUX, C. PAIN 151

The Impact of Social Factors on Acceptable Levels of Safety Integrity
I.H.A. JOHNSTON 157

KEYNOTE ADDRESS

Skills and Technologies for the Development and Evaluation of Safety Critical Systems
J.A. McDERMID 163

Author Index 173

Keyword Index 175

A FORMAL MODEL FOR SAFETY-CRITICAL COMPUTING SYSTEMS

A. Saeed, T. Anderson and M. Koutny

Computing Laboratory, The University, Newcastle upon Tyne NE1 7RU, UK

Abstract. The paper treats a safety-critical computing system as a component of a larger system which could cause or allow the overall system to enter into a hazardous state. It is argued that to gain a complete understanding of such systems, the requirements of the overall system and the properties of the environment must be analysed in a common formal framework. A system development model based on the separation of safety and mission issues is discussed. A formal model for the representation of the specifications produced during the analysis is presented. The semantics of the formal model are based on the notion of a system history. To overcome some of the problems associated with an unstructured specification the concept of a mode is introduced. To illustrate the strategy a simple example is presented.

Keywords. Process control; Safety; Computer control; Requirements analysis and Formal Methods.

1. INTRODUCTION

The general class of systems considered in this paper are process control systems (Smith, 1972); typical application areas include chemical plants and avionic systems. Though each application area has its unique problems, process control systems have significant properties in common. In particular, there is a general relationship between the main components of a process control system (see Fig. 1.1).

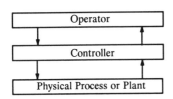

Fig. 1.1 Process control system components

The controller, of a process control system, is constructed to control the behaviour of the physical process. The controller consists of four components: operator console, control system, sensors and actuators. A precise definition of the computing systems considered, in this paper, is given in the context of process control systems.

A safety-critical computing system is a system that is a component of a process control system which could cause or allow the overall system to enter into a hazardous state.

A *hazardous state* is a state which under certain environmental conditions could lead to a disaster. The term *disaster* is usually reserved for calamitous situations such as loss of life, limb, significant revenue or substantial damage to the environment.

In this paper we concentrate on the requirements analysis of safety-critical computing systems. Requirements analysis plays a vital role in the development of systems, since any errors in the identified requirements will corrupt the subsequent stages of system development. Experience has shown that errors in the formulation of requirements are one of the major causes of mishaps (Leveson, 1986). Current methods for requirements elicitation are based on a combination of system safety and software development techniques. A common approach is the use of HAZOPS (Hazard and Operability Studies) and FTA (Fault Tree Analysis) to identify hazards from which the software safety requirements are produced (Leveson, 1989). General guidelines for the integration of system safety and software development techniques are available (HSE, 1987). The main drawback to current methods is that no unified formal framework for requirements analysis is available.

1.1. Formal Framework

To gain a complete understanding of a safety-critical computing system, the requirements of the overall system and the properties of the environment should be analysed in a common formal framework. The benefits of using formal methods during requirements analysis include unambiguity, checks for completeness and consistency, formal verification, and the potential for using automated aids (Jaffe, 1989; Roan, 1986). A unified framework is needed for the analysis because safety is a global issue that can only be addressed by analysing the consequences of system behaviour in the context of the environment (Gorski, 1986; Leveson, 1986). The benefits gained from the adoption of a common framework include: i) improved and precise communication between the members of the analysis team; ii) the ability to perform a rigorous assessment of the effect of the inherent properties of the environment on system behaviour; and iii) the ability to assess the impact of the system on the environment.

The essential attributes that an appropriate formal model must possess are that it must be able to express *physical laws*, *parallelism* and *timing issues* in a *coherent (structured) form*. The necessity for the model to be able to express physical laws stems from the fact that the behaviour of the environment is governed by such laws. The necessity to treat parallelism explicitly in specifications which include the environment has been argued by many researchers (Gorski, 1988). Timing issues will arise in all of the stages of requirements analysis. Timing issues are present in the description of the environment, since physical laws make an explicit reference to time, and in many cases the relationships between the sensors, actuators and the physical process are dependent on time. (The importance of timing issues makes process control systems a subclass of real-time systems.) A structured model is necessary to handle the complexity of these systems. The structure should be present in the techniques used to compose the basic constructs of the model and in the basic constructs themselves.

The remainder of the paper is set out in four sections. In section 2 a development model for requirements analysis is discussed. In section 3 the basic concepts of a specification model are introduced. In section 4 a simple example (reaction vessel) that illustrates the role of the specification model during analysis is presented. Finally, section 5 concludes the paper by discussing how the work presented in the paper relates to an overall approach to requirements analysis.

2. DEVELOPMENT MODEL

Our approach to requirements analysis is concerned with system behaviour at two distinct levels of abstraction: *real world* and *controller*. At the real world level we are concerned with the behaviour exhibited by the physical process. At the controller level, we are concerned with the behaviour exhibited by the sensors and actuators and at the operator console.

For safety-critical systems it has been suggested that for a clear analysis of the safety-related properties, a distinction must be made between the safety-critical and mission-oriented behaviour of a system (Leveson, 1984; Mulazzanni, 1986). In our approach, the distinction between the safety and mission issues is established during the requirements analysis. This distinction allows us to partition the analysis (see Fig. 2.1).

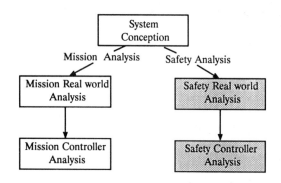

Fig. 2.1. Requirements analysis phases

The decision to make a distinction between the safety and mission issues is reflected in our general structure for safety-critical systems (see Fig. 2.2).

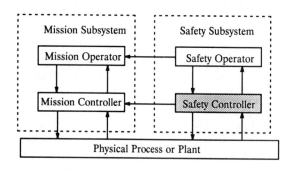

Fig. 2.2. Safety-critical system structure

2.1 Safety Analysis

At the real world level, the safety analysis is performed in three stages. In the first stage the potential *disasters* associated with the mission of the system or the environment are identified. The second stage involves the identification of the *hazards* that can lead to potential disasters and their representation in the formal framework. It also involves the specification of the behaviour of the environment that impinges on the safety-critical behaviour of the system; this specification will be referred to as the *safety real world description (SD)*. The safety real world description and the assumption that the hazards specify all the conditions under which a disaster can occur, comprise the safety (real world)

assumptions of the system. The third stage involves the construction of a *safety constraint* which specifies the behaviour that if exhibited by the system will ensure that no *identified* disaster can occur – under the safety assumptions of the system. At the controller level, the safety analysis is performed in two stages. In the first stage the *relationship* between the *sensors and actuators* of the safety controller and the *physical process* must be specified in the formal framework. This specification together with SD is referred to as the *safety environment description (SED)*. In the second stage, the behaviour that must be exhibited by the sensors and actuators and at the operator console to ensure that the behaviour complies with the safety constraint, under the safety assumptions is specified – as the *safety controller specification (SCS)*. The elicitation of the specification is performed by an analysis of the SED and safety constraint. To guide the analysis general structures for the behaviour of the safety controller have been proposed in terms of phases. The phases for the reaction vessel are illustrated below (see Fig. 2.3).

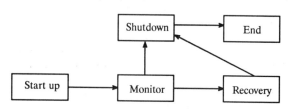

Fig. 2.3. Safety controller structure

2.2. Mission Analysis

At the real world level, the mission analysis is performed in two stages. In the first stage the behaviour of the environment that impinges on system behaviour is specified. The specification will be an extension of SD. It will express the *real world level assumptions* for the mission of the system. The second stage involves the construction of a formal *mission specification* of the system. At the controller level a two stage analysis is performed over the mission controller which is similar to the analysis performed on the safety controller.

2.3. Example System

The concepts described in this paper are illustrated by outlining the safety analysis of a simple chemical plant (the reaction vessel). "The reaction vessel reacts a specified volume of chemical A with a specified volume of chemical B, producing a chemical C. For the reaction between A and B to take place the temperature of the vessel must be raised above a specified activation value."

3. SPECIFICATION MODEL

The behaviour of a process control system is not characterized by a single task, rather as a set of tasks which must be performed in a specific order. In the specification model, each task is defined as a *mode* which specifies the behaviour of the system at the start of the task, during the task and the behaviour which must be exhibited by the system for the task to be completed. A graphical notation, called a *mode graph*, is used to specify the transitions between the modes of a system. The key feature of a mode is the application of system predicates to describe the behaviour of a task. The behaviour specified by modes is defined over a set of system behaviours, that obeys the physical laws of the environment and the construction limitations of the plant. This set of system behaviours is defined by a *history description*. To define modes and history descriptions, the notion of time and system state must be formalized.

3.1. Time

The base time set BT of the specification model will be the set of non-negative real numbers. A time point is an element of BT,

denoted by t, and if more than one time point is required in an expression the time points will be subscripted as: t_0, t_1, ...
Time intervals will be considered as closed finite intervals included in BT, these intervals will be denoted by Int.
For any interval, two boundary points can be defined: the *start* and *end* points. The start point is the first time point of the interval and the end point is the last time point of the interval (i.e., $t_0 \in$ Int is the *start point* of Int iff $\forall t_1 \in$ Int: $t_1 \geq t_0$). The start point of an interval Int will be denoted by s(Int), and the end point by e(Int). For an interval Int we can define its *duration* as e(Int) − s(Int), denoted by dur(Int). The set of all intervals which are included in an interval Int, will be denoted by SI(Int). The system lifetime is the interval of BT denoted by T, where s(T) is the time when the system starts operation and e(T) is the time when the system is finally closed down.

3.2. System State

The state of the system will be given by the values of its state variables. These are (time varying) quantities which either have a significant effect on, or measure factors which affect, the mission or safety of a system.
The vector of all the state variables for a system will be referred to as the state vector. A state vector with n state variables is denoted by: Sv = $\langle p_1, ..., p_n \rangle$, $p_i \neq p_j$, for $i \neq j$, where p_i represents a state variable and i, j $\in \{1, ..., n\}$.
The set of possible values for a state variable are determined by physical laws or construction limitations. This set will be referred to as the variable range, and for variable p_i will be denoted by Vp_i. The state space of a system is denoted as Γ, and is given by the cross product of the variable range set. That is, for a system with n state variables, the state space Γ is given by: $\Gamma = Vp_1 \times Vp_2 \times ... \times Vp_n$.
The state variables of the reaction vessel example, over which the *safety real world (resp. controller) analysis* is performed, are described in Table 1 (resp Table 2). (The contents of the Class column will be explained later.) The safety subsystem of the reaction vessel is illustrated by a schematic diagram (see Fig. 3.1).

TABLE 1 Reaction Vessel Real World Variables

No	Units	Name	Range	Class
p_1	s	Clock	T	Clock
p_2	°K	Temperature	$\{x \in R \mid T_l \leq x \leq T_u\}$	Continuous
p_3	dm³/s	FlowA	$\{x \in R \mid 0 \leq x \leq FA\}$	Continuous
p_4	dm³/s	FlowB	$\{x \in R \mid 0 \leq x \leq FB\}$	Continuous
p_5	dm³/s	Flow	$\{x \in R \mid 0 \leq x \leq Fm\}$	Continuous
p_6	dm³	Volume	$\{x \in R \mid 0 \leq x \leq Vm\}$	Continuous
p_7	dm³/s	OutflowD	$\{x \in R \mid 0 \leq x \leq Om\}$	Continuous
p_8	Ex_Rat	Explosion	{false, true}	Free

TABLE 2 Reaction Vessel Safety Controller Variables

No	Units	Name	Range	Class
p_9	Sa_Set	Safetydial	{on, off}	Free
p_{10}	E_State	EmptySensor	{empty, some}	Free
p_{11}	Lock_Set	LockA	{on, off}	Free
p_{12}	Lock_Set	LockB	{on, off}	Free
p_{13}	mm	ValveD	$\{x \in R \mid 0 \leq x \leq Dm\}$	Continuous
p_{14}	°K	Thermometer	$\{q_1, ..., q_s\}$	Free

3.3. System History

Given the lifetime and state space of a system a model of the behaviour of the system can be constructed as a history. A history

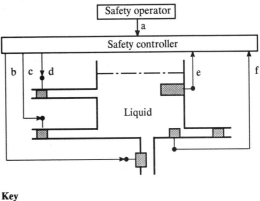

Key
a: Safetydial b: ValveD c: LockB
d: LockA e: Thermometer f: Empty Sensor

Fig. 3.1. Reaction vessel safety subsystem

is a mapping from each time point in the system lifetime to a system state. Process control systems are unpredictable, in the sense that the history is not known in advance. Therefore when reasoning about such systems all "possible" histories must be considered. These form the universal history set of a system.
For every possible history H of the system, the sequence of values taken by a particular state variable p_i can be represented as a function $H.p_i: T \rightarrow Vp_i$. A history itself can be represented as the mapping H: $T \rightarrow \Gamma$, s.t. $H(t) = \langle H.p_1(t), ..., H.p_n(t) \rangle$. The universal history set of a system is denoted as ΓH and is the set of all functions H: $T \rightarrow \Gamma$.

3.4. History Descriptions

The restrictions imposed on the universal history set by the environment are due to the laws of physics and construction of the plant. To specify these restrictions, three formal relations are provided: i) *class relations*, ii) *invariant relations*, and iii) *history relations*. The formal semantics of the relations will be so formed that for every relation we are able to say whether a particular function H: $T \rightarrow \Gamma$ satisfies it or not. The relations will be used in the construction of a history description.
A history description (Desc) is a six-tuple Desc = $\langle T, Sv, VP, CP, IR, HR \rangle$, where T is the system lifetime; Sv is the state vector; VP is the sequence of variable ranges; CP is the sequence of class relations, $\langle Cp_1, ..., Cp_n \rangle$; IR is the sequence of invariant relations, $\langle Ir_1, ..., Ir_m \rangle$; and finally HR is the sequence of history relations, $\langle Hr_1, ..., Hr_k \rangle$.

3.4.1. Class Relations

Class relations are used to impose constraints over the functions which represent the behaviour of a variable. Several useful classes have been defined (Saeed, 1990), such as non-decreasing, continuous, free and perfect clock.
A class relation for a variable p_i (denoted by Cp_i) is a predicate built using relations and logical connectives, and one free function variable p_i. No other free variable may be used.
We will say that a history satisfies a class relation Cp_i if and only if the substitution of $H.p_i$ for p_i results in a well-defined expression which evaluates to true. This satisfaction will be denoted by H **sat** Cp_i.
As an example, consider the definition below.
A perfect clock variable is a variable for which at any time point during the system lifetime the value is equal to the time point.
More precisely, p_i is a prefect clock iff $Cp_i = (\forall t \in T: p_i = t)$.
We will say that a history H of a system satisfies the variable class

relations of a history description if and only if the function $H.p_i$ satisfies the constraints of class Cp_i, $i \in \{1, ..., n\}$. The satisfaction is denoted by H sat CP.

3.4.2. Invariant Relations

Invariant relations are used to express relationships over the state variables which hold for every time point of the system lifetime; these are formulated as system predicates.

A system predicate (SysPred) is a predicate built using standard mathematical functions, relations and logical connectives, and n free value variables $p_1, ..., p_n$ of type $Vp_1, ..., Vp_n$. No other free variables may be used.

The semantics of a system predicate will be defined in terms of the satisfaction condition. A tuple of values $V = \langle x_1, ..., x_n \rangle$, where x_i is of type Vp_i, satisfies a system predicate if and only if substitution of each x_i for p_i within the system predicate results in a well-defined expression which evaluates to true. We denote this satisfaction by writing: V sat SysPred.

A system predicate SysPred is an invariant relation for a history H if and only if $\langle H.p_1(t), ..., H.p_n(t) \rangle$ sat SysPred for all $t \in T$. This satisfaction will be denoted by H sat SysPred.

3.4.3. History Relations

History relations are used to express relationships over the state variables which hold for every interval of the system lifetime, these are formulated as history predicates.

A history predicate is a predicate built using standard mathematical functions, relations and logical connectives, two free time variables T_0, T_1, 2n free value variables $p_{1,0}, ..., p_{n,0}, p_{1,1}, ..., p_{n,1}$ (where $p_{i,j}$ has type Vp_i), and n free function variables $p_1, ..., p_n$ (where p_i is a function of class Cp_i). No other free variables may be used.

We will say that a history H satisfies a history predicate HistPred for an interval Int if and only if the expression resulting from substituting: i) s(Int) for T_0, ii) e(Int) for T_1, iii) $H.p_i(s(Int))$ for $p_{i,0}$ for all i, iv) $H.p_i(e(Int))$ for $p_{i,1}$ for all i, and v) $H.p_i$ for p_i for all i, is a well defined expression which evaluates to true. We will denote this satisfaction by writing: H sat HistPred@Int.

A history predicate HistPred is a history relation for a history H if and only if H sat HistPred@Int for all Int $\in SI(T)$. This satisfaction will be denoted by H sat HistPred.

3.4.4. History Description Set

The history set of a description is the subset of all universal histories which satisfy the relations of the description.
More precisely, for a history description Desc, the history set Set(Desc) is defined as: Set(Desc) = {H$\in\Gamma$H| H sat CP $\wedge$ H sat C(IR) $\wedge$ H sat C(HR)},
where C(IR) = $Ir_1 \wedge ... \wedge Ir_m$ and C(HR) = $Hr_1 \wedge ... \wedge Hr_k$.

3.5. Real-time Satisfaction Conditions

The constraints imposed on histories to specify the behaviour at the real world and controller level will be expressed in terms of three satisfaction conditions - the real-time satisfaction conditions; these are formulated in terms of system predicates.

3.5.1. Point Satisfaction

Point satisfaction is used to impose a constraint over the set of histories which must hold at a specific time point.
A system predicate SysPred is satisfied for a history H at a time point t if and only if $\langle H.p_1(t), ..., H.p_n(t) \rangle$ sat SysPred. This is denoted by H sat SysPred@t.

3.5.2. Interval Satisfaction

Interval satisfaction is used to impose a constraint over the set of histories which must hold during a specific interval.

A system predicate SysPred is satisfied for a history H during an interval Int if and only if H sat SysPred@t for all $t \in$ Int. This is denoted by H sat SysPred@Int.

3.5.3. Events

In our formal framework we capture the notion of an event which occurs (for the first time in an interval) at the end point of an interval. Such an event can be defined in terms of point satisfaction: an event characterized by a system predicate SysPred holds for a history H on an interval Int if and only if SysPred does holds not hold for every time point up to the end point of the interval and holds at the end point of the interval.
The event characterized by SysPred is satisfied for a history H on an interval Int if and only if H sat $\neg$SysPred@t for all $t \in$ Int$-\{e(Int)\}$ $\wedge$ H sat SysPred@e(Int). This is denoted by H sat SysPred$\odot$Int.

3.5.4. Termination Predicate

For most systems it will be necessary to have a predicate that will be true when a system has shut down. This can be achieved by the construction of a system predicate using a perfect clock, called the termination predicate. The termination predicate holds when the value of a perfect clock (say, p_i) is the end point of the system lifetime, i.e., $p_i = e(T)$, abbreviated to Ω.

3.6. Modes

It has been observed that the behaviour exhibited by process control systems can be partitioned into distinct phases. These phases are often referred to as modes (Heninger, 1980; Jahanian, 1988). It is proposed that modes (as defined below) can be used to simplify the construction (and analysis) of formal specifications. It is not suggested that modes are the best way to structure specifications, only that they are a useful method.

A mode is a five-tuple, Mode = $\langle$Start, Inv, End, LB, UB$\rangle$, where Start, Inv and End are system predicates, and LB and UB are time values (or time valued functions).

The semantics will be defined using the **sat**, @ and $\odot$ notation.
Mode = $\langle$Start, Inv, End, LB, UB$\rangle$.
H sat Mode@Int iff the following are satisfied:
H sat Start@s(Int) $\wedge$ H sat Inv@Int $\wedge$ H sat End$\odot$Int $\wedge$ (dur(Int))$\geq$LB $\wedge$ (dur(Int))$\leq$UB.

In other words, a history satisfies a mode over an interval if and only if the history satisfies *Start* at the start point of the interval, satisfies *Inv* during the interval, the event characterized by *End* on the interval and the duration of the interval is within the bounds *LB*, *UB*.

For example, consider the specification of the release mode of the reaction vessel example.

Release Mode

The safety controller is in this mode while it is being set up to monitor the physical process. At the start of the mode the thermometer reading is not greater than tp_1. During the mode valveD must be closed. The safety controller must leave the mode as soon as lockA and lockB are released. The safety controller must not spend more than U_1 seconds in the mode.

Formal specification

Release = $\langle p_{14} \leq tp_1, p_{13} = 0, p_{11} = off \wedge p_{12} = off, 0, U_1 \rangle$.

The start predicate of the release mode stipulates that at the start of the mode the thermometer reading is less than tp_1. The invariant of the mode stipulates that, during the lifetime of the mode, valveD is closed. The event characterized by the end predicate stipulates that lockA and lockB are released (for the first time during the mode) at the end of the mode. The upper bound stipulates that the duration of the mode is less than U_1.

3.7. Mode Sequences

Modes can be used to specify a set of required behaviours of a system, in a structured format, by the construction of a mode sequence.

A mode sequence is a sequence, ModeSeq = ⟨$m_1, ..., m_r$⟩, where m_i is a mode, for $i \in \{1, ..., r\}$.

We will say that a history satisfies a mode sequence if the modes of the sequence are satisfied by the history in the order indicated by the sequence. More precisely, a history H satisfies a mode sequence ⟨$m_1, ..., m_r$⟩ iff $\exists t_0, ..., t_r$: $t_0 \leq ... \leq t_r \wedge t_0 = s(T) \wedge t_r = e(T) \wedge H$ **sat** $m_i@[t_i, t_{i+1}]$, for $i = 0, ..., r-1$. We will denote this satisfaction as: H **sat** ModeSeq.

3.8. Mode Graphs

The requirements of a system can be specified as a set of mode sequences. To visualize such a specification a graphical representation is introduced.

A mode graph (MG) is a four-tuple MG = ⟨M, A, S, E⟩, where M is a (finite) set of modes, A is a set of mode pairs (i.e., $A \subseteq M \times M$), S and E are subsets of M (i.e., S, E $\subseteq$ M). The members of set S are referred to as start modes and the members of set E as end modes. (Thus MG can be viewed as a digraph with nodes M and edges E).

For a mode graph MG the following conditions must hold:
(To simplify the conditions it will be assumed that the mode set (M) is a set of r indexed modes represented as: $\{m_1, ..., m_r\}$.)

i) A must be irreflexive, i.e., $\forall (m_i, m_j) \in A$: $i \neq j$.

In the definition of the following conditions we will make use of the transitive closure set TC of MG. Note: $(m_i, m_j) \in TC$ if and only if there exists a path of length at least one from m_i to m_j.

ii) For any mode which is not a start mode there is a path to it from a start mode, i.e., $\forall m_i \in M-S$: $\exists m_j \in S$: $(m_j, m_i) \in TC$.

iii) For any mode which is not an end mode there is a path from it to an end mode, i.e., $\forall m_i \in M-E$: $\exists m_j \in E$: $(m_i, m_j) \in TC$.

The set of mode sequences specified by a mode graph is given by the set of paths which start at a start mode and end at an end mode. For a mode graph MG this set will be denoted by Seq(MG). We will say that a history H satisfies a mode graph MG iff $\exists ms \in Seq(MG)$: H **sat** ms, this is denoted by H **sat** MG.

Pictorial Representation

A picture of the mode graph MG is a diagram of nodes (annotated ellipses) corresponding to the members of M and arrows corresponding to the members of A, such that if (m_i, m_j) is a member of A then there is an arrow from the node labelled m_i to the node labelled m_j. The start modes will have a broken boundary and the end modes will be shaded (see Fig. 4.1).

4. EXAMPLE SYSTEM

In this section, an outline of the safety analysis of the reaction vessel is presented.

4.1. Safety Real World Analysis

Disaster Identification

Let us suppose that there is a risk of an explosion during the reaction. In the formal framework the occurrence or non-occurrence of an explosion is represented by the elementary system predicate: p_8.

Hazards

Let us suppose that the hazard analysis identifies the following hazard. An explosion may occur if the temperature rises above a specified value (Eact °K) and the vessel is not empty. This hazard can be specified as: $p_2 \geq$ Eact $\wedge p_6 \neq 0$.

Safety real world description

The safety real world description, SD, is constructed by an analysis of the real world to identify the properties which impinge on the critical behaviour of the system. Firstly the analysis identifies the variables which influence the variables in the hazard, and secondly the relationships over the variables are identified. For the reaction vessel some of the identified relations are given in Table 3.

TABLE 3 Reaction Vessel Real World Relations

No	Relationship	Comments
Ir_1	$p_5 = p_3 + p_4$	The flow rate into the vessel is the sum of flowA and flowB.
Hr_1	$p_{2,1} - p_{2,0} \leq \Delta Tm.(T_1 - T_0)$	ΔTm is the maximum change in the temperature per second.
Hr_2	$p_{6,0} = 0 \wedge \forall t$: $p_5 = 0 \Rightarrow p_6 = 0$.	If the vessel is empty at the start of an interval and the flow rate is zero during the interval then the vessel will remain empty during the interval.

Safety constraint

The SC of the reaction vessel is simply the negation of the hazard, that is, SC = ($p_2 <$ Eact $\vee p_6 = 0$). Hence the set of safe histories: SH = {H $\in$ Set(SD): H **sat** SC@T}.

4.2. Safety Controller Analysis

The safety controller must ensure that the system does not enter into a hazardous state.

Safety environment description

The SED is constructed by an analysis of the safety controller. Firstly the analysis identifies the sensors and actuators related to variables in the safety real world description; secondly the relationships over them are identified. For the reaction vessel some relations are given in table 4.

TABLE 4 Reaction Vessel Controller Relations

No	Relationship	Comments
Ir_2	$p_{11} =$ off $\Rightarrow p_3 = 0$	If lockA is on, flowA is zero.
Ir_3	$p_{12} =$ off $\Rightarrow p_4 = 0$	If lockB is on, flowB is zero.
Ir_4	$|p_{14} - p_2| \leq \Delta Tp$	The thermometer reading is always within ΔTp of the temperature.
Hr_3	$T_0 = s(T) \Rightarrow p_6 = 0 \wedge p_{11,0} =$ on $\wedge p_{12,0} =$ on $\wedge p_{13,0} = 0$	At the start of the system lifetime, the vessel is empty, lockA and lockB are on; and valveD is closed.
Hr_4	$T_1 - T_0 < \Delta Sp \Rightarrow p_{14,1} - p_{14,0} < \Delta Te$	ΔTe is the maximum difference between two consecutive temperature readings.

Safety controller Specification

The structure of the safety controller specification is illustrated as a mode graph (see Fig. 4.1).

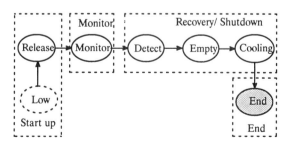

Fig. 4.1. Safety controller mode graph

In the safety controller specification some constants (tp_1, tp_2, tp_3 and U_1) will be used to express system predicates over the temperature and duration of some modes. At the end of a mode specification, where appropriate, conditions (denoted by Ci) will be imposed over these constants. Due to space limitations, only the formal definition of the start up and monitor phases are presented.

Start up phase
The start up phase is specified by the Low and Release modes.

Low mode
The safety controller is in this mode at the start of the system lifetime. While the safety controller is in this mode lockA, lockB are on and valveD is closed. The safety controller leaves this mode as soon as safetydial is at on and the thermometer reading is less than tp_1.

Formal specification
Low = $\langle$true, Init, p_9 = on $\wedge$ p_{14} < tp_1, 0, dur(T)$\rangle$,
where Init = $(p_{11}$ = on $\wedge$ p_{12} = on $\wedge$ p_{13} = 0).

Release mode
See mode example (section 3.6).
C1: $tp_1 + \Delta Tm.U_1 + \Delta Te < tp_2$, $tp_2 < Eact - \Delta Tp$.

Monitor phase
The monitor phase is specified by the monitor mode.

Monitor mode
At the start of the mode the thermometer reading must be less than tp_2. While the safety controller resides in this mode lockA and lockB are off and valveD is closed. The safety controller must leave this mode as soon as the thermometer reading is not less than p_3 or safety dial is at off.

Formal specification
Monitor = $\langle p_{14} < tp_2$, Mval, $p_{14} \geq tp_3 \vee p_9$ = off, 0, dur(T)$\rangle$,
where Mval = $(p_{11}$ = off $\wedge$ p_{12} = off $\wedge$ p_{13} = 0).
C2: $tp_2 < tp_3$, $tp_3 + \Delta Te < Eact - \Delta Tp$.

4.3. Safety Verification

In this section an outline of an illustrative portion of the safety verification is presented. The formal condition that must be proven is: $\forall H \in Set(SED): H$ sat $SCS \Rightarrow H$ sat SC.
The safety condition is proven for the start up phase, the proofs for the other phases can be constructed in a similar way.

Start up phase
The safety condition is proven by showing that the vessel is empty during the Low mode and the temperature is below Eact during the release mode.

Low mode
$\forall H \in Set(SED): \forall Int: s(Int) = s(T) \wedge H$ sat Low@Int $\Rightarrow$
H sat $p_6 = 0$ @Int.
The above condition holds since the vessel is empty at the start of the history (Hr_3), and since lockA, lockB are on and valveD is closed no liquid can flow into the vessel (Ir_1, Ir_2, Ir_3, Hr_2).

Release mode
$\forall H \in Set(SED): \forall Int: H$ sat Low@Int $\Rightarrow H$ sat $p_2 < Eact$@Int.
The above condition holds since the thermometer reading is less than Eact-ΔTp during the mode (C1, Hr_1, Hr_4), which implies that the temperature is less than Eact (Ir_4).

5. SUMMARY AND CONCLUSIONS

This paper has introduced a development model and specification model for requirements analysis. The development model outlined the main stages during requirements analysis, by partitioning the analysis into two levels: the real world and controller. The stages were further refined by maintaining a clear distinction between the safety and mission issues during the analysis. The specification model provided a set of constructs which enable the behaviour of a system and the restrictions imposed by the environment to be expressed in a common formal framework.

This paper gives a flavour of an overall approach to requirements analysis of safety-critical systems. The overall approach (Saeed, 1990) includes: i) formal definitions of consistency and completeness; ii) a set of tools and techniques for the analysis of the specifications produced during the requirements analysis; iii) methodologies for construction and verification of specifications; and iv) a development programme which defines a relationship between the specifications and the roles of the members of the analysis team.

The benefits of the work presented in this paper include: i) the two level approach leads to a full understanding of system behaviour at the real world level, before any complexities are introduced from the controller; ii) the formal model enables a precise specification of timing constraints and provides a means of relating the timing constraints at the controller level to the behaviour exhibited at the real world level; iii) the separation of safety and mission issues allows the formal analysis to be targeted at the safety-critical issues of the system; and iv) the uniform model allows a rigorous analysis of the relationship between a system and environment.

Limitations include: i) the approach only considers requirements analysis – the analysis should be extended to the subsequent phases of system development; and ii) automated tools for analysis (and simulation) would be required for the scheme to be practical for large complex systems.

REFERENCES

Gorski, J. (1986). Design for safety using temporal logic. *Proceedings IFAC Workshop SAFECOMP' 86*, Sarlat, France, 149-155.

Gorski, J. (1988). Formal specification of real time systems, *Computer Physics Communications*, Vol. 50, no. 1-2, 71-88.

Heninger, K.L. (1980). Specifying Software Requirements for Complex Systems: New techniques and their applications. *IEEE Transactions on Software Engineering*, Vol. 6, no. 1, 2-13.

HSE. (1987). *Guidelines on Programmable Electronic Control Systems in Safety-Related Applications, PARTS 1 & 2*. Her Majesty's Stationery Office, London.

Jaffe, M.S. and Leveson, N.G. (1989). Completeness, Robustness, and Safety in Real-Time Software Requirements Specification. *Proc 11th International Conference on Software Engineering*, Pittsburgh, PA, 302-311.

Jahanian, F., and Stuart, D.A. (1988), A Method for Verifying Properties of Modechart Specifications. *Proceedings of IEEE Real-Time Systems Symposium*, Huntsville, Alabama, 12-21.

Leveson, N.G. (1984). Software safety in computer-controlled systems. *IEEE Computer*, Vol. 17, no. 2, 48-55.

Leveson, N.G. (1986). Software Safety: Why, what and how. *ACM Computing Surveys*, Vol. 18, no. 2, 125-163.

Leveson, N.G. (1989). Safety-Critical Software Development. In T. Anderson (Ed.), *Safe and Secure Computing Systems*, Blackwell Scientific Publications, London. pp. 155-162.

Mulazzanni, M. (1985), Reliability versus Safety. *Proceedings IFAC Workshop SAFECOMP '85*, Como, Italy, 149-155.

Roan, A., and Troy, R. (1986). Requirements Modelling of Industrial and Real-Time Systems by Automata and Structured Analysis. *Proceedings IFAC Workshop SAFECOMP '86*, Sarlat, France, 137-142.

Saeed, A. (1990). *A Framework for the Requirements Analysis of Safety-Critical Computing Systems*. Ph.D thesis, University of Newcastle (in preparation).

Smith, C. L. (1972). *Digital Process Control*. Intext Educational Publishers. Scranton, Pennsylvania.

FORMAL APPROACH TO FAULTS IN SAFETY RELATED APPLICATIONS

J. Górski

Institute of Informatics, Technical University of Gdansk, 80-952 Gdansk, Poland

ABSTRACT Hierarchical modelling is necessary in safety analysis - because safety is a higher level, system-wide property. Safety properties are more "visible" if defined with respect to the general model which covers a proper part of the application domain. Investigation of effects of faults which are identifiable and expressible in lower level models and evaluation of their impact on safety properties is an important problem. The paper gives fundamentals for hierarchical verification of specifications built using the temporal logic formalism. It also shows how faults can be included in the analysis.

KEYWORDS safety; formal specification; verification; validation; faults

INTRODUCTION

Formal safety analysis preassumes the existence of formal specifications. Such specifications refer to different models of the application problem and its solution. In [Gorski'89; Gorski'90] it has been argued that, from the safety point of view, a wide spectrum of formal specifications is of interest, starting with a general, conceptual description of the problem and ending with a computer program which implements logical components of system architecture. The temporal logic formalism has been proposed as a "glue" which interfaces the specifications one to another. The safety analysis process involves not only the analysis of system properties expressed in terms of one model but also requires that the specifications expressed in terms of different models are interrelated. Theorem 1 formulated in this paper forms a basis for such type of analysis. Formal analysis is not enough, however, to guarantee safety. If the models, and hence the formal specifications, are not strongly enough connected to the real world of the application, the whole enterprise loses its application oriented sense and becomes merely an exercise in logics and mathematics. Theorem 2 aims on identifying the necessary validation obligations to make sure that a higher level (safety related) system property is guaranteed.

Modelling the reality necessarily involves idealization - some aspects are not included in the model. For instance, while describing the system architecture we concentrate on the positive (expected) behaviours of components. Safety however is a property which must be maintained in any circumstances, also if the system behaviour deviates from the "perfect world" assumption. Therefore the safety analysis must be extended to cover faults and their possible impact. In this paper we show how faulty behaviours can be covered by the formal analysis process. The approach is illustrated by an example which is borrowed from the previous works of this author.

Because of the space limitations we do not include here any introduction to temporal logic, the reader can refer to e.g. [Kroger'87]. The notational conventions in this paper are as follows. The operators of temporal logic refered to in this paper are $\Box$ (henceforth), $\Diamond$ (sometimes), $\ominus$ (next) and $\mathcal{U}$ (until). If σ is a sequence then $\sigma[i]$ denotes the i-th element of σ and $\sigma(i)$ denotes the i-tail of σ, i.e. the sequence $\sigma[i]\sigma[i+1],\ldots$
If $R \subseteq D \times G$ is a relation then $DOM(R)$ denotes a set of all elements of D, for which R is defined, and $COD(R)$ denotes a set of all elements of G for which R is defined. If σ is a sequence of states and F is a formula in the associated temporal language then $\sigma \models F$ denotes that σ satisfies F. If Q and P are formulae then $Q \models P$ denotes that P is a logical consequence of Q.

HIERERCHICAL VERIFICATION

Let be given temporal languages L1 and L2 with fixed interpretations. In particular, there is given a set of state sequences M1 (a set of behaviours) and the satisfaction relation $\models_1$ relates elements of M1 to formulae of L1. Similarly, the satisfaction relation $\models_{L2}$ relates elements of a given set of behaviours M2 to formulae of L2.

Let TRANS be a set consisting of expressions of the form
$P_i \longleftrightarrow W_i$, $i=1,..,n$,
where W_i is a formula of L1, and P_i is a formula of L2. It is assumed that the formulae do not contain free global variables and do not contain the temporal operator Θ (next state).

Let Q be a formula of L1 and Q is composed of subformulae W_i, $i=1,..,n$, connected by logical and temporal operators. In such case Q is <u>convertible</u> through TRANS.

Let $t \in TRANS$ and $t = 'P \longleftrightarrow W'$, and let $\sigma \in M2$. We will introduce the following definitions:

$MAP_t : M2 \longrightarrow 2^{M1}$,

$MAP_t(\sigma) = \{\delta \mid$ for each j, $j \geq 0$,
    $\sigma(j) \models_{L2} P_i$ iff $\delta(j) \models_{L1} W_i\}$,

$MAP_{TRANS}(\sigma) = \bigcap_{t \in TRANS} MAP_t(\sigma)$.

In the sequel, if it does not lead to ambiguities, we will write $MAP(\sigma)$ instead of $MAP_{TRANS}(\sigma)$.

Let Q[TRANS] denotes the formula which is obtained from a convertible formula Q of language L1 by simultaneous substitution of all occurences of subformula W_i by the occurences of subformula P_i, for $i=1,..n$. The resulting formula Q[TRANS] belongs to the language L2.

<u>Lemma 1.</u>
If $\delta \in MAP(\sigma)$, then for $k \geq 0$, we have $\delta(k) \in MAP(\sigma(k))$, where $\delta(k)$ denotes the k-tail of δ, i.e. $\delta(k) = \delta[k] \delta[k+1]..$.
<u>Proof.</u>
Let $\delta \in MAP(\sigma)$. Then, from the definition of MAP we obtain that for each $j=1,..,n$ and for each $i \geq 0$, $\sigma(i) \models_{L2} P_j$ iff $\delta(i) \models_{L1} W_j$. Hence, in particular, the above is true for each $i \geq k$.

<u>Lemma 2.</u>
Let $TRANS=\{..,(P \longleftrightarrow W),..\}$ and let
    $TRANS'=TRANS \cup \{\neg P \longleftrightarrow \neg W\}$.
Then, for each σ,
    $MAP_{TRANS}(\sigma) = MAP_{TRANS'}(\sigma)$,
i.e. TRANS and TRANS' are not distinguishable by the mapping MAP.
<u>Proof.</u>
The proof results immediately from the definition of the mapping MAP if we notice that the condition
    for each $j \geq 0$, $\sigma(j) \models_{L2} P$ iff $\delta(j) \models_{L1} W$
is equivalent to
    for each $j \geq 0$, $\sigma(j) \models_{L2} \neg P$ iff $\delta(j) \models_{L1} \neg W$.

<u>Theorem 1.</u>
Behaviour $\sigma \in M2$, satisfies the formula Q[TRANS] iff each behaviour $\delta \in MAP(\sigma)$ satisfies formula Q, i.e.
    $\sigma \models_{L2} Q[TRANS]$ iff
        for each $\delta \in MAP(\sigma)$, $\delta \models_{L1} Q$.

<u>Proof.</u>
The proof is by induction on the structure of the formula Q.

(1) Let $Q=W$, where $(P \longleftrightarrow W)$ is in TRANS. Then Q[TRANS]=P and the proof results immediately from the definition of $MAP(\sigma)$.

For logical connectives:

(2) Let $Q=R1 \vee R2$, where R1 and R2 are convertible through TRANS and Theorem 1 holds for R1 and R2.

If $\sigma \models_{L2} Q[TRANS]$ then $\sigma \models_{L2} R1[TRANS]$ or $\sigma \models_{L2} R2[TRANS]$ must hold. Let $\delta \in MAP(\sigma)$. Then we have $\delta \models_{L1} R1$ or $\delta \models_{L2} R2$ and from this $\delta \models_{L2} (R1 \vee R2)$.

If $\delta \in MAP(\sigma)$ and $\delta \models_{L1} R1$ or $\delta \models_{L1} R2$ holds then $\sigma \models_{L2} R1[TRANS]$ or $\sigma \models_{L2} R2[TRANS]$ must hold and from this we have $\sigma \models_{L2} (R1[TRANS] \vee R2[TRANS])$.

(3) Let $Q=\neg R$, where R is convertible and Theorem 1 holds for R.

If $\sigma \models_{L2} Q[TRANS]$ then $\sigma \models_{L2} R[TRANS]$ does not hold. Let $\delta \in MAP(\sigma)$. Then, because Theorem 1 holds for R, $\delta \models_{L1} R$ does not hold, and from this we have that $\delta \models_{L1} \neg R$ holds.

If $\delta \in MAP(\sigma)$ and $\delta \models_{L1} \neg R$, then $\delta \models_{L1} R$ does not hold. But then $\sigma \models_{L2} R[TRANS]$ does not hold, which means that $\sigma \models_{L2} \neg R[TRANS]$ must hold.

Other logical connectives ($\wedge$, $\Rightarrow$, $\Leftrightarrow$) can be proven in an analogous way.

For quantifiers:

(4) Let $Q=(\forall x)R$, where x is a global variable and R is convertible and Theorem 1 holds for R.

Because R is convertible it does not contain free global variables. Consequently, $Q[TRANS]=(\forall x)R[TRANS]$ and x is not free in R[TRANS]. From this we have that $\sigma \models_{L2} Q[TRANS]$ holds iff $\sigma \models_{L2} R[TRANS]$ holds, i.e. for each $\delta \in MAP(\sigma)$ must be $\delta \models_{L1} R$. And from this we have immediately $\delta \models_{L1} (\forall x)R$.

Existential quantifier can be proven in an analogous way.

For temporal operators:

(5) Let $Q=\Box R$, where R is convertible and Theorem 1 holds for R.

If $\sigma \models_{L2} Q[TRANS]$ then $\sigma \models_{L2} \Box R[TRANS]$, which means that for each $i \geq 0$, $\sigma(i) \models_{L2} R[TRANS]$ holds and for each behaviour $\alpha \in MAP(\sigma(i))$, $\alpha \models_{L1} R$ holds. Let us take a behaviour $\delta \in MAP(\sigma)$. Then, for each $i \geq 0$, we have $\delta(i) \in MAP(\sigma(i))$ (from Lemma 1), which means $\delta(i) \models_{L1} R$, and consequently $\delta \models_{L1} \Box R$.

If $\delta \in MAP(\sigma)$ and $\delta \models_{L1} \Box R$ then for each $i \geq 0$, $\delta(i) \models_{L1} R$ holds. But then, from Lemma 1, we have that for each $i \geq 0$, $\sigma(i) \models_{L2} R[TRANS]$ holds, and from this $\sigma \models_{L2} \Box R[TRANS]$ must hold.

(6) Let $Q=\Diamond R$, where R is convertible and Theorem 1 holds for R.

If $\sigma \models_{L2} Q[TRANS]$, then for some $i \geq 0$, $\sigma(i) \models_{L2} R[TRANS]$ holds. Let $\alpha \in MAP(\sigma(i))$. Then there must be $\alpha \models_{L1} R$. Let us consider

$\delta \in MAP(\sigma)$. Then, for each $k \geq 0$, we have $\delta(k) \in MAP(\sigma(k))$ (from Lemma 1), which means that $\delta(k) \models_{L_1} R$ holds, and consequently $\delta \models_{L_1} \Diamond R$ also holds.

If $\delta \in MAP(\sigma)$ and $\delta \models_{L_1} \Diamond R$, then for some $i \geq 0$, $\delta(i) \models_{L_1} R$ holds. But then, from Lemma 1, $\sigma(i) \models_{L_2} R[TRANS]$ holds, and from this we have that $\sigma \models_{L_2} \Diamond R[TRANS]$ holds.

(7) Let $Q = (S \cup R)$, where R, S are convertible and Theorem 1 holds for them.

If $\sigma \models_{L_2} Q[TRANS]$ then for some $i \geq 0$, $\sigma(i) \models_{L_2} R[TRANS]$ holds and for each j, $0 \leq j \leq i$, we have $\sigma(j) \models_{L_2} S[TRANS]$. Let $\alpha \in MAP(\sigma(i))$. Then $\alpha \models_{L_1} R$ must hold. Also, for each j, $0 \leq j \leq i$, if $\beta \in MAP(\sigma(j))$ then $\beta \models_{L_1} S$ holds. Let $\delta \in (MAP(\sigma)$. Then (from Lemma 1), $\delta(i) \in MAP(\sigma(i))$ and for each j, $0 \leq j \leq i$, $\delta(j) \in MAP(\sigma(j))$. From this we have that $\delta(i) \models_{L_1} R$ and for each j, $0 \leq j \leq i$, $\delta(j) \models_{L_1} S$ holds, which means that $\delta \models_{L_1} (S \cup R)$ holds.

Let us take $\delta \in MAP(\sigma)$ such that $\delta \models_{L_1} (S \cup R)$ holds. Then, for some $i \geq 0$, we have $\delta(i) \in MAP(\sigma(i))$ (from Lemma 1) and $\delta(i) \models_{L_1} R$, and for each j, $0 \leq j \leq i$, $\delta(j) \in MAP(\sigma(j))$ and $\delta(j) \models_{L_1} S$ holds. But from this we have that $\sigma(i) \models_{L_2} R[TRANS]$ and for each j, if $0 \leq j \leq i$ then $\sigma(j) \models_{L_2} S[TRANS]$ holds, which means that $\sigma \models_{L_2} (S \cup R)[TRANS]$ holds.

The proof for the remaining temporal operators is in an analogous way.

Theorem 1 forms the basis for the algorithm which enables verification of properties expressed in different temporal languages and interpreted in different system models.
The skeleton of this algorithm is given below.

Let SPEC1 be a formula of the language L1 and SPEC2 be a formula of the language L2.

   Verification of SPEC2 against SPEC1.
Step 1
Define a set TRANS such that SPEC1 is convertible through TRANS.
Step 2
Demonstrate that SPEC2 $\models_{L_2}$ SPEC1[TRANS].

If the algorithm is successful then for every behaviour σ which satisfies SPEC2 we have that each corresponding (through MAP_{TRANS}) behaviour δ satisfies SPEC1. If SPEC1 and SPEC2 are specifications of the same system, expressed in two different specification languages, the algorithm shows that for a given translation set TRANS, SPEC2 is consistent with SPEC1.

VALID SPECIFICATIONS

Let be given a relation $OBS \subseteq M2 \times M1$. We assume that the sets M2 and M1 represent behaviours of the same fragment of the real world (the same application problem), expressed in terms of two different models. The OBS relation is a formal expression of this assumption. If $(\sigma, \delta) \in OBS$ then σ and δ are possible observations of the same behaviour of the application under consideration, perceived from the perspective of two different models.

We will say that TRANS is <u>relevant</u> for OBS if the following condition holds:
for each pair (σ, δ),
   if $(\sigma, \delta) \in OBS$ then $\delta \in MAP(\sigma)$, (*)
i.e. if δ and σ denote the same behaviour of the application then they are associated through the mapping MAP.

Let SPEC2 be a formula of L2. We will say that SPEC2 is a <u>valid</u> specification with respect to OBS, if for every $\sigma \in DOM(OBS)$, $\sigma \models_{L_2} SPEC2$, i.e. each σ which models a real behaviour of the application satisfies SPEC2.

Theorem 2.
If SPEC2 is a valid specification, TRANS is relevant and the following condition holds
   SPEC2 $\models_{L_2}$ SPEC1[TRANS]   (**)
then SPEC1 is a valid specification, i.e. each $\delta \in COD(OBS)$ satisfies SPEC1.
Proof.
Let $(\sigma, \delta) \in OBS$. Then because SPEC2 is valid, we have $\sigma \models SPEC2$. And from the relevance of TRANS we have $\delta \in MAP(\sigma)$. Then, from (**) and from Theorem 1 we obtain that $\delta \models SPEC1$.

From Theorem 2 we have that the validity of the verification algorithm presented in the previous section depends on the two critical points:
(1) SPEC2 must be a valid specification of the application, and
(2) TRANS must be relevant.

The requirement for validity of SPEC2 is obvious – if SPEC2 is not valid then the verification is simply an exercise in logic – there is no real-life problem behind it. The requirement on TRANS is not so obvious. From the formal point of view the only restriction is that SPEC1 must be convertible through TRANS. However, in the light of a particular application, the proper choice of TRANS has the principal meaning – it represents the interrelation between two different models of the same reality perceived from the two different viewpoints. The relevance of this choice is formally described by the condition (*). For instance, the interrelation between the conceptual and the architectural models of the "crossing world" described in [Gorski'89] could be expressed informally by the statement:
   "the numbers of signals generated by sensor TI (the entry sensor) and sensor TO (the exit sensor) are equal iff there is no train inside the crossing area".
The above expression defines how the state of the conceptual model of the application (where such objects like cars and trains are explicitly represented) is related to the state of the architectural model (which includes sensors, actuators, computers, etc.). The relevance of this relation can be validated by ensuring that the sensor TI is situated on the railway, at the entry to the crossing area, the sensor TO is situated on the railway, at the exit from the crossing area, there is only one railway which crosses the area, and initially the area is empty, i.e. there are no trains inside. There is also an implicit assumption that the sensors

are fault free, i.e. the signal is generated iff a train is passing through. The extent of such validation is not formalized.

Because the choice of TRANS is restricted by the the condition that TRANS must be relevant, one can expect that SPEC1 will not necessarily be convertible. If so, then if SPEC1 is a conjunction of more elementary formulae (which is the case in most practical situations), we can consider each subformula separately, and concentrate on those which are convertible. This means that the verification coveres only part of the properties specified by SPEC1. An example of such partial verification for the railway crossing application is given in [Gorski'90]. The properties which are not subjected to verification are treated as assumptions which must hold in order to guarantee SPEC1. Their validity must be established directly, without referring to SPEC2.

FORMALIZATION OF FAULTY BEHAVIOURS

Let OBS ⊆ M2 x M1 represents a relation between two models of a given application. If $(\sigma,\delta) \in$ OBS then we will say that it is a <u>fault-free</u> <u>reception</u> of the application.

Let $OBS_F \subseteq$ M2 x M1 denotes another relation between the two models. If $(\sigma,\delta) \in OBS_F$ and $(\sigma,\delta) \in$ OBS then we will say that it is a <u>faulty</u> <u>reception</u> of the application.

The relation OBS_F −OBS is called the <u>fail</u> <u>space</u>.

Let us take a given set TRANS such that TRANS is relevant for OBS and a specification SPEC1 is convertible through TRANS. Let SPEC2 be a valid specification and the condition (**) holds. Then, from Theorem 2 we obtain that SPEC1 is a valid specification of the application.

We will say that SPEC1 is <u>tolerant</u> to the fail space OBS_F − OBS if SPEC1 is valid with respect to OBS_F. From Theorems 1 and 2 we have that SPEC1 is tolerant to the fail space OBS_F − OBS if TRANS is relevant with respect to OBS_F and SPEC2 is a valid specification with respect to OBS_F.

The tolerance of SPEC1 means that SPEC1 remains valid despite of possible faulty receptions of the application's behaviour.

Let us assume that we have two specifications SPEC1 and SPEC2 of an application problem. The specifications are expressed in terms of two different models. SPEC1 directly describes the required safety properties of the application. SPEC2 describes the system from a different perspective, where safety is not so easily expressible, e.g. describes components and their interconnections. The validation analysis shows that SPEC2 is valid for positive (i.e. fault-free) behaviour of components. A relevant translation set TRANS provides for translation and expression of SPEC1 in terms of the model related to SPEC2. Then, through formal analysis we demonstrate the validity of SPEC1. Now a question arises if SPEC1 is still valid if we admit faulty behaviours of some components. From the material presented in this paper it results that in order to investigate if a given (safety related) property SPEC1 of the system is tolerant to a specific fault, we have to fulfill the following obligations:

(1) make sure that SPEC2 is a valid specification of the set of behaviours COD(OBS_F) and if it is not the case, modify SPEC2 accordingly,
(2) make sure that TRANS is relevant with respect to OBS_F, and if it is not the case, modify TRANS accordingly,
(3) if SPEC2 and/or TRANS have been modified, verify that the condition (**) still holds.

EXAMPLE

In this section we will give an example which illustrates how the above approach can be applied in practice. We will refer to the railway crossing application problem which has been studied in detail in the previous papers [Gorski'89; Gorski'89a; Gorski'90].

A hierarchy of specifications have been built, which describe the problem from different viewpoints. In particular, the conceptual model of the problem distinguishes the domain of TRAINS and the CROSSING entity, and the TINSIDE relationship says which trains are presently inside the crossing area. The RLSEMAPHORE is a conceptual railway semaphore and its attribute RLPOSITION distinguishes between two possible values: "closed" and "open". The cars passing through the crossing area are controlled by the GATE entity which attribute GPOSITION distinguishes between three possible values: "down", "up" and "moving". One of the requirements expressed in the associated temporal language says that
(C1)   □(¬GPOSITION(down) ==>
    (RLPOSITION(closed) ∧ ∀t¬TINSIDE(t)),
which reads: if the gate is not down then the railway is closed and there is no train inside the crossing area (the global variable t assumes values from the TRAINS domain).

Another level of the specification hierarchy relates to the system architecture. It is assumed that three sensors TA, TI and TO are situated along the railway in such a way that TA indicates approaching trains, TI indicate that a train enters the crossing area and TO indicates that the train leaves the area out. The sensors are represented by the D-ZONE component of system architecture. The ILIGHT component represents the lights situated at the entry to the crossing area. The lights are controlled by the I signal line which transfers "r" (red) and "g" (green) commands. The GATE component is controlled by the M signal line which transfers "u" (up) and "d" (down) commands. After the gates reach the required position, the acknowledge signal repeating the last accepted command is sent through the P line (before this signal is being sent, the gates are moving). The specification of components states their behavioural properties. For instance,

D-ZONE is specified by
$\Box(0 \le |TO| \le |TI| \le |TA|)$
which states that the number of signals sent by the TA sensor is greater equal the number of signals sent by TI which in turn is greater equal the number of signals sent by TO;

GATE is specified by
$\Box(P \prec M)$ and $\Box(0 \le |M| - |P| \le 1)$
which states that the gate acknowledges the received commands ("$P \prec M$" means that the sequence of commands sent through P is an initial subsequence of commands sent through M), and the gate does not accept the next command before executing and acknowledging the previous one.

ILIGHT is specified by
$\Box$ In(I)
which means that the lights are always ready to accept the next command through the input line I ("In" is a predefined predicate which characterizes if a given line is ready for input). Thus the lights can be switched without any delay.

The complete conceptual and architectural specifications of the crossing problem can be found in [Gorski'89; Gorski'90].

Thus, we are in a position where we have two different descriptions of the same application problem. The real behaviours of the application will be represented by the corresponding behaviours of the associated (conceptual and architectural) models. Formally, it can be characterized by the OBS relation which distinguishes pairs of modelled behaviours which correspond to the same real behaviour of the application. Each model has the associated specification. The validity of such specification means that it adequately characterizes all possible behaviours of the application. For instance, the validity of the specification of the architectural components means that the sensors will not generate spurious signals (otherwise the number of signals generated by e.g. TO could easily exceed the number of signals generated by TA), the gates obey commands and generate proper acknowledgement signals, etc. Each of those statements must be carefully related to the application world because they characterize the strength of the link between the model and the real world. Having decided upon the validity of the architectural model we aim at establishing the link between this model and the more general, conceptual model of the application. Formal description of this link is by building the proper TRANS set. In our case TRANS includes:

(E1) $|TI| = |TO| \longleftrightarrow \forall t \neg TINSIDE(t)$,
(E2) $|TA| = |TI| \lor I\char`^r \longleftrightarrow$
                              RLPOSITION(closed),
(E3) $M\char`^d \land P\char`^d \longleftrightarrow$ GPOSITION(down).

The relevance of TRANS can be established by referring to the common application underlying the two models. The (E1) expression says that the equal numbers of signals sent by TI and TO sensors mean that there is no train inside the crossing area. The relevance of this statement can be validated by ensuring that the sensor TI is situated on the railway, at the entry to the crossing area, the sensor TO is situated on the railway, at the exit from the crossing area, there is only one railway which crosses the area, and initially the area is empty, i.e. there are no trains inside. There is also an implicit assumption that the sensors are fault free, i.e. the signal is generated iff a train is passing through. (E2) means that if the numbers of signals from the TA and TI sensors are equal or the last command to the ILIGHT component was "r" (red) then the railway is closed. To validate the relevance of this statement we have to recall that TA and TI are situated at the entry and at the exit of the approaching zone of the railway and the ILIGHT lights are at the entry of the crossing. So, if there is no train in the approaching zone or the railway light signal is "red", no train can enter the crossing, i.e. the railway is closed. Again, here the implicit assumption is that there is only one railway, the sensors and the lights work correctly and the trains obey the light signals. (E3) means that the situation where the last command sent to the gates is "d" (down) and the last acknowledgement received is "d" is interpreted that the road gates are in the "down" position. The relevance of this requires assuming that the gate commands are obeyed without failure and the acknowledgement signals are generated accordingly.

The (C1) requirement expressed in terms of the conceptual model can be translated (by means of E1, E2, E3) and expressed in terms of the architectural model, in the following way

$\Box(\neg(M\char`^d \land P\char`^d) \Longrightarrow$
                $(|TA| = |TI| \lor I\char`^r) \land |TI| = |TO|)$.

The above formula can be subjected to the formal verification which shows its consistency with the architectural specification. If it is demonstrated, then from Theorem 1 and Theorem 2 we have that (C1) is valid, i.e. each real behaviour will satisfy this property.

Let us now assume that the reality is not as perfect as it has been assumed before: namely, the TO sensor can generate spurious signals. The only assumption which we make is that after a train enters the crossing (i.e. a signal from TI occurs) then TO sensor works correctly, i.e. the first signal from TO denotes that the train leaves the crossing out. If there is no train inside, the TO sensor can generate spurious signals (note that the analogous situation has been analysed [Leveson and Stolzy'85] using the Petri net formalism). To maintain the validity of the architectural specification, the specification of the D-ZONE component is changed as follows:
$\Box(0 \le |TI| \le |TA|)$,
i.e. the specification asserts that TI and TA sensors are fault-free; no assumptions on the behaviour of TO sensor are included.

Our assumption destroys the relevance of the (E1) expression of TRANS: equal numbers of signals received from the TI and TO sensors no longer mean that there is no train inside the crossing. We modify

(E1) to the following form:
$$E1': \exists(n)(n=|TI| \land (\forall k,k' \leq n)(\exists l,l')$$
$$(k=k' \Longrightarrow l=l' \land (TI,k) \ll (TO,l) \land$$
$$(TI,k') \ll (TO,l'))$$
$$\longleftrightarrow \forall t \neg TINSIDE(t).$$

In the above (TI,k) denotes the k-th signal received from TI sensor, (TO,l) denotes the l-th signal received from TO sensor, and $\ll$ is a total ordering of signals (messages) within the architectural model (for more detail see [Gorski'90; NGDO'86]).

The above expression says that if for each signal sent by TI there is a corresponding signal sent by TO which has occured afterwards then there is no train inside the crossing.

The relevance of (E1') can be justified if we recall our assumption that although TO sensor can generate spurious signals, the first signal sent after a train enters the crossing (i.e. after a signal is being sent by TI sensor) means that the train leaves the crossing out (i.e. there are no spurious signals while a train is inside the crossing.

The (C1) requirement can be translated (by means of E1', E2, E3) and expressed in terms of the architectural model, in the following way
$$\Box(\neg(M^d \land P^d) \Longrightarrow$$
$$(|TA|=|TI| \lor I^r) \land$$
$$\exists(n)(n=|TI| \land (\forall k,k' \leq n)(\exists l,l')$$
$$(k=k' \Longrightarrow l=l' \land (TI,k) \ll (TO,l) \land$$
$$(TI,k') \ll (TO,l'))).$$

The above formula is the goal of formal verification. If it can be proven from the architectural specification, then the validity of the property (C1) is demonstrated, i.e. (C1) is tolerant to spurious signals generated by the TO sensor.

CONCLUSIONS

The paper identifies the verification and validation obligations which are to be fulfilled in order to check fault tolerance of safety properties. The attention focuses on the passage from general, conceptual model of the application (where safety properties can be directly formulated) to the architectural model, where component faults are possible to define. The validation obligations to be fulfilled are split into two main parts: validation of the lower level (architectural) specification, and validation of the mapping between the system behaviours perceived at the architectural level and the system behaviours perceived through the conceptual model. The validation can result in changes of specifications of components and in changes of the mapping. If the validation obligations are fulfilled then the validity of the safety properties can be established through formal verification.

The approach proposed here is complementary to the fault tree analysis [FT'81]. The approach is deductive, i.e. goes from elementary properties to their consequences. The fault tree approach is inductive and goes from consequences to the possible causes. In practice, a proper combination of both approaches should be applied.

REFERENCES

[FT'81] Fault Tree Handbook. NUREG-0942, 1981.

[Górski'85] Górski J.: Specification and design of reliable systems in terms of unreliable components. SAFECOMP'85, W.J. Quirk Ed., Pergamon Press, 1985, str.135-140.

[Górski'89] Górski J.: Formal Approach to Development of Critical Computer Applications. Proc. 22-th Hawaii International Conference on System Sciences, Bruce T. Shriver (Ed.), IEEE Computer Society Press, 1989, str. 243-251.

[Górski'89a] Górski J.: Deriving safety monitors from formal specifications. Proc. SAFECOMP'89, Vienna, Austria, 1989, Pergamon Press, str.123-128.

[Górski'90] Górski J.: Problems of specification and analysis of safety related systems - application of temporal logic. Technical University of Gdansk Press, No. 68, 1990.

[Kroger'87] Kroger F.: Temporal logic of programs. Springer-Verlag, 1987.

[Leveson and Stolzy'85] Leveson N.G., Stolzy J.L.: Safety analysis using Petri nets. 15-th Conference on Fault-Tolerant Computing, 1985, str.358-363.

[NDGO'86] Nguyen V., Demers A., Gries D., Owicki S.: A model and temporal proof system for networks of processes. Distributed Computing, vol. 1, No. 1, 1986, str.7-25.

PROVABLY CORRECT SAFETY CRITICAL SOFTWARE

A. P. Ravn*, H. Rischel* and V. Stavridou**†

*Department of Computer Science, Technical University of Denmark,
DK-2800 Lyngby, Denmark
**Department of Computer Science, RHBNC, University of London,
Egham, Surrey TW20 0EX, UK

Abstract

We present the approach to development of provably correct safety critical software emerging through the case studies activity of the ProCoS project. We envisage the development of a safety critical system through six major stages: Control objectives and safety criteria will be captured in **requirements capture languages (RLs)** (formal mathematical models of the problem domain) supporting the notions of durations, events and states. A specification expressed in a **specification language (SL)** satisfying the requirements is derived via requirements transformations. The specification is refined and finally transformed into a **programming language (PL)** program which is refined and mapped by a **verified PL compiler** onto the instruction set of an **abstract hardware machine (AHM)** and is executed via an AHM computer supported by a trusted **kernel** operating system. The aim of this paper is to show how the coordinated development activities fit together once an informal specification of the desired system behaviour has been delivered.

Keywords: Requirements capture, formal specification, program transformation, compiler verification, program validation, MOD 00-55.

1 Introduction

ProCoS (Provably Correct Systems)[1] is an ESPRIT Basic Research Action project whose aim is to select and integrate mathematically based notations and techniques for software engineering and demonstrate their use in the development of provably correct programs for safety critical computing systems. ProCoS is a 3 country, 7 university, 30 month long collaboration between the Technical University of Denmark, Kiel University, Oldenburg University, Manchester University, Oxford University, RHBNC and Aarhus University. The backbone of the ProCoS project is the integration of various research activities into formal system development in an effort to yield a framework enabling the construction of provably correct systems. The integration theme is central to the project and distinguishes it from other formal treatments of safety critical systems. In order to achieve this integration, the stages of software development and the interfaces between the stages must be rigorously specified.

ProCoS project is now half-way complete, so there is still work to do in getting all the interfaces clarified. The following account describes the current views of the Case Studies team within the project, based on some experimentation with some carefully prepared requirements for concrete safety critical systems, i.e. a simple Auto Pilot [10], a Gas Burner [15], and a Railway Crossing [7], and workshop discussions of the issues [14].

*This work is partially supported by the Commission of the European Communities (CEC) under the ESPRIT programme in the field of Basic Research Action proj. no. 3104:"ProCoS: Provably Correct Systems"
†To whom all correspondence should be addressed. Tel (+44) (784) 443429, Fax (+44) (784) 437520.

In order to clarify our perception of the development phases, we postulate that a piece of safety critical software (a program) is described by texts of the following kinds:

R: The requirements, which capture the purpose of the program.

S: A specification, which describes computations prescribed by R.

P: A conventional program satisfying S.

M: Machine code compiled from P.

B: The executable binary image of M in a processor.

If such texts are to be used in formal development, they must be given meaning by mathematical models of computation [25]. Such models are formulated in terms of sets of observable properties of the entity to be modelled, e.g. it may be convenient to postulate that a real-time system is observed through its state, event sequence, time, and termination status. The set of observable properties that concern us is: 1) a set of states, 2) a set of event sequences, 3) a time domain, and 4) a set of status values. A particular implemented entity is then denoted by a subset of observations. A *specification* for an entity is a presumably larger subset containing all acceptable observations for an implementation of the entity. Development of an implementation from a specification consists in selecting an implementable subset of the specification. In general a specification will be *non-deterministic* in that it will allow a choice between several implementations. This allows a developer to *refine* a specification through a number of steps, where each step reduces the amount of non-determinism. (A refinement calculus for sequential program specifications is described in a recent book by Morgan [6]. A calculus for occam is described in [11]). This concept of program development can be illustrated by the following refinement chain[1]:

Environment

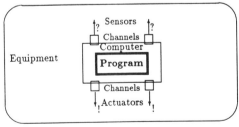

Figure 1: Embedded software system

...	'⊑'	'R'	requirements capture
	'⊑'	S	transformation to a specification
S	⊑	S_1	specification refinement

[1] $X \sqsubseteq Y$ denotes that Y is the same as X or better, i.e. more deterministic. Quoted symbols are informal

$$\begin{array}{rcll}
\vdots & & & \\
S_n & \sqsubseteq & P & \text{transformation to a program} \\
P & \sqsubseteq & P_m & \text{program refinement} \\
\vdots & & & \\
P_m & \sqsubseteq & M & \text{compilation} \\
M & \text{'}\sqsubseteq\text{'} & \text{'}B\text{'} & \text{execution}
\end{array}$$

Organization of this paper: The following sections discuss the stages from Requirements Capture to Compilation, introducing relevant models, notations, and transformations. A section just before the conclusion discusses aspects of the MOD draft standard 00–55 for the procurement of safety critical software [2].

2 Requirements Capture

The above development chain immediately raises the question: Where does the formalisation start? To be able to answer that, we outline the kind of computation and reality we have in mind. This allows us to be more specific on how requirements for a system arise and what parts we consider relevant for program development.

2.1 Environment, equipment, program

An embedded, real-time software system is a *program* placed in a digital computer, which is electronically connected through *channels* to sensors and actuators that measure or change quantities in a piece of *equipment*. The equipment usually being a physical system that evolves over time (Fig. 1), and used in a certain *environment*.

A note on Safety Critical and Correct: The term "safety critical software" is defined in 00-55 as "software used to implement a safety-critical function", i.e. as a property of the *use* of the software, *not* as an intrinsic property of the software. Any safety critical use of equipment must of course be assessed carefully, cf. the MOD Draft Standard 00-56 [3]. We are in agreement with this definition. As for correctness, the goal of ProCoS is a software development method where the *correctness* of the program is stated relative to *precise assumptions about the environment*, and proved to hold for these assumptions.

2.2 System Requirements

Requirements place constraints on a piece of Equipment and its Environment, which together form the total System. The constraints in general specify that the system shall perform certain *Functions*, and will be *Safe* to use. The observations of a system is its *State* evolving over real *Time*, i.e. a subset of the set of functions from Time to State $\{X \mid X : Time \to State\}$. In the following we use the less cumbersome convention of the Z-notation [13] and denote sets by their characteristic predicate, i.e. we denote the set above by:

$$X : Time \to State$$

In real time systems, *Time* is usually modelled by the real numbers $\mathbf{R}$. For continuous systems the state is modelled by a product of state variables which take real values. A state is **safe** when it is bounded. Functional constraints are limits of the state values as time increases.

2.2.1 Example: Auto Pilot

The development of a system including an embedded computer is illustrated by the following example of a simple *autopilot* for a ship. The environment, i.e. the ship will depend on the autopilot to keep it on (a straight line) course, thus the autopilot is a safety critical system. The system requirements can be stated rigorously in the form of a constraint on a mathematical model of the course of the ship. The state variables are continuous, differentiable functions from *Time* to $\mathbf{R}$, which describe the movement of the ship:

y: The deviation from the course.

w: The velocity (speed) of the drift perpendicular to the course, caused by wind, tide etc.

v: The compensating velocity caused by the rudder.

Controlling the course of a ship relies on the interdependence of these variables. In order to have a simple model we assume that elementary mechanics apply, such that the sum of w and v gives the change in y. This *environment assumption* can be stated by,

$$Ship \,\hat{=}\, [y, v, w : Time \to \mathbf{R} \mid \tfrac{dy}{dt} = v + w]$$

The safety requirement for an autopilot quite simple: y and v should stay *bounded* (definitions of simple predicates like *bounded* are omitted for the sake of brevity.). This is stated relative to our assumption:

$$Bounded \,\hat{=}\, [Ship \mid bounded(y) \land bounded(v)]$$

The functional requirement is that y comes *close* to zero shortly after w becomes *constant*.

$$Correcting \,\hat{=}\,$$
$$[Ship \mid \exists \delta t . constant(w, t..t+\delta t) \Rightarrow close(y(t+\delta t), 0)]$$

Development of continuous systems uses *control theory*, based on mathematical and numerical analysis. Analysis of the system above will show that in order to satisfy the requirements we have to: 1) introduce a model state z used to predict the next position of the ship, 2) assume w to be bounded, and 3) assume the change in w ($\frac{dw}{dt}$) to be bounded. The resulting system, assumes,

$$Ship1 \,\hat{=}\, [Ship \mid bounded(w) \land bounded(\tfrac{dw}{dt})]$$

2.2.2 Discrete Systems

For discrete systems, which include computing systems, state variables take values from discrete domains, and are observed only at an enumerable set of time instants, where state transitions happen. State transitions are denoted by a set of *Events*. A discrete model is thus a subset of functions from Time to State and finite sets of events ($\mathbf{F}\,Event$),

$$X : Time \to (State \times \mathbf{F}\,Event)$$

A discrete system is constrained by:

- Classifying states (unsafe, static-safe, dynamic-safe)
- Defining events
- Defining required functions by event sequences.

The safety requirements are to avoid unsafe states, and ensure that dynamic states have limited duration.

2.2.3 Auto Pilot continued

A discrete version of the auto pilot will sample y and v at discrete times ty and tv. Sampling is done periodically with period $p \pm ep$,

$$Sample \,\hat{=}\, [ty, tv : \mathbf{N} \to Time;$$
$$p, ep : \mathbf{R}_+ \mid periodic(ty, p, ep) \land periodic(tv, p, ep)]$$

Using sy (sensor) and av (actuator) to denote integer approximations to sampled values of y and v, we have,

$$Interface \,\hat{=}\, [Ship1;\ Sample;$$
$$sy, av, zy : \mathbf{N} \to \mathbf{Z} \mid approx(sy, y \circ ty) \land approx(av, v \circ tv)]$$

where $\circ$ denotes functional composition and the *AutoPilot* is specified by,

```
┌─ Autopilot ─────────────────────────────
│ Interface
│ ├─────────────────────────────────────
│ │ ∀i • av(i + 1) = av(i) − sy(i + 1)/p−
│ │ (sy(i + 1) − z(i + 1))/p ∧
│ │ ∀i • z(i + 1) = sy(i) ∧
│ │ av(0) = 0 ∧ sy(0) = 0
```

The events of this model are sensor readings, denoted $s.y$ and actuator updates, denoted $a.v$, or formally:

| $Event \triangleq \{s.y | y \in \mathbf{Z}\} \cup \{a.v | v \in \mathbf{Z}\}$

Using ordinary mathematics it is possible to prove that *Autopilot* implies *Bounded* ∧ *Correcting*, i.e. *Autopilot* is indeed a refinement of the requirements, and thus can be taken as a specification for the total system.

In practice, system requirements are not presented as one formula within a single mathematical theory because the equipment is constructed of *components* from different technologies (chemistry, mechanics, electronics, etc.) each with separate theoretical underpinnings. What we can expect, is a formulation of the requirements as a *conjunction* of local constraints on parts of the state and event space. Requirements Capture restates these in a mathematical *domain model* in order to specify overall safety and functional requirements precisely.

2.3 System Design

Requirements also delineate an overall *System design*: A selection of sensors, actuators, and control computers, and their interconnection. Components are modelled as systems, while interconnection is modelled by conjunction (corresponding to parallel composition of models, or communication links in the equipment). In embedded systems a single global *Time* for all components is accepted.

2.3.1 Trace models

When decomposing a system it may be convenient to use a slightly different formulation of models for discrete systems. The following description can be seen as a tentative, outline of a requirements capture language, RL, but much work remains to be done. This formulation specifies a system in terms of finite sequences of events, the *Traces*, tr, of the system. State values are specified by an initial state, s_0, and a final state, s', for each initial segment of a trace.

| $X : [tr : Event^*; s_0, s' : State]$

When finite traces are used to approximate infinite system behaviours the model must include some observation of the possible continuations of a trace. This is done by sets of events accepted as continuations *ready sets*, R [24]. Under reasonable assumptions, ready sets can be derived from the full set of traces [9, 4]. An approach to specifying timing properties by constraints on the duration of specific subtraces, i.e. segments of traces, is explored in a duration calculus [16].

2.3.2 Autopilot Design

A design for the Auto Pilot would specify a sensor component, with a trace consisting of $s.y$ events. I.e. the trace is a *prefix* of the regular expression $<s>^*$, where s denotes an arbitrary $s.y$ value. The current value corresponding to the *last* event of the trace approximates the continuous signal at time t'. And between two events (subtrace $<s, s'>$) the duration is constrained (|) such that the the duration of the state where the sensor is not ready to communicate a value ($s' \notin R$) is at most 0.1 seconds. The sensor has no discrete state.

```
┌─ Sensor ────────────────────────────────
│ Ship1; Event
│ tr : seq{s}
│ ├─────────────────────────────────────
│ │ tr prefix $<s>^*$ ∧
│ │ last(tr) = s.x ⇒ approx(s.x, y(t')) ∧
│ │ $<s, s'> | d(s' \notin R) < 0.1$
```

An *Actuator* is similar. The control program for the computer has a state given by variables sy, av, z. In the specification we use these to denote the final values of the preceding trace. I.e. the state values just before the *last* event was added to the trace. The trace is an alternating sequence of sensor (s) and actuator (a) events. The state values are given by the control algorithm, and durations are constrained by the periodicity.

```
┌─ Control ───────────────────────────────
│ Sensor; Actuator
│ tr : seq Event
│ $sy_0, av_0, z_0, sy, av, z, sy', av', z' : \mathbf{Z}$
│ ├─────────────────────────────────────
│ │ tr prefix $<s, a>^*$ ∧
│ │ last(tr) = a.v ⇒ (sy' = sy ∧ v = av' = av − sy/p−
│ │ (sy' − z')/p ∧ z' = z) ∧
│ │ last(tr) = s.y ⇒ (sy' = y ∧ av' = av ∧ z' = sy) ∧
│ │ $sy_0 = 0 \land av_0 = 0$ ∧
│ │ $<s, a> | d(a \notin R) < ep$ ∧
│ │ $<s, a, s'> | p − ep < d(s' \notin R) < p + ep$
```

The notation above can be seen as a tentative outline of a requirements capture language, RL, but much work remains to be done, especially on the formulation of timing constraints and on combinators for models.

3 Program Specification

The result of Requirements Capture has been: 1) Specification of safety requirements, S, which constrain values and duration of system states. 2) Specification of functional requirements, F, which define initial states, event sequences, and final states of the system. 3) Interface specifications, I, which constrain interface component states and events (these can probably be subdivided into safety and functional requirements).

Program requirements, P, are the requirements for the control components, and the criterion for system correctness is $(P \land I) \Rightarrow (F \land S)$. The component design above will be correct, when it is established that *Control* ∧ (*Sensor* ∧ *Actuator*) implies a trace specification corresponding to *Autopilot*.

The specification language should primarily allow other engineers to read, understand, and certify the specification. It should also allow transformations leading to a program, which **can be proven to be a refinement of the specification**. In development of the SL languages we assume that the following aspects of a computation are separated in the requirements:

Behaviour: The possible communication sequences.

Calculation: The relationship between state, input and output values.

Timing: Duration of program states.

The initial proposal for a specification language SL_0 gives a program specification, S, in two major parts: 1) The trace specification part *tracespec(S)* of S defines the behaviour by assertions, $TA(S)$, about the structure of traces over a finite set of logical channel identifiers (the alphabet α_S). 2) The state specification part *statespec(S)*, defines the effects of a single communication on the statespace (*states[S]*). It defines a unique initial state (σ_S) for a set $VAR(S)$ of declared state variables, and introduces the set $CA(S)$ of communication assertions.

The trace specification part for the Auto Pilot defines the alphabet to contain two channels s and a, and gives one trace assertion:

```
INPUT s : -128..127
OUTPUT a : -128..127

TRACE ON { s, a } IN PREF (s.a)*
```

where PREF (s.a)* is the set of all prefixes of the regular expression (s.a)* and . denotes concatenation. The state specification part defines three variables av, sy, and z, two of which are initialized to 0. There are two communication assertions, one for each of the channels s and a:

```
STATE  av : INT = 0
       sy: INT = 0
       z : INT

COM s?y ON {z,sy}:
       sy' = y AND z' = sy

COM a!v ON {av,z,sy}:
       av' = av - 2*sy + z AND sy' = sy AND v = av'
```

where the COM construct denotes the effect of a communication on a given state.

Duration assertions: These have not yet been incorporated in the program specification language. The relation between time units and physical time is a parameter to the specification, and so is the latency of channels and the scheduling discipline.

4 Program Development

As dictated by the layered approach of ProCoS, we are selecting a small number of gradually more ambitious programming languages ($PL_0, PL_1, \ldots, PL_N$), complete with abstract and operational semantics which are essential for the production of the **verified PL compiler**. The languages range from a minimum roughly corresponding to the Dijkstra/Gries nondeterministic language, to a maximum superset of occam-2. We have so far identified the first two layers, PL_0 and PL_1.

PL_0 is used for the initial pilot studies and therefore the level of ambition is such that we are confident of being able to demonstrate the correctness of its compiler. Its primary features are:

- Assignable variables
- A single data type (integer) with the standard operators
- Simple sequential control structures
- Interaction with the environment through an input and an output channel.

Our approach to the design process does not involve conventional programming; it is based on provably correct transformations from SL specifications to PL programs. The use of manual transformations probably means that proofs will not be formal in the strict logical sense; but they will be rigorous in the sense of conventional mathematical proofs. This approach stems from our desire to deliver correct systems. It is also based on the standard practice of providing (mechanised) compiler transformations (as embodied in compilers) for programs, as opposed to merely instituting a rigorous process of manually producing machine code by looking at the program!

With the present specification language, the work by Olderog and Rossig [4, 12] gives a solid foundation for developing transformation rules that allow the control structure of a program to be derived from trace assertions.

The transformation of an SL_0 specification S into a PL program P satisfying S is done in a top down manner. Each of the four semantic sections, system alphabet α_S, trace assertions $TA(S)$, variable declarations $VA(S)$ and communication assertions $CA(S)$ of the specification, contribute to the final program. The static properties α_S and $VAR(S)$ can be mapped directly onto corresponding PL declarations of channels and variables. Dynamic system behaviour is dealt with using the following transformation rules. For each channel $b \in \alpha_S$, we construct a boolean expression $guard(b)$, a communication process $comm(b)$ and a list of processes $trans(b)$ on the basis of the communication assertion ca_b. These three program parts are used to build up a guarded choice construct $gc(b)$ and a guarded alternative construct $ga(b)$. These are then put together in alternation or conditional processes. The appropriate control structure around these processes can be deduced from nondeterministic automata for the language $L[S]$ defined over α_S.

Although we do not envisage development of transformation tools within ProCoS(if we were to produce a tool to automate the process we would have to prove its correctness), the long term goal has to be the provision of automatic rather than manual transformations. We have identified a set of automatically transformable SL_0 transformations, ATS_0, which is a syntactic subset of SL_0 specifications. ATS_0 specifications contain **a single trace assertion and syntactically constrained regular expressions**. Experiments undertaken within the project indicate that term rewriting techniques may be useful in providing automated transformation tools in the future.

The transformation techniques above relate to a generalised PL. The current common denominator for all project partners is PL_0. However, two additional PL variants have been designed; PL_0^R [17] which enhances PL_0 with recursion and procedure calls (we aim to use this to bootstrap the compiler) and $PL_\|$ [26] which is PL_0 extended with parallelism and nondeterminism.

Refinement of sequential operations, and data types is considered a minor problem, as there is a considerable literature on the subject, c.f. [6, 27].

5 The compiler

A thorny issue in much formal development work is the possibility of introducing errors in the final program by using an unproven compiler. Compiler correctness is therefore one of the key aspects of ProCoS. Producing a verified compiler is not a trivial proposition and the majority of the project resources are dedicated to this activity. We distinguish two verification aspects, namely the correctness of the PL to ML transformation rules (*compiling verification*) and the correctness of the compiler program embodying these transformations (*compiler verification*).

The project is currently exploring two styles of compiling specifications:

- Inductive function definitions defining a mapping of abstract source language trees to abstract target language trees [23], and

- predicative specifications defining a relation between abstract source and target language trees [5].

The latter is an elegant idea which avoids the need to reconcile the semantics of PL and ML and is based on the notion of refinement which is imposed by the compiler on the source program. Refinement is modelled by a partial order $\sqsubseteq$ between PL programs, where $p \sqsubseteq q$ means that q is better (more deterministic) than p. All program constructors maintain this order when applied (they are monotonic).

The target machine language ML_0 is a subset of the transputer instruction set. In the refinement approach the machine code is interpreted by a program $IsfmT$ where s and f are the start and finish addresses for an ML_0 program in memory m (ROM) with data space T (RAM). RAM and ROM are disjoint. Correct compilation of a PL_0 program p is specified by a predicate $Cpsfm\Psi\Omega$ where Ψ is a symbol table mapping p-identifiers to RAM addresses and Ω denotes the as yet unallocated RAM space ($alloc_\Psi \cap \Omega = \emptyset$ and $alloc_\Psi \cup \Omega = RAM$). The machine code of p is therefore interpreted by $Isfm(alloc_\Psi + \Omega)$. The compilation is correct when the interpretation of the machine code with the appropriate translation between data spaces is a refinement of p, i.e.

$$Cpsfm\Psi\Omega \triangleq SEQ[\hat{\Psi}_\Omega, p] \sqsubseteq SEQ[Isfm(alloc_\Psi + \Omega), \hat{\Psi}_\Omega]$$

where $\hat{\Psi}_\Omega$ translates from machine state to program state and then forgets the machine state.

The correctly compiled source program will run on a transputer architecture under a trusted kernel which is also being developed by ProCoS. We expect that each processor will support a single, occam–like process. Programs resulting in several processes will be handled by a hardware configuration involving at least that number of processors.

The ProCoS project has demonstrated that such verified compilers are possible. [18, 17] show the proof of correctness of the compiling specification for languages PL_0 and PL_0^R.

6 Program dependability

Software validation is usually perceived as a mix of formal verification, static code analysis or testing and it is striving to establish correctness and dependability properties. The ProCoS approach guarantees that once the program has been derived is provably correct wrt the specification. To the extend that dependability requirements enter the specification, they are also provably satisfied. But clearly a requirements specification can only take into account some of a host of environmental factors which may affect program execution in situ. Therefore, a *computer system* with provably correct programs may still fail in some circumstances. Questions of the overall reliability can perhaps best be answered by techniques known from hardware fault estimation [19]. This is however, outside the scope of ProCoS.

It is particularly important to complement the ProCoS program derivation process with interface validation and testing (possibly by deriving test harnesses and data from the formal specification) and fault tolerance mechanisms (such as recovery blocks or shelf checking code). It is equally important to accurately state which properties have *not* been included in the specification so that reliance cannot justifiably be placed on such features.

7 ProCoS and the draft 00–55

The growing concerns about the dependability of computer controlled high integrity systems are reflected in the emergence of industry wide standards such as the HSE guidelines on the use of programmable electronic systems in safety related applications [20], the MOD interim draft standard 00-55 on requirements for the procurement of safety critical software in defence equipment [2] and the SafeIT standardisation framework [21] to name but a few. 00-55 in particular, has provided valuable insight in the ProCoS software development activity. Although here we present a summary of our views, a full ProCoS assessment of 00-55 can be found in [22]. That assessment, as well as the discussion here, solely relates to the draft published for comment in the spring of 1989. It may well be the case, that, as a result of the public consultation, the revised draft will redress existing problems.

00-55 has implications both for the organisation procedures of software production and the use of technology. Although we are mainly concerned with the latter, we would like to make some remarks on the former. The goal of organisational procedures is to effect an orderly and controllable software development process with the required degree of management visibility which forms the basis of approvals and budgeting. Such procedures must reflect the technology used whereas the responsibility of each development agent should be expressed and understood in technology independent terms. It is quite clear that a chaotic development process cannot produce any kind of dependable software, regardless of the technology employed. Sound organisational procedures are a prerequisite for developing high quality software. We feel that the draft 00-55 fails to address this matter adequately as it, for instance, requires the production of perhaps inconsistent and certainly superfluous documentation.

Our feeling towards the technology advocated by the standard is, overall, positive, as it embodies existing best practice and looks to the future via its incursion in the domain of formal methods. There is however, an unfortunate tendency to blur the distinction between electromechanical components of the equipment and software. Viewing a program as a physical object runs contrary to all attempts to use formal methods in specifying and developing software.

A central question to ProCoS is, whether a program developed using the approach presented in this paper would comply with the 00–55 requirements. We fear that the answer is no, mainly because of implementation restrictions such as the use of floating point arithmetic and multiprocessing. We do not see any reason why these practices have to be banned if formal reasoning can be used to establish their correctness as we propose to do in ProCoS. Another factor is that we do not envisage to perform static analysis on our programs since they are axiomatically correct wrt the specification. In this context, it seems unreasonable to blindly require the use of inappropriate techniques when they clearly serve no useful purpose. This is a case where the future is prescribed strictly in terms of the present. Our feeling is therefore that the standard is overly restrictive in some practical aspects and not flexible enough to allow technological progress to migrate into the development of safety critical software.

8 Summary

We have presented a ProCoS approach for developing provably correct safety critical software starting from system requirements. We show how to derive discrete systems requirements which are stated in a formal program specification based on the concepts of traces and states. This specification is then mapped via proven transformation rules onto a program which is compiled by a verified compiler onto a machine language supported on a transputer configuration via a trusted kernel. We have touched upon software development procedures and the views expressed in the draft 00-55 standard. We have presented arguments that provable correctness is a *currently* realistic proposition for compact embedded safety critical systems. The development is based on the traditional engineering use of mathematics to model systems (such as control theory). Our contribution lies in the study of discrete models, their compositionality and their link to program development.

Acknowledgements

Many of the ideas in this paper are the results of the work of other ProCoS project members. In particular, we wish to acknowledge the contribution of Tony Hoare, Zhou Chao Chen, He Jifeng, Ernst Rudiger Olderog and Stefan Rossig.

References

[1] D Bjørner, *A ProCoS Project Description*, ESPRIT BRA 3104, EATCS Bull., No 39, October 1989.

[2] Requirements for the procurement of safety critical software in defence equipment, *Interim Defence Standard 00-55*, Ministry of Defence, Directorate of Standardisation, Kentigern House, 65 Brown St., Glasgow G2 8EX, May 1989.

[3] Requirements for the analysis of safety critical hazards, *Interim Defence Standard 00-56/1 Draft* Directorate of Standardisation, Kentigern House, 65 Brown St., Glasgow G2 8EX, May 1989.

[4] E-R Olderog, *Nets, Terms and Formulas: Three Views of Concurrent Processes and Their Relationship*, Institut für Informatik und Praktische Mathematik, Christian-Albrechts-Universität, Kiel, W. Germany, December 1988.

[5] CAR Hoare, *Algebra and Refinement in Compiler Design*, Procs of BCS-FACS Workshop on Program Refinement, January 1990.

[6] C Morgan, *Programming from Specifications*, Prentice-Hall Int. Series in Comp. Sc., Prentice-Hall 1990.

[7] J Nordahl, *Requirements Specification for a Railway Level Crossing*, ProCoS Note ID/DTH JNO 2, Feb. 1990.

[8] J Nordahl, *Dependability in a Process Algebraic Framework*, ProCoS Rep. ID/DTH JNO 4, Apr. 1990.

[9] E-R Olderog, CAR Hoare, *Specification Oriented Semantics for Communicating Processes*, Acta Informatica, 23, 9-66 (1986).

[10] AP Ravn, *Control Program for an Autopilot: Requirements*, ProCoS Case Study 0, ProCoS Note ID/DTH APR 3, Nov. 1989.

[11] AW Roscoe, CAR Hoare, *The Laws of Occam Programming* Theoretical Comp. Sc. 60 (1988), 177-229.

[12] S Rošsig, *Transformation of SL_0 specifications into PL programs*, ProCoS Rep. OLD SR 1, Mar. 1990.

[13] JM Spivey, *The Z Notation, A reference manual*, Prentice-Hall Int. Series in Comp. Sc., Prentice-Hall 1989.

[14] V Stavridou, AP Ravn, *Specification and Development of Safety Critical Software*, ProCoS Rep. RHC VS 1, June 1990.

[15] EV Sørensen, AP Ravn, H Rischel, *Control Program for a Gas Burner: Part 1: Informal Requirements*, ProCoS Case Study 1, ProCoS Rep. ID/DTH EVS2, March 1990.

[16] CC Zhou, CAR Hoare, AP Ravn, *A duration calculus for real-time requirements in embedded software systems*, ProCoS Rep. OU ZCC 2, June 1990.

[17] J He, J Bowen, *Compiling Specification for ProCoS language PL_0^R*, ProCos Report No OU HJF 4, May 1990.

[18] M Franzle, H Langmaack, M Muller-Olm, *Development of Proven Correct Compilers in ProCoS*, 1990.

[19] BW Jonson, *Design and Analysis of Fault Tolerant Digital Systems*, Addison Wesley, 1989.

[20] *Programmable Electronic Systems in Safety Related Applications – General Technical Guidelines*, UK Health and Safety Executive, 1987.

[21] RE Bloomfield, J Brazendale, *SafeIT: A Framework for Safety Standards*, Department of Trade and Industry, Kingsgate House, 66/74 Victoria Street, London SW1E 6SW, May 1990.

[22] AP Ravn, V Stavridou, *Specification and Development of Safety Critical Software: An Assessment of MOD Draft Standard 00–55*, Proceedings of ICSE 12 Workshop on *Industrial Use of Formal Methods*, March 1990, submitted for further publication.

[23] M Franzle, *Compiling Specification for ProCoS Programming Language Level 0*, ProCoS Rep. Kiel MF 4, April 1990.

[24] CAR Hoare, *A calculus of total correctness for communicating processes*, Science of Computer Programming, 1, 44-72, 1981.

[25] CAR Hoare, *Let's Make Models*, Note, PRG, Oxford University, April 1990.

[26] KM Jensen, HH Lovengreen, *Definition of the ProCoS Programming Language $PL_{||}$*, ProCoS Report ID/DTH KMJ 8, June 1990.

[27] CB Jones, *Systematic Software Development using VDM*, Prentice-Hall Int. Series in Computer Science, 1986.

ASPECTS OF PROVING COMPILER CORRECTNESS

B. v. Karger

*Christian-Albrechts-Universität, Institut für Informatik, Preußerstraße 1-9,
D-2300 Kiel, FRG*

Abstract

The ProCoS project aims at constructing a development environment for safety critical software for embedded systems, containing a specification language, a parallel programming language, a transputer machine language, and appropriate tools. All parts of this system shall be provably correct. It will thus become possible to specify, write, compile and run programs whose correct behaviour will solely depend on mathematical arguments (and correct functioning of the underlying hardware). A most crucial part of the project is the construction of a provably correct compiler. The aim of this paper is to show how the introduction of an intermediate language has helped to reduce the complexity of this correctness proof.

Keywords: Assembly Language, Compilers, Computer Software, Software Tools, Verification

1 Introduction

The ProCoS project aims among other things at a provably correct compiler from a subset of occam 2 to the code of a transputer machine. The compiler is constructed in two stages: First it is written in the language of recursive function definitions or a similar language (the compiler specification) and then the recursive functions are translated into transputer code (the compiler implementation). This paper focusses on the correctness proof for the specification. Transition systems are used for defining the semantics of both source and target programs, but, unfortunately, different styles are used. A structured operational semantics (SOS) in the style of Plotkin [Plotkin 81] is natural for the source language. Unfortunately, a SOS is only useful for structured programs and not for the flat sequences of instructions we have in machine programs, nor is it possible to impose structure on arbitrary machine programs, because they may contain jumps. For this reason an instruction oriented operational semantics (IOS) has been chosen for the machine language. Another reason for this decision was the ProCoS connection to the safemos project, where transputer hardware is proven correct with respect to the Z specification of its instructions (it is relatively easy to derive an IOS from a Z specification).

Now, proving correctness of the specification means showing that translation into machine code does not change the semantics of a given source program. It would be much easier to do such a proof if the same kind of semantics definition were used for both languages. This difficulty could be removed if it were possible to introduce an intermediate (assembly) language, the semantics of which could be described both ways and then proving that the resulting semantic functions are equal. It is not, however, possible to do exactly that, because such a language would have to be structured in order to allow for a SOS and at the same time be sequential in order to allow for an IOS.

Therefore we use two intermediate languages AL (assembly language) and TAL (tree structured assembly language), that are almost identical. AL arises from TAL by coding the tree structure into a sequence and there is an obvious one-to-one correspondence between TAL programs and well-formed AL-programs. We then give a SOS for TAL, an IOS for AL and prove their equivalence.

The paper is organized as follows: First we report the definitions of the languages TAL and AL as used in the first phase of the ProCoS project (sections 2–5). In the next section we collect the definitions and results from the theory of block structured sequences we need for making precise the semantics definition for AL and for the equivalence proof. Section 7 introduces the notion of wellformedness and section 8 gives the compiling function. The next two sections deal with relating AL configurations to TAL configurations. The equivalence theorem is presented and proved in section 11. Finally, section 12 offers a short review and some conclusions.

Due to lack of space only a few typical proofs could be included. For more details, see [vK 1].

2 Syntax

The languages TAL and AL are assembly languages for a slight modifications of a subset of the inmos transputer instruction set (see for example the transputer manual [inmos 88] or [Farr 87] for a Z specification of a slightly bigger subset. We define the abstract syntax of TAL and of AL by

$$
\begin{array}{rcl}
\textit{Simple} & ::= & \text{ldc } \textit{Con} \mid \text{eqc } \textit{Con} \mid \text{adc } \textit{Con} \\
 & & \text{def } \textit{Lab} \mid \text{j } \textit{Lab} \mid \text{cj } \textit{Lab} \mid \\
 & & \text{ldl } \textit{Adr} \mid \text{stl } \textit{Adr} \mid \text{in} \mid \text{out} \mid \\
 & & \text{stopp} \mid \text{seterr} \mid \text{testerr} \mid \text{stoperr} \mid \\
 & & \text{rev} \mid \text{not} \mid \text{gt} \mid \text{add} \mid \text{mul} \mid \text{div} \mid \text{rem} \\
\textit{Simple}' & ::= & \text{exit} \mid \text{cexit} \mid \textit{Simple} \\
\textit{Ins}_{AL} & ::= & \textit{Simple}' \mid \text{begin} \mid \text{end} \mid \text{do} \mid \text{od} \\
\textit{Ins}_{Block} & ::= & BLOCK \mid LOOP \mid \textit{Simple} \\
\textit{Ins}_{Loop} & ::= & BLOCK \mid LOOP \mid \textit{Simple}' \\
BLOCK & ::= & \underline{\text{block}}\ \textit{Ins}_{Block}^* \\
LOOP & ::= & \underline{\text{loop}}\ \textit{Ins}_{Loop}^* \\
TAL & ::= & BLOCK \mid LOOP \\
AL & ::= & \textit{Ins}_{AL}^* \\
\textit{Con} & ::= & \textit{Bit32} \\
\textit{Lab} & ::= & \mathbb{N} \\
\textit{Adr} & ::= & \mathbb{N} \\
\textit{Bit32} & ::= & \{-2^{31}, \ldots, 2^{31} - 1\} \\
\mathbb{N} & ::= & \{0, 1, \ldots\}
\end{array}
$$

3 Transition Rules for *Simple*

In this section we report the part of the semantics that is shared by **AL** and **TAL**. We do this by defining a transition system $\xrightarrow{\lambda}_{Ins}$ that describes the effect of a simple instruction on the machine state. Because of the way $\xrightarrow{\lambda}_{Ins}$ is used in the semantics definitions both of **AL** and of **TAL**, most of its features will not be relevant for the equivalence proof. This is a desirable effect because it means that we can add many features (e.g. indirect addressing, new arithmetic instructions) to the languages without having to reconsider the proof. We use the following notations:

$$State := \{continue, stopped, abnormal\}$$
$$\Sigma := \{(wsp, A, B, C, err, state) \mid$$
$$wsp \in Wspace, A, B, C \in Bit32,$$
$$err \in \{0,1\}, state \in State\}$$

The components of an element $\sigma \in \Sigma$ are denoted by wsp_σ, A_σ, B_σ, C_σ, err_σ, and $state_\sigma$. If σ is clear from context we will drop the subscript.

Variation is denoted with square brackets and the (simultaneous) assignment operator $:=$. For example

$$\sigma[B := C, A := B] = (wsp_\sigma, B_\sigma, C_\sigma, C_\sigma, err_\sigma, state_\sigma).$$

The symbol $\bot$ represents an arbitrary element of *Bit32*. A transition rule in which $\bot$ occurs is really a schema for a large number of transition rules (namely $|Bit32|^n$ rules if there are n occurrences of $\bot$). Similarly $\bot_b$ represents an arbitrary element of $\{1,0\}$. We use the following abbreviations:

"push(x)" for "$A := x, B := A, C := B$".
"pop"   for "$A := B, B := C, C := \bot$".
"clear"   for "$A := \bot, B := \bot, C := \bot, err := \bot_b$".

We assume that the following arithmetic operations, errordetecting functions and predicates are already defined:

$$+_s, -_s, *_s, \div_s, |_s, +_u, -_u, *_u : Bit32 \times Bit32 \to Bit32$$

$error_{op} : Bit32 \times Bit32 \to \{0,1\}$ for each of the above op's

$$=_s, <_s, >_s, \leq_s, \geq_s, =_u, <_u, >_u, \leq_u, \geq_u : Bit32 \to Bit32$$

There is no need for us to know the exact definitions of these functions and predicates, because they are part of both languages and are not affected by the compilation.

The set of start configurations is $Simple' \times \Sigma$ and the set of terminal configurations is $\Sigma \cup \Sigma \times Lab$. Every transition goes in one step from a start configuration to a terminal configuration. Transitions are labelled with communications τ stands for the empty communication.

$$(\text{ldc } c, \sigma) \xrightarrow{\tau}_{Ins} \sigma[\text{push}(c)]$$
$$(\text{eqc } c, \sigma) \xrightarrow{\tau}_{Ins} \sigma[A := (A =_u c)]$$
$$(\text{ldl } a, \sigma) \xrightarrow{\tau}_{Ins} \begin{cases} \sigma[\text{push}(wsp(a))] \\ \quad \text{if } a \in \text{dom}(wsp) \\ \sigma[state := \text{abnormal}] \\ \quad \text{if } a \notin \text{dom}(wsp) \end{cases}$$
$$(\text{stl } a, \sigma) \xrightarrow{\tau}_{Ins} \begin{cases} \sigma[\text{pop}, wsp(a) := A] \\ \quad \text{if } a \in \text{dom}(wsp) \\ \sigma[state := \text{abnormal}] \\ \quad \text{if } a \notin \text{dom}(wsp) \end{cases}$$
$$(\text{cj } l, \sigma) \xrightarrow{\tau}_{Ins} \begin{cases} \sigma & \text{if } A \neq 0 \\ (\sigma, l) & \text{if } A = 0 \end{cases}$$
$$(\text{j } l, \sigma) \xrightarrow{\tau}_{Ins} (\sigma[\text{clear}], l)$$
$$(\text{def } l, \sigma) \xrightarrow{\tau}_{Ins} \sigma$$
$$(\text{gt}, \sigma) \xrightarrow{\tau}_{Ins} \sigma[\text{pop}, A := (B > A)]$$
$$(\text{not}, \sigma) \xrightarrow{\tau}_{Ins} \begin{cases} \sigma[A := 0] & \text{if } A \neq 0 \\ \sigma[A := 1] & \text{if } A = 0 \end{cases}$$
$$(\text{stopp}, \sigma) \xrightarrow{\tau}_{Ins} \sigma[state := \text{stopped}]$$
$$(\text{exit}, \sigma) \xrightarrow{\tau}_{Ins} (\sigma[\text{clear}], 0)$$
$$(\text{cexit}, \sigma) \xrightarrow{\tau}_{Ins} \begin{cases} \sigma & \text{if } A = 1 \\ (\sigma, 0) & \text{if } A \neq 1 \end{cases}$$
$$(\text{seterr}, \sigma) \xrightarrow{\tau}_{Ins} \sigma[err := 1]$$
$$(\text{testerr}, \sigma) \xrightarrow{\tau}_{Ins} \sigma[\text{push}(err), err := 0]$$
$$(\text{stoperr}, \sigma) \xrightarrow{\tau}_{Ins} \begin{cases} \sigma \\ \quad \text{if } err = 0 \\ \sigma[state := \text{stopped}] \\ \quad \text{if } err = 1 \end{cases}$$
$$(\text{adc } c, \sigma) \xrightarrow{\tau}_{Ins} \sigma[A := A +_s c,$$
$$err := (err \lor \text{error}_{+_s}(A, c))]$$
$$(\text{add}, \sigma) \xrightarrow{\tau}_{Ins} \sigma[\text{pop}, A := A +_s B,$$
$$err := (err \lor \text{error}_{+_s}(A, B))]$$
$$(\text{mul}, \sigma) \xrightarrow{\tau}_{Ins} \sigma[\text{pop}, A := A *_s B,$$
$$err := (err \lor \text{error}_{*_s}(A, B))]$$
$$(\text{sub}, \sigma) \xrightarrow{\tau}_{Ins} \sigma[\text{pop}, A := A -_s B,$$
$$err := (err \lor \text{error}_{-_s}(A, B))]$$
$$(\text{div}, \sigma) \xrightarrow{\tau}_{Ins} \sigma[\text{pop}, A := A \div_s B,$$
$$err := (err \lor \text{error}_{\div_s}(A, B))]$$
$$(\text{rem}, \sigma) \xrightarrow{\tau}_{Ins} \sigma[\text{pop}, A := A |_s B,$$
$$err := (err \lor \text{error}_{|_s}(A, B))]$$
$$(\text{in}, \sigma) \xrightarrow{?a}_{Ins} \begin{cases} \sigma[\text{clear}, wsp(C) := a] \\ \quad \text{if } A = wordsize \land \\ \quad C \in \text{dom}(wsp) \\ \sigma[state := \text{abnormal}] \\ \quad \text{if } A \neq wordsize \lor \\ \quad C \notin \text{dom}(wsp) \end{cases}$$
$$(\text{out}, \sigma) \xrightarrow{!a}_{Ins} \begin{cases} \sigma[\text{clear}] \\ \quad \text{if } A = wordsize \land \\ \quad C \in \text{dom}(wsp) \land \\ \quad wsp\{B\} = A \\ \sigma[state := \text{abnormal}] \\ \quad \text{if } A \neq wordsize \lor \\ \quad C \notin \text{dom}(wsp) \end{cases}$$

4 The Transition system for *AL*

An **AL** configuration consists of an instruction pointer iptr $\in \mathbf{N}$ and a machine state σ. A configuration (iptr, σ) is terminal iff $state_\sigma = $ stopped. None of the below rules may fire if the configuration is terminal (this is a meta rule).

A few things in this section are formally defined in later sections. These are

- *SWF* is the set of semi well structured **AL** programs (cf. section 7)
- $B(\pi, i)$ and $E(\pi, i)$ give the index of the beginning resp./ the end of the smallest block or loop of π that contains the i'th instruction (cf. section 6)
- dest(π, i, l) calculates the destination of a j l instruction that occurs at position i in the program π. (cf. section 6)

Elementary Instruction Rule

$$\frac{\pi(\text{iptr}) = i \ \land \ (i, \sigma) \xrightarrow{\lambda}_{Ins} \sigma'}{\pi \vdash (\text{iptr}, \sigma) \xrightarrow{\lambda}_{AL} (\text{iptr}+1, \sigma')}$$

Block Rules

$$\frac{\pi(\text{iptr}) = \text{begin}}{\pi \vdash (\text{iptr}, \sigma) \xrightarrow{\tau}_{AL} (\text{iptr}+1, \sigma)}$$

$$\frac{\pi(\text{iptr}) = \text{end}}{\pi \vdash (\text{iptr}, \sigma) \xrightarrow{\tau}_{AL} (\text{iptr}+1, \sigma)}$$

Loop Rules

$$\frac{\pi(\text{iptr}) = \text{do}}{\pi \vdash (\text{iptr}, \sigma) \xrightarrow{\tau}_{AL} (\text{iptr}+1, \sigma)}$$

$$\frac{\pi(\text{iptr}) = \text{od}}{\pi \vdash (\text{iptr}, \sigma) \xrightarrow{\tau}_{AL} (B(\pi, \text{iptr})+1, \sigma[clear])}$$

Exit Rule
$$\frac{\pi(\text{iptr}) = i \ \wedge \ (i, \ \sigma) \xrightarrow{\tau}_{Ins} (\sigma', \ 0)}{\pi \vdash (\text{iptr}, \ \sigma) \xrightarrow{\tau}_{AL} (E(\pi, \ \text{iptr}) + 1, \ \sigma')}$$

Forward Jump Rule
$$\frac{\pi(\text{iptr}) = i \ \wedge \ (i, \ \sigma) \xrightarrow{\tau}_{Ins} (\sigma', \ l) \ \wedge \ l > 0}{\pi \vdash (\text{iptr}, \ \sigma) \xrightarrow{\tau}_{AL} (\text{dest}(\pi, \ \text{iptr}, \ l) + 1, \ \sigma')},$$

Abortion Rule
$$\frac{\text{iptr} \notin \text{dom}(\pi) \ \vee \ state_\sigma = \text{abnormal}}{\pi \vdash (\text{iptr}, \ \sigma) \xrightarrow{\lambda}_{AL} (\text{iptr}', \ \sigma')}$$

The last rule requires some explanation. In safety critical systems, the programmer has to assume the responsibility for the correctness of his program and the compiler writer is responsible for the correct compilation of correct programs. There is no safe way to recover from an address error in midaction and it is the duty of the programmer and the compiler writer to prevent this from ever happening. They cannot blame the hardware designer for any consequences of such an error. We express this responsibility by allowing the machine to do anything whatsoever, once it has reached an abnormal state. Thus, the hardware designer is free in his decision how to handle such exceptions; any error handler (or none at all) will be correct with respect to the above abortion rule, but the programmer will not be allowed to rely on any properties of the error handler. (We are following here the philosophy of nondeterminism introduced in [Hoare 87]).

5 Transition Rules for TAL

An **TAL** configuration consists of an program fragment $s \in SWF$ (the part of the original program that has still to be processed) and a machine state σ. A configuration (s, σ) is terminal iff $state_\sigma = $ stopped. None of the below rules may fire if the configuration is terminal (this is a meta rule).

Sequence Rule
$$\frac{(i, \ \sigma) \xrightarrow{\lambda}_{Ins} \sigma'}{(i; s, \ \sigma) \xrightarrow{\lambda}_{TAL} (s, \ \sigma')}$$

Forward Jump Rule
$$\frac{(i, \ \sigma) \xrightarrow{\tau}_{Ins} (\sigma', \ l) \ \wedge \ \forall k \in \text{dom}(a) : a(k) \neq \text{def } l}{(i; a; \text{def } l; s, \ \sigma) \xrightarrow{\tau}_{TAL} (s, \ \sigma')}$$

Block Rule
$$\overline{(\underline{\text{block}} \ s; t, \ \sigma) \xrightarrow{\tau}_{TAL} (s; t, \ \sigma)}$$

Loop Rules
$$\frac{t(0) \neq \text{def } 0}{(\underline{\text{loop}} \ s; t, \ \sigma) \xrightarrow{\tau}_{TAL} (s; \underline{\text{loop}} \ s; \text{def } 0; t, \ \sigma)}$$

$$\frac{t(0) = (\text{def } 0)}{(\underline{\text{loop}} \ s; t, \ \sigma) \xrightarrow{\tau}_{TAL} (s; \underline{\text{loop}} \ s; t, \ \sigma[\text{clear}])}$$

Abortion Rules
$$\overline{(\epsilon, \ \sigma) \xrightarrow{\lambda}_{TAL} (\epsilon, \ \sigma[state := \text{abnormal}])}$$

$$\frac{state_\sigma = \text{abnormal}}{(s, \ \sigma) \xrightarrow{\lambda}_{TAL} (s', \ \sigma')}$$

6 Block Structured Sequences

6.1 Sequences

Finite sequences over a set M are represented as total functions from some interval $\{0, ..., n\} \subset \mathbf{N}$ to M, where $n \geq 1$ (we allow $n = -1$, because we have to allow the empty interval). The set of all finite sequences over a set M is denoted by M^*. For convenience, we identify the subset containing all sequences of length 1 with M itself. The empty sequence is denoted by ϵ. Concatenation of the sequences a and b is written as $a; b$ and

$a(n)$ denotes the nth element of the sequence a (counting from 0). The length of a sequence a is denoted by $|a|$. For example, if $a = 4; 7; 1; 1$ then $|a| = 4$, $\text{dom}(a) = \{0, 1, 2, 3\}$, and $a(1) = 7$. Furthermore we write $a[i..j]$ to denote the subsequence of a beginning at index i and ending at index j (including the borders). We abbreviate $a[0..j]$ by $a[..j]$ and $a[i..|a| - 1]$ by $a[i..]$. Thus, if $a = 1; 2; 3; 4; 5$, then $a[2..4] = a[2..] = 3; 4; 5$. If $i < 0$ or $j \geq |a|$, then $a[i..j]$ is undefined. If $i = j$, then $a[i..j] = a[i]$. If $i < j$, then $a[i] = \epsilon$ (the empty sequence).

6.2 Block Structure

Definition 1 (level) For $\pi \in Ins^*_{AL}$ and $i \in \mathbf{Z}$ let
$$\begin{aligned} level \ := \ & |\{j \in dom(\pi) \ | \ j \leq i \ \wedge \ \pi(j) \in \{\text{begin}, \text{do}\}\}| \\ & - \ |\{j \in dom(\pi) \ | \ j < i \ \wedge \ \pi(j) \in \{\text{end}, \text{od}\} \ \}| \end{aligned}$$

Note the asymmetry that arises from using $j \leq i$ in the first term and $j < i$ in the second one. It stems from the wish that the level of a begin instruction be equal to the level of the corresponding end instruction — and that means increasing the level <u>at</u> every begin (or do) but decreasing it <u>after</u> every end (or od).

Definition 2 (block structure) $\pi \in Ins^*_{AL} - \{\epsilon\}$ is called block structured iff

- $level(\pi, |\pi|) = 0$.
- $\pi(0) \in \{\text{begin}, \text{do}\} \ \wedge \ \pi(|\pi| - 1) \in \{\text{end}, \text{od}\}$.
- $\forall i \in \{1, ..., |\pi| - 2\} : level(\pi, i) = 1 \Rightarrow \pi(i) \notin \{\text{end}, \text{od}\}$.

Let BS be the set of all block structured elements of Ins^*_{AL} and let SBS be the smallest subset of Ins^*_{AL} closed under concatenation and containing BS, $Simple'$, and ϵ.

6.1 Properties of BS and SBS:

- *The decomposition of a semi block structured sequence into elements of $BS \cup Simple'$ is unique (blocks do not overlap).*
- *No proper prefix or postfix of a block structured sequence is block structured.*
- *If $\pi_1; \pi_2 \in SBS$, then $\pi_1 \in SBS$ iff $\pi_2 \in SBS$.*
- *Let $\pi \in Ins^*_{AL}$. Then $\pi \in SBS$ iff $\text{begin}; \pi; \text{end} \in BS$ iff $\text{do}; \pi; \text{od} \in BS$.*

6.3 Locator Functions

In order to formulate the transition rules for **AL**, we must be able to construct from an arbitrary position in a program the beginning and the end of the smallest block (or loop) surrounding it. We formally define such functions B and E as follows:

Definition 3 ($B(\pi, i)$ and $E(\pi, i)$)
Let $B(\pi, i) := max(M_B(\pi, i))$ and $E(\pi, i) := min(M_E(\pi, i))$, where
$$M_B(\pi, i) := \{j \in dom(\pi) \ | \ j \leq i \ \wedge \ level(\pi, j) \leq level(\pi, i) \ \wedge \ \pi(j) \in \{\text{begin}, \text{do}\}\}$$
$$M_E(\pi, i) := \{j \in dom(\pi) \ | \ j \geq i \ \wedge \ level(\pi, j) \leq level(\pi, i) \ \wedge \ \pi(j) \in \{\text{end}, \text{od}\}\}$$

If $M_B(\pi, i) = \emptyset$ resp. $M_E(\pi, i) = \emptyset$, then $B(\pi, i)$ resp. $E(\pi, i)$ is undefined. It is possible that one of these is defined while the other is not. However, this will not happen if $\pi \in BS$ (both B and E are always defined) or $\pi \in SBS$ (for a given position, B and E are either both defined or both undefined).

6.2 Properties of B and E: Let $0 \leq i \leq j \leq k < |\pi|$. Then:

- *If $\pi(i) \in \{\text{begin}, \text{do}\}$ and $\pi(j) \notin \{\text{begin}, \text{do}\}$, then $i = B(\pi, j) \Leftrightarrow \pi[i + 1..j - 1] \in SBS$*
- *If $\pi(i) \in \{\text{end}, \text{od}\}$ and $\pi(j) \notin \{\text{end}, \text{od}\}$, then $j = E(\pi, i) \Leftrightarrow \pi[i + 1..j - 1] \in SBS$*
- $(\pi, i) \in dom(B) \cap dom(E) \ \Rightarrow \ \pi[B(\pi, i)..E(\pi, i)] \in BS$.

6.3 Idempotency Lemma Let $H_1, H_2 \in \{B, E\}$.
Then $(\pi, i) \in dom(H_1) \ \Rightarrow \ H_1(\pi, H_2(\pi, i)) = H_1(\pi, i)$.

Next we define a (partial) function dest : $SBS \times \mathbf{N} \times \mathbf{N} \to \mathbf{N}$ that maps a program $\pi \in SBS$, a position i and a label l to the corresponding label definition:

Definition 4 ($dest(\pi, i)$)
Let $\pi \in SBS$, $i \in dom(\pi)$ and $l \in \mathbf{N}$ and let $\pi' := \mathtt{begin}; \pi; \mathtt{end}$. Then set
$$M(\pi, i, l) := \{j \in \mathbf{N} \mid i \leq j < |\pi| \wedge \pi(j) = \mathtt{def}\, l \wedge$$
$$E(\pi', i+1) = E(\pi', j+1)\}$$
and $dest(\pi, i, l) := min(M(\pi, i, l))$.

Note that $dest(\pi, i, l)$ is undefined if $M(\pi, i, l) = \emptyset$.

The above definition is designed to handle forward jumps. We could make it work for backward jumps as well by dropping the condition $i \leq j$ in the definition.

6.4 Properties of dest: Let $\pi \in SBS$ and $i \in dom(\pi)$ with $\pi(i) \in Simple'$. Then:
(dest-1) $(\pi, i, l) \in dom(dest)$ $\Leftrightarrow$
          $dest(\pi; \psi, i, l) = dest(\pi, i, l)$ $\Leftrightarrow$
          $dest(\pi; \psi, i, l) < |\pi|$.
(dest-2) $dest(\phi; \pi, i + |\phi|, l) = dest(\pi, i, l) + |\phi|$.
(dest-3)   If $i < j < |\pi|$ and $l \in \mathbf{N}$ such that $\pi(j) = \mathtt{def}\, l$ and $\pi[i..j] \in SBS$, then $dest(\pi, i, l) \leq j$.
(dest-4) $(\pi, i, l) \in dom(dest) \Rightarrow \pi[i..dest(\pi, i, l)] \in SBS$.

7 Wellformedness

Definition 5 $\pi \in Ins_{AL}^*$ is called well-formed iff:
(wf-1) $\pi \in BS$.
(wf-2) $\forall i \in dom(\pi) : \pi(i) \in \{\mathtt{cexit}, \mathtt{exit}\} \wedge (\pi, i) \in dom(B) \Rightarrow \pi(B(\pi, i)) = \mathtt{do}$.
(wf-3) $\forall i \in dom(\pi) : \pi(i) = \mathtt{begin} \Rightarrow \pi(E(\pi, i)) = \mathtt{end}$.
(wf-4) $\forall i \in dom(\pi) : \pi(i) = \mathtt{do} \Rightarrow \pi(E(\pi, i)) = \mathtt{od}$.
π is called semi-well-formed iff $\pi \in SBS$ and (wf-2) - (wf-4) hold. The set of all well-formed (resp. semi-well-formed) elements of Ins_{AL}^* is called WF (resp. SWF).

For $\pi \in BS$ the proviso "$(\pi, i) \in dom(B)$" in (wf-2) is redundant, but not for $\pi \in SBS$. In a semi-well-formed program there may be exit instructions that are not part of a loop, provided they are at the outermost level. This situation must be permitted, because it arises, when the body of a loop is transferred out of the loop as a consequence of one of the TAL Loop Rules.

7.1 Characterization of SWF
SWF is the smallest subset of Ins_{AL}^* that is closed under concatenation and contains WF, $Simple'$ and ϵ. Every element of SWF can be decomposed uniquely into elements of $WF \cup Simple'$.

7.2 Subprogram Lemma: Let $\pi \in SWF$, $0 \leq i \leq j < |\pi|$ and $\pi' = \pi[i..j]$. Then:
(sub-1) $\pi' \in WF \Leftrightarrow \pi' \in BS$.
(sub-2) $\pi' \in SWF \Leftrightarrow \pi' \in SBS$.

7.3 Splitting Lemma: Let $\pi_1; \pi_2 \in SWF$. Then $\pi_1 \in SWF$ iff $\pi_2 \in SWF$.

Definition 6 $\pi \in SBS$ is called exit free if there are no exit or cexit instructions at the outermost level of π. This property can be formally expressed by saying that $\pi(i) \in \{\mathtt{cexit}, \mathtt{exit}\}$ implies $(\pi, i) \in dom(B)$ for all $i \in dom(\pi)$.

7.4 Enclosure Lemma: Let $\pi \in Ins_{AL}^*$. Then we have:
$\mathtt{do}; \pi; \mathtt{od} \in WF$ $\Leftrightarrow$ $\pi \in SWF$.
$\mathtt{begin}; \pi; \mathtt{end} \in WF$ $\Leftrightarrow$ $\pi \in SWF \wedge \pi$ is exit free.

8 Translating TAL into AL

Definition 7   Define a map $\mathcal{C} : Ins_{Loop}^* \to Ins_{AL}^*$ by
$\mathcal{C}(\epsilon) \quad := \quad \epsilon$
$\mathcal{C}(i) \quad := \quad i \quad (\text{for } i \in Simple')$
$\mathcal{C}(\underline{\mathtt{block}}\, s) \quad := \quad \mathtt{begin}; \mathcal{C}(s); \mathtt{end}$
$\mathcal{C}(\underline{\mathtt{loop}}\, s) \quad := \quad \mathtt{do}; \mathcal{C}(s); \mathtt{od}$
$\mathcal{C}(s; t) \quad := \quad \mathcal{C}(s); \mathcal{C}(t)$

Obviously, $\mathcal{C}$ is a function. We show

Theorem 1 (Syntactical Equivalence)
$\mathcal{C}$ is a bijective mapping from Ins_{Loop}^* onto SWF. The restriction of $\mathcal{C}$ to $\mathbf{TAL}$ is a bijection onto WF.

Proof: We start by showing that $\mathcal{C}(Ins_{Loop}^*) \subseteq SWF$ and $\mathcal{C}(\mathbf{TAL}) \subseteq WF$. Thus, let $\pi \in Ins_{Loop}^*$. We proceed by induction on the structure of π. If $|\pi| = 0$, then $\mathcal{C}(\pi) = \mathcal{C}(\epsilon) \in SWF$ by definition. If $|\pi| > 1$ we can write $\pi = \pi_1; \pi_2$ with $|\pi_i| < |\pi|$ for $i = 1, 2$. By induction hypothesis, $\mathcal{C}(\pi_i) \in SWF$ for $i = 1, 2$. Since SWF is closed under concatenation, we conclude $\mathcal{C}(\pi) = \mathcal{C}(\pi_1); \mathcal{C}(\pi_2) \in SWF$. So we can assume $|\pi| = 1$. If $\pi \in Simple'$ then $\mathcal{C}(\pi) \in SWF$ by definition of SWF. Two cases remain:

Case 1: $\pi = \underline{\mathtt{loop}}\, s$ with $s \in Ins_{Loop}^*$
Then $\mathcal{C}(\pi) = \mathtt{do}; \mathcal{C}(s); \mathtt{od}$ and $\mathcal{C}(s) \in SWF$ by induction hypothesis. Hence $\mathcal{C}(\pi) \in WF$ by the Enclosure Lemma.

Case 2: $\pi = \underline{\mathtt{block}}\, s$ with $s \in Ins_{Block}^*$
Then $\mathcal{C}(\pi) = \mathtt{begin}; \mathcal{C}(s); \mathtt{end}$ and $\mathcal{C}(s) \in SWF$ by induction hypothesis. Since $s \in Ins_{Block}^*$, there are no (c)exits in s, hence $\mathcal{C}(s)$ is exit free. Hence $\mathcal{C}(\pi) \in WF$ by the Enclosure Lemma.

It remains to show the following surjectivity conditions:
(1) $\forall \pi \in SWF: \exists \pi' \in Ins_{Loop}^*: \mathcal{C}(\pi') = \pi$.
(2) $\forall \pi \in WF: \exists \pi' \in \mathbf{TAL}: \mathcal{C}(\pi') = \pi$.
To show (1) let $\pi \in SWF$. If $|\pi| \leq 1$, then $\pi = \epsilon$ or $\pi \in Simple'$ and we can take $\pi' = \pi$. If $\pi = \pi_1; \pi_2$ with $\pi_1, \pi_2 \in SWF - \{\epsilon\}$ we obtain π_1' and π_2' by induction and take $\pi' = \pi_1'; \pi_2'$. The only remaining case is where $\pi \in WF$. Then either $\pi = \mathtt{begin}; \psi; \mathtt{end}$ or $\pi = \mathtt{do}; \psi; \mathtt{od}$ for some $\psi \in SWF$. Again we obtain a suitable ψ' from the induction hypothesis and we take $\pi' = \underline{\mathtt{block}}\, \psi'$ resp. $\pi' = \underline{\mathtt{loop}}\, \psi'$. Since this π' lies in $\mathbf{TAL}$, the argumentation for the $\pi \in \overline{WF}$ case proves (2) as well. $\square$

9 Computing the Program Remainder

In order to relate **TAL** configurations to **AL** configurations, it is necessary to determine from an **AL** program π and an instruction pointer i the corresponding program remainder that appears in the first component of the corresponding **TAL** configuration. To achieve this, we will first define a function $\mathcal{D}$ that computes the program remainder within SWF, and then translate the result with $\mathcal{C}^{-1}$ to Ins_{Loop}^*.

Definition 8   We define a function $\mathcal{D} : SWF \times \mathbf{N} \to SWF$. If $(\pi, i) \in SWF \times \mathbf{N}$, let $b = B(\pi, i)$, $e = E(\pi, i)$ and set

$$\mathcal{D}(\pi, i) := \begin{cases} \epsilon & \text{if } i \notin dom(\pi) \\ \pi(i); \mathcal{D}(\pi, i+1) & \text{if } \pi(i) \in Simple' \\ \pi[b..e]; \mathcal{D}(\pi, e+1) & \text{if } i = b \\ \mathcal{D}(\pi, i+1) & \text{if } \pi(i) = \mathtt{end} \\ \pi[b..e]; \mathtt{def}\, 0; \mathcal{D}(\pi, i+1) & \text{if } \pi(i) = \mathtt{od} \end{cases}$$

9.1 Lemma: $\mathcal{D}$ is a well-defined total function from $SWF \times \mathbf{N}$ to SWF.

Definition 9   Define $\mathcal{E} : SWF \times \mathbf{N} \to Ins_{Loop}^*$ by $\mathcal{E}(\pi, i) := \mathcal{C}^{-1}(\mathcal{D}(\pi, i))$.

Taking Theorem 1 together with the preceding lemma, we obtain

9.2 Corollary: $\mathcal{E}$ is a total function from $SWF \times \mathbf{N}$ to Ins_{Loop}^*.

10 Moving the instruction pointer

10.1 Induction Lemma
If $\pi = a; b; c \in SWF$ with $b \in SWF$ then $\mathcal{D}(\pi, |a|) = b; \mathcal{D}(\pi, |a; b|)$.

Proof: We use induction on $|b|$. If $b = \epsilon$, there is nothing to prove. So assume $|b| > 0$.

Case 1: $\pi(|a|) \in \textit{Simple}'$
By definition of $\mathcal{D}$ we have:
(1) $\mathcal{D}(\pi, |a|) = \pi(|a|); \mathcal{D}(\pi, |a|+1)$.
Let $a' := a; b(0)$ and $b' := b[1..]$. Obviously, $b' \in SBS$ and $a; b = a'; b'$. The induction hypothesis yields

$$\mathcal{D}(\pi, |a|+1) = \mathcal{D}(\pi, |a'|) = b'; \mathcal{D}(\pi, |a'; b'|) = b'; \mathcal{D}(\pi, |a; b|)$$

Since $\pi(|a|) = b(0)$ the desired conclusion follows from (1).

Case 2: $\pi(|a|) \in \{\text{begin}, \text{do}\}$
Let $\psi := \pi[|a|..E(\pi, |a|)] = \pi[B(\pi, |a|)..E(\pi, |a|)]$. By definition of $\mathcal{D}$,
(2) $\mathcal{D}(\pi, |a|) = \psi; \mathcal{D}(\pi, |a; \psi|)$.
Since $b \in SWF$, ψ is a prefix of b. Define $a', b' \in \textit{Ins}_{AL}^*$ by $a' := a; \psi$ and $b = \psi; b'$. Then $a; b = a'; b'$ and we obtain the desired result by applying the induction hypothesis to $\pi = a'; b'; c$.

Because of $b \in SBS$ and $\pi(|a|) = b(0)$ we cannot have $\pi(a) \in \{\text{end}, \text{od}\}$. Hence the above case distinction is complete. □

The following result is the key to the proof of the Correctness Theorem:

10.2 Movement Lemma: *Let* $\pi \in WF$ *such that* $\pi(j) \neq \text{def } 0$ *for all* $j \in dom(\pi)$ *and let* $i \in dom(\pi)$. *Then we have:*
(move-1) $\pi(i) \in \textit{Simple}' \Rightarrow$
     $\mathcal{D}(\pi, i) = \pi(i); \mathcal{D}(\pi, i+1)$.
(move-2) $\pi(i) = \text{begin} \Rightarrow \exists a, b \in SWF:$
     $\mathcal{D}(\pi, i) = \text{begin}; a; \text{end}; b$
     $\wedge \; \mathcal{D}(\pi, i+1) = a; b$.
(move-3) $\pi(i) = \text{end} \Rightarrow \mathcal{D}(\pi, i+1) = \mathcal{D}(\pi, i)$.
(move-4) $\pi(i) = \text{do} \Rightarrow \exists a, b \in SWF:$
     $\mathcal{D}(\pi, i) = \text{do}; a; \text{od}; b$
     $\wedge \; \mathcal{D}(\pi, i+1) = a; \text{do}; a; \text{od}; \text{def } 0; b$
     $\wedge \; b(0) \neq \text{def } 0$.
(move-5) $\pi(i) = \text{od} \Rightarrow \exists a, b \in SWF:$
     $\mathcal{D}(\pi, i) = \text{do}; a; \text{od}; \text{def } 0; b$
     $\wedge \; \mathcal{D}(\pi, B(\pi, i)+1) = a; \mathcal{D}(\pi, i)$.
(move-6) $\pi(i) \in \{\text{j } l, \text{cj } l\} \wedge (\pi, i, l) \in dom(dest) \Rightarrow$
     $\mathcal{D}(\pi, i) = \pi[i..d]; \mathcal{D}(\pi, d+1)$,
     where $d = dest(\pi, i, l)$.
(move-7) $\pi(i) \in \{\text{cexit}, \text{exit}\} \Rightarrow \exists a \in SWF:$
     $\mathcal{D}(\pi, i) = \pi(i); a; \text{def } 0; \mathcal{D}(\pi, E(\pi, i)+1)$
     $\wedge \; dest(\mathcal{D}(\pi, i), 0, 0) = |a|+1$.

(move-1) and (move-3) are obvious from the definition of $\mathcal{D}$.

Proof of (move-2): Let $j := E(\pi, i)$. Then $a' := \pi[i..j] \in BS$ by the properties of E. The wellformedness condition (wf-3) implies $\pi(j) = \text{end}$. Let $a := \pi[i+1..j-1]$. The Subprogram Lemma shows that $a' \in WF$, whence $a \in SWF$ by the Enclosure Lemma. Letting $b := \mathcal{D}(\pi, j+1)$, we have $\mathcal{D}(\pi, i) = a'; b = \text{begin}; a; \text{end}; b$ by definition of $\mathcal{D}$. Applying the Induction Lemma to the decomposition $\pi = \pi[..i]; a; \pi[j..]$, we obtain $\mathcal{D}(\pi, i+1) = a; \mathcal{D}(\pi, j)$. Since $\pi(j) = \text{end}$, we have $\mathcal{D}(\pi, j) = \mathcal{D}(\pi, j+1) = b$. (move-2) is now established. □

Proof of (move-5): Let $h := B(\pi, i)$. (wf-3) implies $\pi(h) = \text{do}$. With $a := \pi[h+1..i-1]$ and $b := \mathcal{D}(\pi, i+1)$ we get $\mathcal{D}(\pi, i) = \text{do}; a; \text{od}; \text{def } 0; b$ from the definition of $\mathcal{D}$. As in the proof of (move-2), the Induction Lemma can be used to show that $\mathcal{D}(\pi, h+1)$ has the desired form. □

Proof of (move-4): Let $j := E(\pi, i)$. (wf-4) implies $\pi(j) = \text{od}$. With $a := \pi[i+1..j-1]$ and $b := \mathcal{D}(\pi, j+1)$ we get $\mathcal{D}(\pi, i) = \text{do}; a; \text{od}; b$ from the definition of $\mathcal{D}$. Since $a \in SWF$, the Induction Lemma implies, as in the proof of (move-2), $\mathcal{D}(\pi, i+1) = a; \mathcal{D}(\pi, j)$. Because of $\pi(j) = \text{od}$, (move-5) yields $\mathcal{D}(\pi, j) = \text{do}; a'; \text{od}; \text{def } 0; b'$ and $\mathcal{D}(\pi, B(\pi, j)+1) = a'; \mathcal{D}(\pi, j)$. But $B(\pi, j)+1 = i+1$ and we conclude $a = a'$. A look at the proof of (move-5) reveals that also $b = b' = \mathcal{D}(\pi, j+1)$. Checking the definition of $\mathcal{D}$, we see that $b(0) \neq \text{def } 0$ and (move-4) is established. □

Proof of (move-6): We have $a := \pi[i..dest(\pi, i, l)] \in SWF$ by (dest-4). Hence (move-6) follows from the Induction Lemma. □

Proof of (move-7): Set $b := \pi[i..E(\pi, i)-1]$. Then $b \in SBS$ by the properties of E, and the Induction Lemma yields:
(2) $\mathcal{D}(\pi, i) = b; \mathcal{D}(\pi, E(\pi, i))$.
We infer from $\pi(i) \in \{\text{cexit}, \text{exit}\}$ and (wf-2) that $\pi(B(\pi, i)) = \text{do}$. The Idempotency Lemma and (wf-4) imply: $\pi(E(\pi, i)) = \pi(E(\pi, B(\pi, i))) = \text{od}$. Looking up the definition of $\mathcal{D}$ and using the Idempotency Lemma, we get:
(3) $\mathcal{D}(\pi, E(\pi, i)) = \pi[B(\pi, i)..E(\pi, i)]; \text{def } 0; \mathcal{D}(\pi, E(\pi, i)+1)$.
Let $b' := b[1..]$ and $a := b'; \pi[B(\pi, i)..E(\pi, i)]$. Then we have $\pi(i); a = b; \pi[B(\pi, i)..E(\pi, i)]$ and we can rewrite (3) as:
(4) $b; \mathcal{D}(\pi, E(\pi, i)) = \pi(i); a; \text{def } 0; \mathcal{D}(\pi, E(\pi, i)+1)$
Since $b \in SWF$ and $b(0) \in SWF$, we have $b' \in SWF$, whence $a \in SWF$ by the properties of E. Taking (2) and (4) together, we have:
(5) $\mathcal{D}(\pi, i) = \pi(i); a; \text{def } 0; \mathcal{D}(\pi, E(\pi, i)+1)$.
Since a consists of substrings of π, we have $a(j) \neq \text{def } 0$ for all $j \in dom(a)$ by hypothesis. We conclude $dest(\pi(i); a; \text{def } 0, 0, 0) = |a|+1$. Using (5) and (dest-1), we obtain $dest(\mathcal{D}(\pi, i), 0, 0) = |a|+1$. The proof of (move-7) is now complete. □

11 The Equivalence Theorem

Theorem 2 (Semantic Equivalence)
Let $\pi \in WF$ *such that* $\pi(j) \neq \text{def } 0$ *for all* $j \in dom(\pi)$. *Suppose* $\pi \vdash (i, \sigma) \xrightarrow{\lambda}_{AL} (i', \sigma')$. *Then we have:*
$(\mathcal{E}(\pi, i), \sigma) = (\mathcal{E}(\pi, i'), \sigma')$ *or* $(\mathcal{E}(\pi, i), \sigma) \xrightarrow{\lambda}_{TAL} (\mathcal{E}(\pi, i'), \sigma')$.

Proof (Sketch): Given the Movement Lemma, the proof of the Equivalence Theorem is quite straightforward. The overall strategy is as follows: Suppose we have some transition
(1) $\pi \vdash (i, \sigma) \xrightarrow{\lambda}_{AL} (i', \sigma')$.
Then there must be an **AL** rule (R) that justifies it. Every rule in question gives rise to a distinct case. In each case, we proceed along the following steps:

1. Use (R) to relate i to i' and σ to σ'.

2. Use the Movement Lemma to translate the relation between i and i' into a relation between $\mathcal{D}(\pi, i)$ and $\mathcal{D}(\pi, i')$.

3. Retranslate with $\mathcal{C}^{-1}$ to get a relation between $\mathcal{E}(\pi, i)$ and $\mathcal{E}(\pi, i')$.

4. Check for equality or else a **TAL** rule that allows the transition from $\mathcal{E}(\pi, i)$ to $\mathcal{E}(\pi, i')$.

We will exemplify this method by proving theorem 2 by carrying out the analysis for the First Loop Rule and for the Forward Jump Rule.

Case 1: First Loop Rule
If (1) is allowed by the First Loop Rule, then its precondition and conclusion must hold. The precondition yields $\pi(i) = \text{do}$ and the conclusion forces $\sigma = \sigma'$, $i' = i+1$, and $\lambda = \tau$. With (move-4) we get $a, b \in SWF$ such that $\mathcal{D}(\pi, i) = \text{do}; a; \text{od}; b$ and $\mathcal{D}(\pi, i') = a; \text{do}; a; \text{od}; \text{def } 0; b$. Translating with $\mathcal{C}^{-1}$ gives:
$$\mathcal{E}(\pi, i) = \underline{\text{loop }} \mathcal{C}^{-1}(a); \mathcal{C}^{-1}(b)$$
and
$$\mathcal{E}(\pi, i') = \mathcal{C}^{-1}(a); \underline{\text{loop }} \mathcal{C}^{-1}(a); \text{def } 0; \mathcal{C}^{-1}(b).$$
Moreover $b(0) \neq \text{def } 0$ by (move-4), whence also $\mathcal{C}^{-1}(b)(0) \neq \text{def } 0$. **TAL**'s First Loop Rule gives now the desired conclusion:
$$(\mathcal{E}(\pi, i), \sigma) \xrightarrow{\lambda}_{TAL} (\mathcal{E}(\pi, i'), \sigma).$$

The analysis is similar for the Elementary Instruction Rule, the Block Rules, the Second Loop Rule, and the Abortion Rule. However, the cases involving jumps are a bit more tricky. Essentially, we need to know that application of $\mathcal{C}$ does not hide any label declarations. This is guaranteed by the following little lemma:

11.1 *Let* $a, b \in SWF$ *and* $l \in \mathbb{N}$ *such that* $dest(a; b, 0, l) = |a|$. *Then* $\mathcal{C}^{-1}(a)(j) \neq \text{def } l$ *for all* $j \in dom(\mathcal{C}^{-1}(a))$.

Proof: Let $x := C^{-1}(a)$, $j \in \text{dom}(x)$ and assume $x(j) = \text{def } l$. Then
$$\begin{aligned} a &= C(x) = C(x[..j-1]); C(\text{def } l); C(x[j+1..]) \\ &= C(x[..j-1]); \text{def } l; C(x[j+1..]) \end{aligned}$$

By Theorem 1, $C(x[..j-1]) \in SWF$. Now (dest-1) and (dest-3) give $\text{dest}(a; b, 0, l) \leq |C(x[..j-1])| < |a|$, contradicting the assumption. □

Case 6: Forward Jump Rule

The precondition of the Forward Jump Rule yields
$$(\pi(i), \sigma) \xrightarrow{\tau}_{Ins} (\sigma', l)$$
and $l > 0$. Its conclusion implies $i' = \text{dest}(\pi, i, l) + 1$. Checking through the transition rules for $\rightarrow_{Ins}$, we find that $\pi(j) \in \{j\,l, cj\,l\}$. Let $a = \pi[i..\text{dest}(\pi, i, l)]$ and $a' = a[..|a|-2]$. Then $a = a'; \text{def } l$ and (move-6) yields
$$D(\pi, i) = a'; \text{def } l; D(\pi, i')$$
Translating with C^{-1}, gives
$$\mathcal{E}(\pi, i) = C^{-1}(a'); \text{def } l; \mathcal{E}(\pi, i').$$
Since we have
$$\begin{aligned} & \text{dest}(a'; \text{def } l, 0, l) \\ (\text{dest-2}) =\ & \text{dest}(\pi[..\text{dest}(\pi, i, l)], i, l) - i \\ (\text{dest-1}) =\ & \text{dest}(\pi, i, l) - i \\ \text{def. of } a =\ & |a| - 1 \\ =\ & |a'|, \end{aligned}$$
The above lemma yields $C^{-1}(a')(j) \neq \text{def } l$ for all $j \in \text{dom}(C^{-1}(a'))$. Hence application of **TAL**'s Forward Jump Rule gives
$$(\mathcal{E}(\pi, i), \sigma) \xrightarrow{\tau}_{TAL} (\mathcal{E}(\pi, i'), \sigma').$$

The analysis for the Exit Rule is similar. □

Strictly speaking, Theorem 2 does not really say that the languages **AL** and **TAL** are equivalent. It only claims that, for $\pi \in$ **TAL**, any behaviour exhibited by $C(\pi)$ can also be exhibited by π itself. The compilation C is therefore correct in the sense that the compiled program $C(\pi)$ satisfies any specification satisfied by π. However, Theorem 2 does not exclude the possibility that $C(\pi)$ might be more deterministic than π. Actually, this is not the case. To prove this assertion, we would have to show the following

Theorem 3 *Let $\pi \in WF$ such that $\pi(j) \neq \text{def } 0$ for all $j \in \text{dom}(\pi)$ and let $i \in \mathbb{N}$. Suppose $(\mathcal{E}(\pi, i), \sigma) \xrightarrow{\lambda}_{TAL} (\pi', \sigma')$. Then there exist $i', n \in \mathbb{N}$ and $k \in \{1, \ldots, n\}$ such that*
$$(i, \sigma) \xrightarrow{\lambda_1}_{AL} \ldots \xrightarrow{\lambda_n}_{AL} (i', \sigma'),$$
where $E(\pi, i') = \pi'$, $\lambda_k = \lambda$ and $\lambda_j = \tau$ for all $j \neq k$.

This theorem can be proved in a similar way. There is one case now for every **TAL** rule. To do them, a kind of "Reverse Movement Lemma" is required: Instead of determining $\mathcal{D}(\pi, i')$ from $\mathcal{D}(\pi, i)$ and i', we would have to calculate i' from $\mathcal{D}(\pi, i)$ and $\mathcal{D}(\pi, i')$. Checking through the **TAL** rules, one sees that is not difficult.

12 Conclusions

We have demonstrated how the structural operational semantics can be related to a more instruction oriented operational semantics. An equivalence proof has been given for a simple assembly language containing blocks, loops and (possibly conditional) jump instructions.

We have seen that the equivalence proof consists of the following parts:

1) a general theory of block structured sequences

2) a proof of the syntactical equivalence (bijectivity of the compilation function)

3) for each transition rule R of **AL**, a (short) proof that the effect R has on an **AL** state can be simulated by the effect of zero or more **TAL** rules on the corresponding **TAL** state.

This structure suggests that it will be very easy to add new constructs to the languages and extend the equivalence proof:

- Because of the modularity 3), no changes to the old parts of the proof are necessary.
- Any new "simple" instructions (an instruction is simple if it does not change the flow of control) does not even affect the proof, because it generates only a transition rule for $\rightarrow_{Ins}$, but not for $\rightarrow_{AL}$.
- The general theory 1) is already there and may be applied to any new language constructs.

Moreover, the case analysis 3) tends to use similar arguments over and again. It should therefore be possible to apply automatic theorem proving. At Kiel, Soo Woo Lee is currently working on redoing the proof with the aid of the Boyer-Moore prover. Note that we do not use mechanical proof systems in order to arrive at proofs more quickly or with less effort. Our primary motive is the increased confidence that can be placed into a proof that has been done by hand and by machine. Meanwhile, ProCoS is moving to the next level of ambition and correctness proofs have to be done for richer language. For the assembly languages this means inclusion of indirect addressing, parallel processes, (possibly recursive) procedure calls and non-deterministic choice of guarded alternatives. These languages have been defined both in **AL** and in **TAL** style [vK 2], [vK 3]. The equivalence proof looks feasible, but has not yet been done.

References

[Farr 87] J. R. Farr *A Formal Specification of the Transputer Instruction Set* M.Sc.thesis, Progr. Research group, Oxford University, UK, 1987

[Fränzle 89] M. Fränzle, B. von Karger, Y. Lakhneche, *Compiling Specification and Verification* ProCoS Report, 1989

[Hoare 87] C.A.R. Hoare, I.J. Hayes, He Jifeng, C. C. Morgan, A.W. Roscoe, J.W. Sanders, I.H. Sorensen, J.M. Spivey, B.A. Sufrin *Laws of Programming* Comm./ of the ACM, 87, Vol 30, Number 8

[inmos 88] inmos limited *The Transputer Instruction Set: A Compiler Writer's Guide* Prentice-Hall International, UK, 1988

[vK 1] B. v. Karger *On the Equivalence of the ProCoS AL and TAL languages* ProCoS Report, 1990

[vK 2] B. v. Karger *Definition of the ProCoS Level 1 Assembly Language* ProCoS Report, 1990

[vK 3] B. v. Karger *Definition of the ProCoS Block Structured Level 1 Assembly Language* ProCoS Report, 1990

[Plotkin 81] G.D. Plotkin, *A Structural Approach to Operational Semantics* Monograph DAIMI FN-19, Computer Science Department, Denmark, 1981

A CONCEPT OF A COMPUTER SYSTEM FOR THE EXECUTION OF SAFETY CRITICAL LICENSABLE SOFTWARE PROGRAMMED IN A HIGH LEVEL LANGUAGE

W. A. Halang and Soon-Key Jung

Department of Computing Science, University of Groningen, P.O. Box 800, 9700 AV Groningen, The Netherlands

Abstract. There are already a number of established methods and guidelines, which have proven their usefulness for the development of high integrity software employed for the control of safety critical technical processes. Prior to its application, such software is still subjected to appropriate measures for its verification and validation. However, according to the present state of the art, these measures cannot guarantee the correctness of larger programs with mathematical rigour. Therefore, the licencing authorities do not approve safety relevant systems yet, whose behaviour is exclusively program controlled. In order to provide a remedy for this unsatisfactory situation, the concept of a special computer system is developed, which can carry out safety related functions within the framework of distributed process control systems or programmable logic controllers. It explicitly supports sequence controls, since many automation programs including safety relevant tasks are of that kind. The architecture features full temporal predictability, determinism, and supervision of the program execution and of all other activities of the computer system and supports the software verification method of diverse inverse documentation. The system can be programmed in a high level language and is based on a library of function modules, whose correctness can be mathematically proved. The concept utilises an operating system and a compiler only in a very rudimentary form. Hence, it enables the safety licencing of the software running on the computer system. In the microprogram of this computer, which may be based on the VIPER chip, a minimum operating system is provided, whose only task is to start the execution of runnable subroutines, which are marked in a ready list. As the elementary units of application programming, the set of basic function modules is provided in ROMs. These modules are of application specific nature and generally different for each application area. For the formulation of safety related automation programs these basic functions are only interconnected with each other. The prototype of a tool has been developed allowing to carry through this kind of programming in graphical form.

Keywords. Computer architecture; computer control; high integrity software; high level languages; programmable controllers; programming environments; real time computer systems; safety critical automation; safety licencing; software engineering.

INTRODUCTION

In the literature (Clutterbuck and Carré, 1988; DGQ-NTG, 1986; Ehrenberger, 1983; EWICS, 1985; Faller, 1988; Grimm, 1985; Grimm, 1988; Hausen, Müllerburg, and Schmidt, 1987; Hölscher and Rader, 1984; IEC, 1986; ANSI/IEEE, 1983; ANSI/IEEE, 1984a; ANSI/IEEE, 1984b; Jülly, 1987; Krebs, 1984; Krebs and Haspel, 1984; Redmill, 1988; Schmidt, 1988; Traverse, 1987; VDI, 1985; Voges, 1986) measures and guidelines have been compiled, which have proven their usefulness for the development of (almost) error free computer programs employed for the control of safety critical technical processes and automation systems. Prior to its application, such software is still subjected to appropriate methods for its verification and validation. According to the present state of the art, with mathematical rigour the correctness of only rather small program modules can be established. For the correctness proofs object code must be considered, since compilers — or even assemblers — are themselves far too complex software systems, as that their correct operation could be verified. Therefore, the licencing authorities do not approve safety relevant systems yet, whose behaviour is exclusively program controlled. A prominent example for this policy is the case of a Canadian nuclear power plant, whose construction was recently completed. It was designed in a way that its systems, including the safety relevant ones, relied on computer control only. David L. Parnas was charged with the safety licencing of the real time software written for utilisation in this power plant. Owing to their complexity, the correctness of these programs could not be proven. Consequently, a licence for putting the nuclear power plant into service was not granted, which is very costly for its owner. For these reasons, the only accepted method of applying computer control for safety critical functions in automation systems is to let programmable electronic systems run in parallel with hardwired ones, which are realised in proven technologies such as relais or SSI/MSI TTL logic and for which long established verification techniques exist. In the case of discrepancies between the generated results, the output of the hardwired logic will always override the ones yielded by computers.

Naturally, owing to their greater flexibility and higher information processing capability, it is desirable to let programmable electronic systems also take care of safety relevant functions. This requires that the corresponding control software and its time behaviour can be safety licenced. The greatest obstacle which needs to be overcome for reaching this goal is to cope with complexity. In one of his recent notes, Dijkstra (1989) has also identified this problem and provides some guidance towards its solution:

> Computing's core challenge is how *not* to make a mess of it.

> ... so we better learn how not to introduce the complexity in the first place.

> The moral is clear: prevention is better than cure, in particular if the illness is unmastered complexity, for which no cure exits.

In order to provide a remedy for the above described unsatisfactory situation with respect to the safety licencing of programmable electronic systems and their software, in the sequel the concept of a special computer system is presented in a constructive way, which can carry out safety related functions within the framework of programmable logic controllers or distributed process control systems. The leading idea followed throughout this design is to combine already existing software engineering and verification methods with novel architectural support. Thus, the semantic gap between software requirements and hardware capabilities is closed, relinquishing the need for not safety licensable compilers and operating systems. By keeping the complexity of each component in the system as low as possible, the safety licencing of the hardware in combination with application software is enabled on the basis of well-established and proven techniques.

According to the quest of preventing complexity throughout, the concept is restricted to the implementation of rather simple, but nevertheless for industrial control applications very useful and widely employed, programmable electronic systems, viz. programmable logic controllers

(PLC). Therefore, in the subsequent section, we take a closer look at the present situation of developing software for PLCs, which promises to improve considerably and, at the same time, to facilitate safety licencing. As a special feature, our architecture provides explicit support for sequence controls, since many automation programs including safety relevant tasks are of this nature. Correspondingly, we shall briefly summarise the theory of sequential function charts, which are used for the formulation of sequence control applications. Then, the concept of the hardware architecture will be presented. It features full temporal predictability, determinism, and supervision of the program execution process and of all other activities of the computer system. Furthermore, and very importantly, it supports the software verification method of diverse inverse documentation (Krebs and Haspel, 1984). The computer could be based on the VIPER chip (Kershaw, 1987), which is the only available microprocessor the correctness of whose design has been formally proven. It would be more advantageous, however, to design a specific processor implementing the new features in hardware and preventing inappropriate software behaviour as far as possible. In the microprogram of this computer, an execution control program is provided, whose only task is to activate runable subroutines, which are marked in a ready list. Such rudimentary form of an operating system is feasible, since its low complexity allows for safety licencing. The application software is formulated in a high level graphical language, which is based on a library of function modules. By their nature, these elementary units are specific to and generally different for each application area. They are well-defined and compact in size so that their correctness can be proved with mathematical rigour. When the latter has been carried through, the object code of the basic function modules in provided in ROMs. Now, for the formulation of safety related automation programs, incarnations of these basic functions only need to be interconnected with each other. As described in the final section of this paper, the prototype of a tool has been developed allowing to perform this kind of programming entirely in graphical form. The tool also contains a compiler for the interpretation of the source drawings, from which it generates object code mainly consisting of sequences of procedure calls. However, the compiler does not need to be safety licenced, since the loaded application software can be verified with the method of diverse inverse documentation. Owing to the applied programming paradigm and the architectural support provided, the utilisation of the mentioned safety licencing technique turns out to be very easy, economical, and time-efficient.

SOFTWARE DEVELOPMENT FOR PROGRAMMABLE LOGIC CONTROLLERS

Computer aided software development tools for programmable logic controllers are presently only available in vendor specific form. The commonly used programming methods can be subdivided into two main groups: a textual one based on instruction lists, i.e. low level machine specific programming languages similar to assembly languages, and a semi-graphical one employing ladder diagrams. The latter representation is a formalisation of electric circuit diagrams to describe relais based binary controls.

In order to improve this situation, the International Electrotechnical Commission (IEC) has worked out for later standardisation a detailed draft defining four compatible languages for the formulation of industrial automation projects (IEC, 1988). Two of them are textual and the other two are graphical. The languages are suitable for all performance classes of PLCs. Since they provide a range of capabilities, which is larger than would be necessary for just covering the classical application area of PLCs, viz. binary processing, they are also apt for the front-end part of distributed process control systems. It is the goal of this standardisation effort to replace the programming in machine, assembly, and procedural languages by employing object oriented languages with graphical user interfaces. Therefore, it emphasises the high level graphical Function Block Diagram language (FBD), that was derived from diagrams of digital circuits, in which each chip represents a certain module of the overall functionality. The direct generalisation of this concept leads to function blocks which may have inputs and outputs of any data type and which may perform arbitrary processing functions (cp. Fig. 1). The schematic description of logical and functional relationships by symbols and connecting lines representing a conceptual signal flow provides easy conceivability. A function diagram is a process oriented representation of a control problem, independent of its realisation. It serves as a means of communication between different interest groups concerned with the engineering and the utilisation of PLCs, which usually represent different technical disciplines. The FBD language is supplemented by another graphical component: the Sequential Function Chart language (SFC). The representation method of sequential function charts (cp. Fig. 2) can be considered as an industrial application of the general Petri net concept, utilised to formulate the co-ordination and co-operation of asynchronous sequential processes. The IEC proposal's second high level language is called Structured Text (ST). Emphasising modularisation, it has a Pascal-like syntax and functionality, but also provides the task concept to handle parallel real time processes.

Based on these two high level IEC languages, a system independent rapid prototyping and CASE tool with graphical user interface for PLCs and other process control systems was developed, which will be outlined later in this paper. The ST language is used for the formulation of project specific software modules in the form of function blocks containing — and at the same time hiding — all implementation details. These modules are then utilised and interconnected in the graphical languages FBD/SFC to express solutions of automation and control problems. Thus, the advantages of graphical programming, viz. orientation at the engineer's way of thinking, inherent documentation value, clearness, and easy conceivability, are combined with the ones of textual programming, viz. unrestricted expressibility of syntactic details, of control structures, of algorithms, and of time behaviour. The guiding principle for the development of the CASE tool was to combine support for rapid prototyping, for structured top-down design, as well as for the engineering of reliable software to be licenced and applied in safety critical environments. The latter is achieved by reducing the number of different ways in which a given problem may be solved.

The system independent software engineering and rapid prototyping for PLCs is carried out in two steps:

1. set-up of a function block library, and
2. interconnection of function block instances (cp. Fig. 3).

As the elementary units of process control application programming, sets of basic function blocks are introduced. A project has been carried through for the identification and definition of such function modules suitable for control purposes in the chemical industry. The project revealed that some 40 functions are sufficient for the formulation of the large majority of the automation problems occurring there. Owing to their simplicity and universality, they can be re-used in many different contexts. When programmed in the ST language, the source code length of these modules does not exceed two pages. Therefore, the possibility is given to prove their correctness with mathematical rigour. For this purpose, a number of already well-established methods and guidelines for the development of high integrity software employed for the control of safety critical technical processes is utilised. Thus, the safety licencing of the generated software is enabled. Examples of such software verification techniques are symbolic program execution, diverse inverse documentation, inspection, simulation and, in some cases, complete test.

In the second of the above mentioned steps, for rapid prototyping or for the formulation of automation applications with safety properties, the solution of a control problem is worked out in the form of a function block diagram determining the interaction between function modules. To this end, the user invokes from his library function block instances, places them, and interconnects them with each other. Besides the provision of constants as external input parameters, the function block instances and the parameter flows between them, which are represented by the connection lines, are the only language elements used on this programming level. In order to provide maximum system independence and expressive power, a compiler transforms the logic contained in the diagrams into the ST language. Owing to the simple structure of this logic, the generated ST programs only contain sequences of procedure calls besides the necessary declarations.

SEQUENTIAL FUNCTION CHARTS

The mathematical model of sequential function charts has been derived from the well-known theory of Petri-nets. A sequential function chart is a directed graph defined as a quadrupel

$$(S, T, L, I)$$

with

$S = (s_1, ..., s_m)$ a finite, non-empty set of steps,
$T = (t_1, ..., t_n)$ a finite, non-empty set of transitions,
$L = (l_1, ..., l_k)$ a finite, non-empty set of links between a step and a transition, or a transition and a step, and finally
$I \subset S$ the set of initial steps.

The sets S and T represent the nodes of the graph. The initial steps are set at the beginning of the process and determine the initial state. With each step actions are associated, which are being executed while a step is set. Each transition is controlled by a Boolean condition: if the preceding step is set and (in the Boolean sense) the condition turns true, the subsequent step is set and the preceding one is reset.

Hence, a sequential function chart directly and precisely provides an answer to the question: "How does the system react if a certain step is set and the subsequent transition condition will be fulfilled?". The end of a step is characterised by the occurrence of the process information, which fulfills the condition for the transition to the following step. Consequently, steps cannot overlap. Actions can be initiated, continued, or terminated during a step.

The representation method of sequential function charts can be considered as an industrial application of the general Petri-net concept, utilised to formulate the co-ordination of asynchronous processes. The main elements of sequential function charts are

- steps,
- transitions,
- actions, and
- connections linking steps and transitions with one another.

They are employed under observation of the following boundary conditions:

- any step may be associated with one or more actions,
- there is a transition condition for any transition.

A control application is statically represented by a sequential function chart. By its interpretation observing certain semantic rules, the dynamic aspect of the described control procedure can be revealed.

A sequential function chart generally consists of steps, which are linked to other steps by connectors and transitions. One or more actions are associated with each step. The transition conditions are Boolean expressions formulated as function charts or in Structured Text.

At any given point in time during the execution of a system,

- a step can be either active or inactive, and
- the status of the PLC is determined by the set of active steps.

With each step a Boolean variable X is associated expressing with the values "1" or "0" its active or inactive state, respectively. A step remains active and causes the associated actions as long as the conditions of the subsequent transitions are not fulfilled. The initial state of a process is characterised by initial steps, which are activated upon commencement of the process. There must be at least one initial step.

The alternating sequence step/transition and transition/step must be observed for any process, i.e.

- two steps may never be linked directly, i.e. they must be separated by a transition, and
- two transitions may never be linked directly, i.e. they must be separated by a step.

A transition is either released or not. It is considered to be released when all steps connected to the transition and directly preceding it are being set. A transition between steps cannot be performed, unless

- it is released, and
- its transition condition has turned true.

The elaboration of a transition causes the simultaneous setting of the directly following step(s) and resetting of the immediately preceding ones. Transitions which are to be executed concurrently must be synchronised.

Actions are graphically represented by rectangles, which are connected to the steps initiating them. Each rectangle is internally subdivided into three areas, the first of which is called action qualifier. It characterises the action and is composed of the following elements:

N non-stored, unconditional
R reset
S set / stored
L time limited
D time delayed
P having pulse form
C conditional

THE SYSTEM CONCEPT

We assume that a technical process is to be controlled by a distributed computer system. First, all safety relevant hardware and software functions and components are clearly separated from the rest of the automation system and, then, also from each other. Under any circumstances it should be the objective in this process to keep the number of these functions as small as possible. Each of the thus identified functions is assigned to a separate special processor in the distributed system (Krebs and Haspel, 1984). In order to cope with intermittent errors and failures caused by wear, the hardware of each of these computers is to be fault tolerant. There is an extensive literature about fault tolerant hardware and corresponding products are already commercially available. Therefore, we shall not discuss this topic here any further. However, it must be emphasised that as hardware platform only safety licenced components such as the VIPER microprocessor are acceptable.

In the microprogram of this processor for safety functions a minimum operating system is provided, which should better be called an execution control program. Its only task is to start the execution of runable subroutines, which are marked in a corresponding list. According to the model of synchronous programming (Lauber, 1989), this program may have the form of a "Cyclic Executive". A further, and even better, possibility for the realisation of the internal execution control would be the implementation of the deadline driven scheduling algorithm (Henn, 1978; Halang, 1990), which also allows the supervision of execution times and the early recognition of overload situations. Both mentioned algorithms are such compact, that their correctness can easily be proved with formal methods.

As the elementary units of application programming, a set of basic function modules is introduced and as well stored in the microprogram, or in a program ROM, respectively. In contrast to the execution control program, the basic functions are of application specific nature. Essentially, the sets of basic functions implemented in each specific case are oriented at the application area, although certain functions like analogue and digital input/output may have general relevance.

A project has been carried through for the identification and definition of basic function modules suitable for process control applications in the chemical industry. The project revealed that about 30 to 50 functions are sufficient for the formulation of the large majority of the occurring automation problems. These functions have been written in the language "Structured Text" (ST) (IEC, 1988). They are relatively short software modules with lengths of one or two pages of source code. Therefore, the possibility is given to prove their correctness mathematically. This can be achieved with the methods described by (Cheheyl and co-workers, 1981; Craigen and co-workers, 1988; Gordon, 1986; Luckham, 1977), but also by symbolic program execution or, in some cases, by complete test.

Now, for the formulation of automation applications with safety properties, these basic functions are only interconnected with each other, i.e. single basic functions are invoked one after the other and, in the course of this, they pass parameters. Besides the provision of constants as external input parameters, the basic functions' instances and the parameter flows between them are the only language elements used on this programming level. The software development is carried out in graphical form: the instances of the basic functions are represented by rectangular symbols and the data flows are depicted as connecting lines. Then, a compiler transforms the graphically represented program logic into the language Structured Text. Owing to the simple structure, this logic is only able to assume, the generated ST-programs contain no other features than sequences of procedure calls and the necessary declarations. Further compilers for the conversion of such ST-programs into low level target languages and corresponding linkage editors will be of comparable simplicity as the first mentioned compiler. Hence, there should be the possibility for their safety licencing.

The above outlined high level programming method is very similar to the programming language LUCOL (O'Neill and co-workers, 1988), which has specifically been developed for safety critical applications. The LUCOL modules correspond with the basic functions and LUCOL programs are sequences of invocations of these modules with data passing as well. This similarity can now be utilised to carry through the correctness proof of programs, which are produced by interconnection of basic functions, since the verification tool SPADE (Clutterbuck and Carré, 1988) has already successfully been applied for this purpose (O'Neill and co-workers, 1988). SPADE works especially well on procedures, which are closed in themselves and possess well defined external interfaces, such as represented by the basic functions.

Although the static analyser SPADE and the comparable Malpas (Malpas) have proven to be highly valuable tools, they cannot establish alone the correctness of software. However, neither their utilisation nor the availability of safety licenced compilers and linkage editors as mentioned above are necessary preconditions to employ the here proposed high level programming method. For the application software may be safety licenced by subjecting its loaded object code to diverse backward documentation, the verification method which was developed in the course of the Halden nuclear power plant project (Krebs and Haspel, 1984). This technique consists of reading the machine programs out of the computer memory and giving them to a number of teams working without any mutual contact. All by hand, these teams disassemble and decompile the code, from which they finally try to regain the specification. The software is granted the safety licence if the original specification agrees with the inversely obtained re-specifications. Of course, the method is generally extremely cumbersome, time-consuming, and expensive. This is due to the semantic gap between a specification formulated in terms of user functions and the machine instructions carrying them out. Applying the programming paradigm of basic function modules, however, the specification is directly mapped onto sequences of subroutine invocations. The object code consists of just these calls and parameter passing. It takes only minimum effort to interpret such code and to redraw graphical program specifications from it. The implementation details of the function modules are part of the architecture. Thus, they are invisible from the application programming point of view and do not require safety licencing in this context.

IMPLEMENTATION DETAILS

After having outlined above the concept for a grapically, i.e. at a high level, programmable computer system, which is designed with safety related applications in mind, we shall now discuss some aspects in more detail.

As hardware for this computer the VIPER chip (Kershaw, 1987) could be employed, which is the only microprocessor yet that has passed a safety licencing procedure. In case of a new development, maximum simplicity and a minimum application specific instruction set ought to be emphasised. If possible, the instruction set should then correspond to the low level programming language "Instruction List" (IL), which will be standardised by the IEC together with Structured Text, and which essentially represents a single address assembly language.

Essentially, all operating system and application programs must be provided in read only memories in order to prevent any modification by a malfunction. For practical reasons, there will generally be two types of these memories. The code of the execution control program and of the basic function modules resides in mask programmed ROMs, which are produced under supervision of and released by the licencing authorities, after the latter have rigorously established the correctness of the modules and the correctness of the translation from Structured Text into target code. The sequences of subprogram invocations together with the corresponding parameter passing, representing the application programs at the architectural level, are written into (E)PROMs by the user. As outlined above, only this part of the software is subject to verification by diverse backward documentation. Again, this is performed by the licencing authorities, which finally still need to install and seal the (E)PROMs in the target process control computers.

Many automation programs including safety relevant applications have the form of sequence controls. Therefore, the language ST also provides specific language features for their programming ("Sequential Function Charts", SFC): step, transition, and action. In the here considered context, for purposes of a clear concept, of easy conceivability and verifiability, and in order not to leave the Petri-net concept of sequence controls, we only permit the utilisation of non-stored actions. All other types of actions can be expressed in terms of non-stored ones and a reformulated sequential control logic. Parallel branches in sequential function charts should either be realised by hardware parallelism or already resolved by the application programmer in the form of explicit serialisation.

To the end of redundant hardware support and supervision of the SFC control structures in the special architecture for safety related applications the following measures should be taken. The identification of the step active at a given time is kept in a separate register, which is especially protected and which cannot be accessed by the software. The register's contents is displayed on an external operator's console. A memory protection mechanism prevents the erroneous access to the program code of the steps being not active. A special subprogram return branch instruction is provided supporting the cyclic operation of programmable logic controllers. It branches to the initial address of the program area corresponding to the active step. With some further special instructions, table processing (Ghassemi, 1983) is implemented, which is frequently employed for realising actions. As described in more detail below, the time consumed for the execution of the actions of each step is supervised.

In programmable logic controllers it is checked after the execution of the active step's actions, whether the Boolean condition for the transition to the subsequent step(s) is fulfilled. If this is not the case, the considered step remains active and the associated actions are executed once more. The execution time for a step varies from one cycle to the next depending upon the program logic of these actions and the external condition evaluated each time. Therefore, the measurement of external signals and the output of values to the process is generally not carried out at equidistantly spaced points in time, although this may be intended in the control software.

A basic cycle is introduced in order to achieve full determinism of the time behaviour of programmable logic controllers. The length of the cycle is selected in a way as to accommodate during its duration the execution of the most time consuming step occurring in an application (class). At the end of the action processing and after the evaluation of the corresponding transition condition, the occurrence of a clock signal is expected, which marks the begin of the next cycle. An overload situation or a run time error, respectively, is encountered when the clock signal interrupts an active application program. In this case a suitable error handling has to be carried through. Although the introduction of the basic cycle exactly determines a priori the cyclic execution of the single steps, the processing instants of the various operations within a cycle, however, may still vary and, thus, remain undetermined. Since a precisely predictable timing behaviour is only important for input and output operations, the problem can be solved as follows. All inputs occurring in a step are performed en bloc at the beginning of the cycle and the thus obtained data are buffered until they will be processed. Likewise, all output data are first buffered and finally sent out together at the end of the cycle.

The here presented architecture features temporal predictability, determinism, and supervision of the program execution and of all other activities of the computer system. This is in distinct contrast to conventional programmable electronic systems which, at best, allow to make approximate predictions about the instants when programmed activities will actually take place. In the course of timing considerations, error recognition times and the duration required for putting a system into the safe state must be taken into account under any circumstances, too.

The software verification method of diverse backward documentation (Krebs and Haspel, 1984) is greatly facilitated by the here introduced problem oriented architecture. Owing to the employment of basic function modules with application specific semantics as the smallest units of software development, the effort for the method's utilisation is by orders of magnitude less than in the cases reported by Dahll, Mainka, and Märtz (1988). Furthermore, the employed principle of software engineering reduces the number of possibilities to solve a given single problem in different ways. Therefore, it is considerably simpler to check the equality of the reversely documented software with the original program. Finally, it is to be mentioned that tools for the graphical reverse documentation of memory resident programs are part of the standard support software of distributed process control systems, and that the application of the verification method (Krebs and Haspel, 1984) is thus facilitated.

PROGRAMMING ENVIRONMENT

A set of tools has been developed supporting the system independent graphical programming of PLCs. In the sequel we shall give a survey on the functionality of these tools, which are aimed at facilitating methodical and structured top-down design, the design in the conceptual phase, and the set-up of a library of well-tested standard modules for a project or a class of projects (cp. Fig. 4).

When a new software is to be developed, the programmer uses a CAD tool to set up his drawings of function blocks and sequential function charts. In particular, he fetches appropriate graphical objects from his library, places them in a work sheet, and links, according to his logic, connectors to these objects. After the drawing process is complete, the work sheet is stored and, then, submitted to a utility program of the CAD tool, which generates lists of all objects and all interconnection nodes occurring in the drawing. These lists constitute a textual representation which is fully equivalent to the contents of the considered worksheet. They are then submitted to a second postprocessor, viz. a compiler generating code in the IEC language Structured Text. In particular, the compiler produces complete program units with the declarations of input, output, and local variables, of tasks, and of instantiated function blocks, with calling sequences of functions, function block instances, and tasks, and with the encoded description of steps, transitions, and actions. The texts of these program units is stored in a library, which, of course, can also be filled by a text editor with hand coded modules. It serves as input for the final postprocessor, i.e. a translator, that generates executable programs for the here described architecture. Owing to the restricted form the source programs are only able to assume, the translator produces calling sequences to and parameter passings between the user invoked functions and function blocks, the object code of which is contained in systems ROM.

For documentation and verification purposes, functions and function blocks are also coded in the language Structured Text and stored in a text library. In order to save a further editing step needed to set up entries for the graphical library of the CAD tool, it is desirable to have a tool available which automatically generates from the modules' source codes the corresponding graphical symbols to be used in the drawing process. Therefore, a further tool was developed, which interprets the name and the input and output declarations of a function or function block in order to generate the textual description of an appropriately sized graphic symbol. The output of this tool is finally subjected to a utility of the CAD system, which translates the descriptions into an internal form and places the latter into its component library.

CONCLUSION

In a constructive way, and only using presently available methods and hardware technology, in this paper for the first time a computer architecture was defined, which enables the safety licencing of entire programmable electronic systems including the software. This goal was achieved by the following measures:

- consequent separation of program and data storage and utilisation of ROMs wherever possible,
- hardware support and supervision of predictable and fully deterministic software behaviour,
- hardware support for the main application area of programmable logic controllers, viz. sequence controls,
- avoiding the need for a complex operating system,
- utilisation of a very high level, graphical software engineering method,
- closing of the semantic gap between architecture and user programming by basing the software development on a set of non-trivial function (s) (blocks) with application specific semantics,
- removal of compilers from the chain of items requiring safety licencing (although their complexity is largely decreased by the architecture), and
- by providing a feasible application level and architectural support for the software verification method of diverse backward documentation.

It is hoped that the concept presented here leads to the breakthrough of allowing to relinquish discrete or relais logic from taking care of safety critical functions in industrial processes by programmable electronic systems executing safety licenced high integrity software.

REFERENCES

Cheheyl, M., and co-workers (1981). Verifying Security. *Computing Surveys*, *13*, 279 – 339.

Clutterbuck, D. L., and B. A. Carré (1988). The verification of low-level code. *IEE Software Engineering Journal*, 97 – 111.

Craigen, D., and co-workers (1988). m-EVES: A tool for Verifying Software. In *Proc. 10th International Conference on Software Engineering*, pp. 324 – 333.

Dahll, G., U. Mainka, and J. Märtz (1988). Tools for the Standardised Software Safety Assessment (The SOSAT Project). In W. D. Ehrenberger (Ed.). *Safety of Computer Control Systems 1988.* IFAC Proceedings Series, 1988, No. 16. Pergamon Press, Oxford. pp. 1 – 6.

DGQ-NTG (1986). *Software-Qualitätssicherung.* Schrift 12-51. Beuth, Berlin.

Dijkstra, E. W. (1989). The next forty years. *EWD 1051*.

Ehrenberger, W. D. (1983). Softwarezuverlässigkeit und Programmiersprache. *Regelungstechnische Praxis rtp*, *25*, 24 – 29.

EWICS TC7 Software Sub-group (1985). Techniques for the Verification and Validation of Safety-Related Software. *Computers & Standards*, *4*, 101 – 112.

Faller, R. (1988). Sicherheitsnachweis für rechnergestützte Steuerung. *Automatisierungstechnische Praxis atp*, *30*, 508 – 516.

Ghassemi, A. (1983). Problemorientierter Entwurf und zuverlässige Realisierung von Prozesssteuerungen. *Regelungstechnische Praxis rtp*, *25*, 478 – 483.

Gordon, M. (1986). Why higher-order logic is a good formalism for specifying and verifying hardware. In G. Milne and P.A. Subrahmanyam (Eds.). *Formal Aspects of VLSI Design.* North Holland, Amsterdam.

Grimm, K. (1985). Klassifizierung und Bewertung von Software-Verifikationsverfahren. In *Technische Zuverlässigkeit — Generalthema: Softwarequalität und Systemzuverlässigkeit.* VDE-Verlag, Berlin, Offenbach. pp. 79 – 90.

Grimm, K. (1988). Methoden und Verfahren zum systematischen Testen von Software. *Automatisierungstechnische Praxis atp*, *30*, 271 – 280.

Halang, W. A. (1990). A Practical Approach to Pre-emptable and Non-pre-emptable Task Scheduling with Resource Constraints Based on Earliest Deadlines. In *Proc. Euromicro '90 Workshop on Real Time.* IEEE Computer Society Press, Washington.

Hausen, H. L., M. Müllerburg, and M. Schmidt (1987). Über das Prüfen, Messen und Bewerten von Software. *Informatik-Spektrum*, *10*, 123 – 144.

Henn, R. (1978). Antwortzeitgesteuerte Prozessorzuteilung unter strengen Zeitbedingungen. *Computing*, *19*, 209 – 220.

Hölscher, H., and J. Rader (1984). *Mikrocomputer in der Sicherheitstechnik.* Verlag TÜV Rheinland, Köln.

IEC (1986). Standard 880 *Software for computers in the safety systems of nuclear power stations.*

IEC (1988). International Electrotechnical Commission, TC 65: Industrial Process Measurement and Control, SC 65A: System Considerations, WG 6: Discontinuous Process Control. Working Draft *Standards for Programmable Controllers*, Part 3: Programming Languages. IEC SC65A/WG6/TF3 (Coordinator)4, 1 November 1988.

ANSI/IEEE (1983). Standard 829-1983 *IEEE Standard for Software Test Documentation.* New York.

ANSI/IEEE (1984a). Standard 730-1984 *IEEE Standard for Software Quality Assurance Plans.* New York.

ANSI/IEEE (1984b). Standard 830-1984 *IEEE Guide to Software Requirements Specifications.* New York.

Jülly, U. (1987). Funktionsweise und Programmierung einer speicherprogrammierten Sicherheitssteuerung. *Automatisierungstechnische Praxis atp*, *29*, 532 – 535.

Kershaw, J. (1987). *The VIPER Microprocessor.* Royal Signals and Radar Establishment, Malvern, England. Report No. 87014. 26 pp. (Cp. also the journal *SafetyNet — VIPER Microprocessors in High Integrity Systems.*)

Krebs, H. (1984). Zum Problem des Entwurfs und der Prüfung sicherheitsrelevanter Software. *Regelungstechnische Praxis rtp*, *26*, 28 – 33.

Krebs, H., and U. Haspel (1984). Ein Verfahren zur Software-Verifikation. *Regelungstechnische Praxis rtp*, *26*, 73 – 78.

Lauber, R. (1989). *Prozessautomatisierung*, Band 1. Springer, Berlin Heidelberg New York London Paris Tokyo.

Luckham, D. C. (1977). Program verification and verification-oriented programming. In B. Gilchrist (Ed.). *Information Processing*, *77*. pp. 783 – 793.

Malpas (Malvern Program Analysis Suite). Rex, Thompson & Partners Ltd., Farnham, England.

O'Neill, I. M., D. L. Clutterbuck, P. F. Farrow, P. G. Summers, and W. C. Dolman (1988). The Formal Verification of Safety-Critical Assembly Code. In W. D. Ehrenberger (Ed.). *Safety of Computer Control Systems 1988*. IFAC Proceedings Series, 1988, No. 16. Pergamon Press, Oxford. pp. 115 – 120.

Redmill, F. J. (Ed.) (1988). *Dependability of Critical Computer Systems - 1*. Elsevier, London New York.

Schmidt, K. P. (1988). *Rahmenprüfplan für Software*. Formblätter und Anleitung für Prüfungen von Software nach den Güte- und Prüfbestimmungen Software RAL-GZ 901 und der Vornorm DIN 66285 "Anwendungssoftware, Prüfgrundsätze". Arbeitspapiere der GMD 312. Gesel- schaft für Mathematik und Datenverarbeitung GmbH, Sankt Augustin.

Traverse, P. (1987). AIRBUS and ATR System Architecture and Specification. In U. Voges (Ed.). *Software Diversity in Computerized Control Systems, Dependable Computing and Fault-Tolerant Systems*, Vol. 2. Springer, Wien New York.

VDI (1985). Richtlinie VDI 2880 Blatt 5 Entwurf *Speicherprogrammierbare Steuerungsgeräte — Sicherheitstechnische Grundsätze*". Beuth, Berlin.

Voges, U. (1986). Anwendung von Software-Diversität in rechnergesteuerten Systemen. *Automatisierungstechnische Praxis atp*, *28*, 583 – 588.

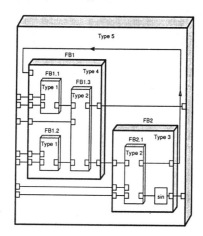

Fig. 1. A function block according to IEC 65

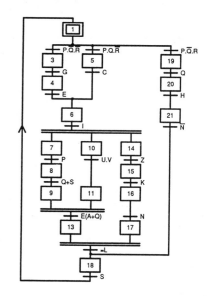

Fig. 2. A sequential function chart

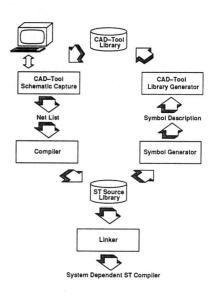

Fig. 4. A CASE tool set

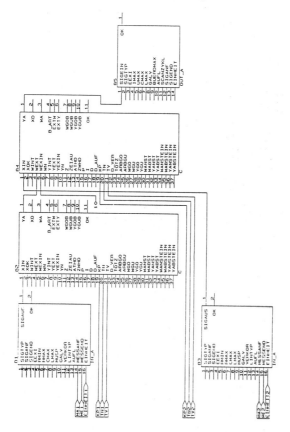

Fig. 3. A graphically formulated program

MODELING AND VERIFYING SYSTEMS AND SOFTWARE IN PROPOSITIONAL LOGIC

G. Stålmarck and M. Säflund

Logikkonsult NP AB, VI-Novembervägen 210, S-125 34 Älvsjö, Sweden

Abstract. We present a new proof procedure for analysing large propositional formulas. A definition of hardness of tautologies is presented and it is stated that a boolean representation of a system, although resulting in a large formula, often is easily verifyed. We also give a detailed example of software verification, using the proof procedure, in the railway signalling area.

Keywords. Hard Tautologies, Proof Procedures, Propositional Logic, Railways, System Verification.

1. BOOLEAN MODELLING OF SYSTEMS

In many cases it is easy to check or evaluate the output of a system given some particular inputs. In general it is much harder to find possible inputs resulting in a given output.

The former problem, here refered to as *simulation*, often turns out to be in P, the class of problems computable in deterministic polynomial time, if so, the latter problem, here refered to as *verification*, by definition belongs to NP, the class of problems for which a given solution can be checked in P, (for an exaustive survey of problems in NP see (Garey and Johnson 1979)).

Cooks theorem (Cook 1971) states that each problem in NP can be represented (in deterministic polynomial time) as a satisfiability problem of a boolean formula. According to Cooks theorem we state:

All systems that can be simulated in P can be verified by checking the satisfiability of some boolean formulas (of size polynomially related to the system).

Boolean representation of so called bi-stable systems such as *combinatorial circuits* and *relay-systems* are studied already by Shannon (1938) and followed, e.g. by Hohn (1960) and Whitesitt (1961).

Examples

Syntax used for the logical connectives in the boolean formulas:

OR	∨
AND	&
IMPLICATION	→
EQUIVALENCE	↔
NOT	¬
FALSITY	⊥
n-ary XOR	! (x1 ! x2 ! ... ! xn)

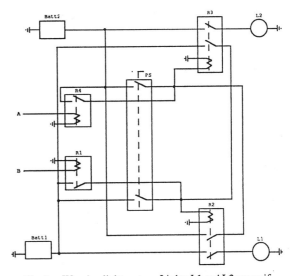

Fig. 1. Warning light system. Lights L1 and L2 go on if either of the signals A or B go on, or if the pressure switch PS goes on.

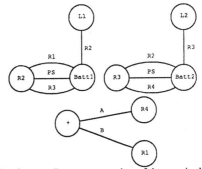

Fig. 2. Intermediate representation of the warninglight system.

Example 1. In Fig. 2 the intermediate representation describing the system in Fig. 1 is shown. (Note that relays can correspond to both a node and an edge.)

Below the boolean representation of the system and the functional specification is shown.

```
((
 (L1 ↔ R2) &            ; Light 1 iff Relay R2 is
                        ; pulled
 (L2 ↔ R3) &            ; Light 2 iff Relay R3 is
                        ; pulled
 (R2 ↔ (PS ∨ R3 ∨ R1)) & ; Relay R2 pulled iff pressure
                        ; switch is closed
                        ;    or Relay R1 is pulled
                        ;    or Relay R3 is pulled
 (R3 ↔ (PS ∨ R2 ∨ R4)) & ; Relay R3 pulled iff pressure
                        ; switch is closed
                        ;    or Relay R2 is pulled
                        ;    or Relay R4 is pulled
 (R4 ↔ A) &             ; Relay R4 is pulled iff
                        ; signal A
 (R1 ↔ B) )             ; Relay R1 is pulled iff
                        ; signal B
 →
 (      ; Functional specification of the system:
    ( L1 & L2) ↔ (A ∨ B ∨ PS))
)
)
```

Applying a proof procedure (introduced in section II) yields the following result (note the counter model showing the state of the system violating the functional specification):

```
##############################################
THE FORMULA IS FALSIFIABLE.
##############################################
COUNTER MODEL NR 1
   L1
   R2
   L2
   R3    ; The system does not satisfy the
         ; functional specification
```

To represent more complex systems, i.e. systems involving time, it is convenient to use a language with more expressive power such as Boyer-Moore Logic, see (Journal of Automated Reasoning, Special issue on verification 1989), 1st order predicate logic or temporal logic, see (Galton Ed. 1987).

However, it is in principle possible to translate temporal logic into 1st order predicate logic (with some extra axioms concerning the accessibility relation see (Shoham 1988)) and predicate logic into boolean formulas, given that the domain of each quantifier is finite. That kind of translation may of course give rise to extremely large formulas.

Example 2. Figure 3 shows a simple flow chart. The boolean representation of the flow chart and all possible paths of length less than or equal to seven steps terminating the computation:

```
// Description of Graph
(Start[0] &
 (Start[0] → Build[1]) &
 (Build[1] → Set[2]) &
 (Set[2] → Terminal_test[3]) &
 (Terminal_test[3] →
    (Valid[4] ! Applicable_rule[4])) &
 (Applicable_rule[4] →
    (Not_provable[5] ! Apply_rule[5])) &
 (Apply_rule[5] → Terminal_test[6]) &
 (Terminal_test[6] →
    (Valid[7] ! Applicable_rule[7])) &

// Exit at stop nodes
 (Not_provable[5] →  ¬(Terminal_test[6] ∨ Valid[7
    ∨ Applicable_rule[7])) &
 (Valid[4] → ¬(Not_provable[5] ∨ Apply_rule[5] ∨
    Terminal_test[6] ∨ Valid[7] ∨ Applicable_rule[7])) &

// Stop nodes
 (Valid[4] ! Not_provable[5] ! Valid[7]));
```

Which would give the following result:

```
##############################################
THE FORMULA IS SATISFIABLE.
##############################################

   MODEL NR 1
Start[0]
Build[1]
Set[2]
Terminal_test[3]
Valid[4]

   MODEL NR 2
Start[0]
Build[1]
Set[2]
Terminal_test[3]
Applicable_rule[4]
Not_provable[5]

   MODEL NR 3
Start[0]
Build[1]
Set[2]
Terminal_test[3]
Applicable_rule[4]
Apply_rule[5]
Terminal_test[6]
Valid[7]
```

II. A NEW METHOD FOR ANALYZING PROPOSITIONAL FORMULAS

In this section we present a new proof procedure for propositional (boolean) logic (developed by Stålmarck (1989a), patent pending). A new technique is used for evaluating so called Natural Deduction rules on subformulas of a given formula.

For definitions of proof theoretic concepts used below we refer to (Prawitz 1965) and (Stålmarck 1991).

The number of computation steps needed to prove a formula with the proposed method is polynomially related to the least number of different free assumptions needed in any subderiva-

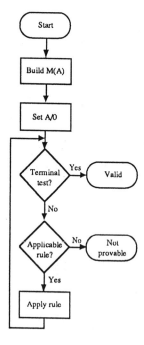

Fig. 3. Flow chart for the algorithm described in section II.

tion of a so called normal derivation of the formula. We state:

Theorem 1.

For each constant k it is decidable in polynomial time if there is a Natural Deduction proof, satisfying the subformula principle for classical logic, of a given formula A with less or equal than k different free assumptions in each subderivation.

The Method

1. Initialization. A first selection is done by the application of the following:

Lemma (from (Stålmarck 1989b)).

Each formula can be reduced (in linear time) to a logically equivalent formula of the shape A or the shape ¬A where A is negation-free, through the following transformations:

$(\neg A \rightarrow B)$	to	$(A \vee B)$
$(A \rightarrow \neg B)$	to	$\neg(A \& B)$
$(\neg A \rightarrow \neg B)$	to	$(B \rightarrow A)$
$(A \vee \neg B)$	to	$(B \rightarrow A)$
$(\neg A \vee B)$	to	$(A \rightarrow B)$
$(\neg A \vee \neg B)$	to	$\neg(A \& B)$
$(A \& \neg B)$	to	$\neg(A \rightarrow B)$
$(\neg A \& B)$	to	$\neg(B \rightarrow A)$
$(\neg A \& \neg B)$	to	$\neg(A \vee B)$
$\neg\neg A$	to	A.

If a formula A reduces to the negation of a negation free formula then A is obviously not logically valid.

If A reduces to a negation free formula, B, then we continue and build the matrix of B, M(B), according to the following:

Definition 1.

Assume that $a_1...a_n$ are all propositional variables in a formula A, $B_1...B_k$ are all compound subformulas of A ($= B_k$) and $B_i = (C_i \Delta D_i)$, (where C_i and D_i are subformulas of A and Δ is &, $\vee$ or $\rightarrow$). $b_1,...b_k$ are "new" variables such that b_i differs from each a_j and f is a function with $f(a_i) = a_i$ and $f(B_i) = b_i$.

$$(b_i \leftrightarrow (f(C_i) \Delta f(D_i)))$$

The *matrix* of A, M(A), is:

$$(b_k \leftrightarrow (f(C_k) \Delta f(D_k)))$$

Definition 2.

Each row $t_i = (b_i \leftrightarrow (f(C_i) \Delta f(D_i)))$ in a matrix is called a *triplet* and b_i is called *the variable representing* the formula B_i.

2. Computation part. We assume that the formula to be proved is false and try to derive a contradiction using the matrix formulation of the Natural Deduction rules below.

To simplify the description we only treat triplets with $\Delta = \rightarrow$, we also write triplets as (x,y,z) as an abbreviation for

$(x \leftrightarrow (y \rightarrow z))$.

m with or without indices is used to represent matrixes or submatrixes. x, y and z are used for terms (1, 0 or variables).

Matrix calculus M.

Simple rules:

	M-rules	Corresponding ND-rules
r1	m1 (0,y,z) m2 ———— y/1 z/0 m1 m2	$\dfrac{\neg (B \rightarrow C)}{B} \quad \dfrac{\neg (B \rightarrow C)}{\neg C}$
r2	m1 (x,y,1) m2 ———— x/1 m1 m2	$\dfrac{C}{(B \rightarrow C)}$
r3	m1 (x,0,z) m2 ———— x/1 m1 m2	$\dfrac{\neg B}{(B \rightarrow C)}$
r4	m1 (x,1,z) m2 ———— x/z m1 m2	$\dfrac{B}{(B \rightarrow C) \leftrightarrow C}$
r5	m1 (x,y,0) m2 ———— x/¬y m1 m2	$\dfrac{\neg C}{(B \rightarrow C) \leftrightarrow \neg B}$
r6	m1 (x,x,z) m2 ———— x/1 m1 m2	$\dfrac{(A \leftrightarrow (A \rightarrow C))}{A}$
r7	m1 (x,y,y) m2 ———— x/1 m1 m2	$\overline{(B \rightarrow B)}$

The triggering triplet is omitted in the conclusion of each rule according to the following properties:

(i) a triplet can only be triggered once

(ii) a triggered triplet cannot be a terminal triplet (see below).

Terminal triplets:

(1,1,0), (0,0,z) and (0,y,1).

Test rule:

M-rule	Corresponding ND-rule
[m x/1] [m x/0] D1 D2 m(S1) m(S2) ———————————— m(S)	[A] [¬A] D1 D2 B B ————————— B

Where Si is the set of variable instantiations from the end matrix in the derivation Di and all possible instantiations if the end matrix contain a terminal triplet, and S is the intersection of S1 and S2.

Suffix rule:

M-rule	Corresponding ND-rule
m1 (x,y,z) m2 (u,y,z) m3	$(A \leftrightarrow (C \rightarrow D))\quad (B \leftrightarrow (C \rightarrow D))$
m1 m2 x/u m3	$(A \leftrightarrow B)$

The standard ND introduction and elimination rules for $\rightarrow$ and the classical *reductio ad absurdum* rule is easily derived in M.

Hence, M is a sound and complete system for the classical implication calculus and via transformations from general boolean formulas to formulas built from variables, $\rightarrow$ and $\perp$ (or to pure implication formulas (Stålmarck 1989b)), also sound and complete for full classical propositional logic.

M is divided in a sequence of subsystems M0, M1... by limiting the test rule t1 in the following way:

M0 is M without the test rule t1

M_{i+1} is M with the premise derivations in the rule t1 restricted to Mi derivations.

We outline the algorithm for the M0 case (omitting the suffix rule):

Program: prove_M0
begin
    {build matrix M(A)}
    {set formula variable to 0}

    while NOT {Terminal} do
    begin
        if {Simple Rule is applicable} then
            {apply Simple Rule}
        else
            exit("A not provable in M0"}
    end

    exit("A is proved")
end.

Example

Let A be the formula $(a1 \rightarrow (a2 \rightarrow a1))$ and b1, b2 new variables.

Build matrix M(A):   (b1,a2,a1)
                              (b2,a1,b1)

Set formula variable to 0: (b1,a2,a1)
                              (0,a1,b1)

No triplet is terminal.

Apply simple rule r1:   (b1,a2,a1)
                           (0,a1,b1)
                           (0,a2,1)

Terminal triplet: (0, a2, 1)

A is valid.

Performance

The algorithm for M0 can be implemented to run in linear time.

The extension of the algorithm to any Mi is straightforward and can be implemented to run within time complexity $O(n^{2i+2})$.

The method is not sensitive to the number of variables or logical connectives of boolean formulas but to the formula structure according to theorem 1 above.

We propose the Mi sequence to be used in a definition of hardness of boolean tautologies (Krishnamurthy 1981) as follows:

Definition 3.

A valid formula A is i-hard if A is provable in Mi but not provable in Mj for any $j<i$.

By experience we know that boolean representations of optimization problems often result in hard formulas whereas system representations often are easy according to our definition of hardness and hence, easily solved by the method. Boolean formulas with more than 100 000 logical signs have been successfully analyzed.

As an example we can mention the STERNOL program treated in part III below. Each of the 500 double value formulas are proved within M1.

To actually obtain (counter-) models and a complete proof procedure we apply backtracking on a given variable ordering to some chosen Mi proposed by Arnborg (1987).

This makes the proof lengths sensitive to the ordering of the variables (=subformulas), similar to methods using so called binary decision diagrams, see (Bryant 1986).

III. A COMPUTER CONTROLLED INTERLOCKING SYSTEM

In this section we will study a computer controlled railway interlocking system used by the Swedish state railways (SJ).

The Computer Model

The computer model is built on the geographical principle, i.e. the (physical) objects in the yard (points, track crossings, signals etc) are represented by logical objects in the computer model. Logical objects interact with neighbouring objects. Besides the interaction with other logical objects, there is also an interaction with the physical objects in the yard and the control- and supervisory system.

The rules for a logical object's state transitions are described by equations.

For each variable there is a set of equations. The equations control the assignment of values to a variable. A set of equations defining a variable is called an Equation Group. All logical objects of the same type have the same set of equation groups.

An equation group is computed by two parallel systems A and B. The A-system computes the equations starting from low addresses and the B-system starts from high addresses. This mechanism ensures that variables are uniquely determined, i.e. should different values be computed by the two systems, the interlocking system will be closed down and all signals set to the aspect "stop".

The equations are written in the language STERNOL.

The STERNOL Language

In STERNOL there are Object Types. Objects contain multivalued Variables that can take any value in a finite domain. The possible values for a variable are determined by boolean equations.

Simulating and Verifying

We identify two important problems, simulating and verifying equation groups.

Simulation is understood as the problem to compute the value of a set of variables given a set of input variables. Simulation has a low degree of computational complexity and can be performed in linear time, using e.g. the method described in section II. Simulation will not be discussed further.

Verification problems are of the kind: "is there an assignment of values to variables which results in a given state?" and "will a given premise hold for every assignment of values to variables?". On the equation group level, questions of the first kind are in NP and questions of the latter kind are in co-NP.

Verification problems:

(i) Double value check, i.e. can two equations in an equation group be simultaneously satisfied.

(ii) Completeness, i.e. does each assignment of truth values satisfy at least one equation in each equation group.

(iii) Equivalence, i.e. do two equations denote the same set of assignments.

We will now study examples on the translation of verification problems into the Satisfiability problem and into the Tautology problem. The examples are taken from a session with the Circuit Verification Tool, CVT, an micro computer based application implementing the method described in section II.

Examples

Syntax used for the logical connectives in the STERNOL programs:
```
OR        +
AND       *
```

(i) Double value check

The equation group for a given variable (Fig. 4.) is translated, in linear time, into a boolean formula, seen below, that is satisfiable only for the assignments of values to variables that satisfy at least two different equations.

```
(DS=3 ↔ (R[1]=1 & B=1 & (I005=1v R[0]=6) v R[1]=2 &
B=2 & (I006=1v R[0]=6) v R[1]=3 & B=3 & (I007=1v
R[0]=6)) & UF=0 & UM=0 & (PK=5 & ¬U[1]=5 & T0=1v PK=3)
& HA[1]=1 & T1=2 & MFI1=0 & MFI2=0 & I019=1) &

(DS=2 ↔ R[1]=1 & B=1 & (I005=1v R[0]=6) & MFI1=0 &
MFI2=0 & UF=0 & UM=0 & (PK=3v PK=5 & ¬U[1]=5 & T0=1) &
(HA[1]=0 v T1=1)) &

(DS=1 ↔ B=4 & R[1]=7 & KO[1]=1 & OT30=16384 & MFI1=0 &
MFI2=0)
```

Information that has to be added to the boolean formula representing the equation group:

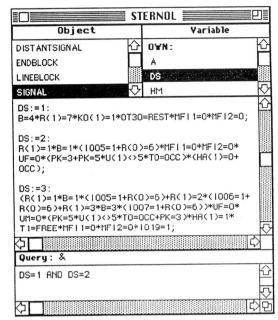

Fig. 1. STERNOL equation group for the variable DS.

a) the assumption that each defining STERNOL variable has at most one of the values occurring in the equation group

b) logical dependencies between relational operators

c) the assumption that the analyzed STERNOL variable has two different values.

The additions stated as a formula (note the syntax used for n-ary XOR - (x1 ! x2 ! ...)):

```
((OT30=16384) v ¬(OT30=16384)) &
((KO[1]=1) v ¬(KO[1]=1)) &
((R[1]=1 ! R[1]=2 ! R[1]=3 ! R[1]=7)v
¬(R[1]=1 v R[1]=2 v R[1]=3 v R[1]=7)) &
((B=1 ! B=2 ! B=3 ! B=4) v
¬(B=1 v B=2 v B=3 v B=4)) &
((T1=2 ! T1=1) v ¬(T1=2 v T1=1)) &
((HA[1]=1 ! HA[1]=0) v¬(HA[1]=1 v HA[1]=0)) &
((I019=1) v ¬(I019=1)) &
((MFI2=0) v ¬(MFI2=0)) &
((MFI1=0) v ¬(MFI1=0)) &
((PK=5 ! PK=3) v ¬(PK=5 v PK=3)) &
((T0=1) v ¬(T0=1)) &
((U[1]=5) v ¬(U[1]=5)) &
((UM=0) v ¬(UM=0)) &
((UF=0) v ¬(UF=0)) &
((I007=1) v ¬(I007=1)) &
((I006=1) v ¬(I006=1)) &
((R[0]=6) v ¬(R[0]=6)) &
((I005=1) v ¬(I005=1))
```

The formulas above are appended to the formula representing the question and the analysis is performed as a satisfiability test. Each model satisfying the boolean formula are value assignments that will satisfy more than one STERNOL equation. Stating the double value question is straightforward, e.g. the question whether DS=1 and DS=2 simultaneously is written:

```
(DS=1 AND DS=2) .
```

Note that the information added in clause a) gives a translation from multivalued into boolean variables.

(ii) Completeness

The equation group for a given variable is translated, in linear time, into a boolean formula that is valid iff every possible assignment of STERNOL variables satisfies at least one equation in the equation group.

Information that has to be added to the boolean formula representing the equation group:

a) and b) as above

c) the assumption that the boolean formula, including the information added in a) and b), implies that the analyzed variable will take one of the values defined.

The analysis is performed as a tautology test. All counter models are value assignments that do not satisfy any of the equations. An example of a counter model is shown in Fig. 5.

(iii) Equivalence

When an equation group has been rewritten it is often of interest whether the new and the old equations are equivalent.

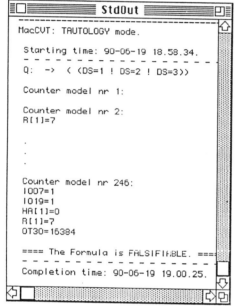

Fig. 5. The result of a completeness test. Note the first counter model where all boolean variables are false.

The equations to be analyzed for equivalence are translated, in linear time, into a boolean formula that is valid iff the the equations are satisfied by exactly the same value assignments.

Information that has to be added to the boolean formula representing the equations:

a) and b) as above

c) the assumption that the boolean formula, including the information added in a) and b), implies that the variable definitions are equivalent.

The analysis is performed as a tautology test. All counter models are value assignments resulting in different equations beeing satisfied.

Interactive analysis

The CVT is an interactive environment, no double value test takes more than five seconds and a majority of the equation groups are each analyzed in less than one second (on a Motorola 68020, 16 MHz micro computer). A complete analysis of the STERNOL program, e.g. with respect to double values, can easily be executed in batch, thus reducing overall time required.

CONCLUSION

A new method makes it possible to prove large boolean formulas.

The method can also be used for defining hardness of boolean tautologies. System-representations often turn out to be easy according to the definition and therefore easy to verify.

The method, implemented in an interactive verification tool, has been used to verify interlocking equations in SJ:s (Swedish state railways) STERNOL-programs.

A hardware implementation of the method forms the base in a generic platform, ASSURE, for various verification applications that is under development.

REFERENCES

Arnborg, S. (1987). Personal communication.

Bryant, (1986) Graph-Based Algorithms for Boolean Function Manipulation. IEEE Trans. on Computers 677-691.

Cook, S.A. (1971). The Complexity of theorem-proving procedures. Proc. 3rd Ann. ACM Symp. on Theory of Computing 151-158.

Galton, A. (1987). Editor Temporal Logics and their Applications, Academic Press.

Garey, M., and Johnson, D. (1979). Computers and Intractability, New York, W.H. Freedman and Company.

Hohn, F. (1960). Applied Boolean Algebra, New York, Macmillan.

Journal of Automated Reasoning (1989). Special issue on System Verification Vol. 5, No. 4.

Krishnamurthy, B. (1981). Examples of Hard Tautologies and Worst-Case Complexity for Propositional Proof Systems Dissertation, Massachusetts.

Prawitz, D. (1965). Natural Deduction, Dissertation, Stockholm.

Shannon, C.E. (1938). A Symbolic Analysis of Relay and Switching Circuits, Trans Am. Inst. Elec. Eng. no. 57, 713-723.

Shoham, Y. (1988). Reasoning About Change: Time and Causation from the Standpoint of Artificial Intelligence, MIT Press.

Stålmarck, G. (1989a). Swedish Patent Application no. 8902191-9.

Stålmarck, G. (1989b). A Note on the Computational Complexity of the Pure Classical Implication Calculus. Information Processing Letters 277-278.

Stålmarck, G. (1991). Normalization Theorems for Full First Order Classical Natural Deduction (to appear March 1991 in) The Journal of Symbolic Logic.

Whitesitt, J.E. (1961). Boolean Algebra with applications, Addison-Wesley.

Copyright © IFAC SAFECOMP'90,
London, UK, 1990

LOGICAL FOUNDATIONS OF A PROBABILISTIC THEORY OF SOFTWARE CORRECTNESS

T. Grams

Fachbereich Angewandte Informatik und Mathematik, Fachhochschule Fulda, Marquardstraße 35, D-6400 Fulda, FRG

Abstract. While statistical theories of probability have some deficiencies with regard to software and on the other hand the pure logical treatment of programs seems to be restricted to small problems, there is a need for a more powerful language and basic theory for both parties. A unified theory for both aspects of program dependability is provided by the theory of *logical probabilities*.

Within this theory the theorems of Eckhardt/Lee and Littlewood/Miller are restated. The theory comprises - among others - the following two theorems on the failure probability and on the probability of incorrectness of diverse systems: $p_{div} \geq p^2$ and $u_{div} \leq u^2$, where p is the failure probability of one program and p_{div} is the failure probability of the diverse system. Corresponding definitions hold for the probabilities of incorrectness u and u_{div} respectively.

The first of the theorems tells us that diversity may be useless with weak software. The second theorem says: diversity may pay off with high quality software. The conclusion from these theorems is, that software diversity should not be used as a substitute for careful software construction or for testing effort. Software diversity should rather be used as an extra measure to fulfill especially high safety demands.

Keywords. Computer software; probabilistic logic; reliability theory; software reliability; software correctness.

INTRODUCTION

There is an ongoing debate on whether probabilistic terminology and thinking can be applied to software. The opponents to probabilistic theories on software dependability argue, that statistics about the correctness of programs are not feasible: Every program exists only once even when it is copied, and programs do not fail in the same sense as hardware does. Programs are either correct forever or are faulty right from the beginning. Consequently probabilistic modelling seems to be inadequate. And there are examples of false modelling resulting from only superficial analogies between hardware and software reliability. The opponents to probabilistic thinking prefer a more logical treatment of software dependability according to the logical character of software.

On the other hand probabilistic thinking has some tradition in the field of software reliability and there are strong arguments for it: programs may contain (pessimists would say: they do contain) some errors, and these errors become apperent in a statistical manner according to the statistics of the environment and input process. This paper aims at a synthesis of the two ways of thinking.

For evaluating software dependability two measures will be introduced: the *failure probability* and the *probability of correctness*.
These notions cannot be founded on a statistical theory of probability: There are no statistics on propositions like "tomorrow the sun will be shining". And there are no statistics on the proposition "this program will ever perform correctly" either.

Therefore a theory of *logical probabilities* has to be developed which is much more adequate to the evaluation of software than statistical probability.

In the following logical expressions and predicates are written in a Pascal like manner (ISO 7185).

LOGICAL PROBABILITY

The Test Predicate

Logical probabilities for programs are introduced as an example of a general theory of probabilistic induction (Carnap, Stegmüller, 1959).

Starting point is the notion of the *test predicate Q*. This predicate is the complete description of a program with regard to its failure behaviour. Its value is TRUE for all input data on which the program always operates correctly. In the theory of program proving the predicate Q is called the *weakest precondition* of the program (Gries, 1981). For large programs the construction of this predicate may be infeasible. But the notion of the test predicate Q is clear and distinct.

Definition: May variable x denote the input data of a program S. The test predicate Q(x) is defined to be TRUE if and only if the program S always performs correctly on x. Otherwise the value of Q(x) is defined to be FALSE.

The possible values of x are $a_1, a_2, ..., a_N$ (input cases). These form the input space X. Under the aspect of software dependability the programm S can be fully characterized by the values $Q(a_1), Q(a_2), ..., Q(a_N)$. If S is correct Q gets the value TRUE in all theses cases.

The evaluation of the test predicate may be difficult for complex programs. But in simple cases the predicate can be constructed by the method of correctness proofs (Baber, 1987; Gries, 1981).

For example: A program should perform the addition of two nonnegative integers A and B (input data). Program variable y should show the result: y=A+B. This expression serves as postcondition of the program. The program S is given by "z:= A; y:= B; y:= z+y". The internal representation of nonnegative integers shall be restricted to the range 0..max. Starting with the postcondition "y=A+B" and by applying the proof rule for the assignment statement the weakest precondition of the program can be constructed:

{A+B ≤ max}
{(A+B=A+B) AND (A+B ≤ max)}
z:= A;
{(z+B=A+B) AND (z+B ≤ max)}
y:= B;
{(z+y=A+B) AND (z+y ≤ max)}
y:= z+y;
{y=A+B}

"A+B ≤ max" is the weakest precondition of the program with respect to the postcondition "y=A+B". The variables A and B are free in both predicates. The weakest precondition can be written in the form Q(A, B). "x" has been introduced to denote the input variable: x = (A, B). Therefore Q is defined by the equivalences Q(x) = Q(A, B) = (A+B ≤ max).

For simplicity let us assume a 2-bit representation of nonnegative integers, i.e. max=3. There are 16 possible input cases:

a_1 = (0, 0) a_2 = (0, 1) a_3 = (0, 2) a_4 = (0, 3)
a_5 = (1, 0) a_6 = (1, 1) a_7 = (1, 2) a_8 = (1, 3)
a_9 = (2, 0) a_{10} = (2, 1) a_{11} = (2, 2) a_{12} = (2, 3)
a_{13} = (3, 0) a_{14} = (3, 1) a_{15} = (3, 2) a_{16} = (3, 3)

These values form the input space X. The values $Q(a_1), Q(a_2), ..., Q(a_N)$ of the test predicate can be evaluated for every input. Examples:

$Q(a_3)$ = Q(0, 2) = (0+2 ≤ 3) = (2 ≤ 3) = TRUE

$Q(a_{12})$ = (2+3 ≤ 3) = (5 ≤ 3) = FALSE

The test predicate can be conceived as a function:

Q: X ⟶ {FALSE, TRUE}

If a system contains only one program, this function contains everything relevant with respect to reliability or safety. The function Q serves as a *case description* of the system.

Let M denote the set of all case descriptions, i.e. the set of all Boolean functions defined on X. The number of elements contained in some set S will be denoted by c(S), where "c" is an abbreviation of "cardinality". The set M contains 2^N elements: $c(M) = 2^N$. This number is generally very large. Let the input data be represented by K bits. The cardinality of the input space takes the value 2^K ($N = c(X) = 2^K$). In very simple cases, where the input data can be represented by 2 bytes only the cardinality of the set of case descriptions is beyond imagination: $N = 2^{2 \cdot 8} = 2^{16}$ = 65536; this yields $c(M) = 2^{65536}$.

A system may contain several (say n) different programs with the same input space and designed to fulfill the same specification. Then a *case description D* is given by the n-tupel of test-predicates of these programs: $D = (Q_1, Q_2, ..., Q_n)$.

Now a formal language can be introduced, containing the following elements:

1. the constants $a_1, a_2, ..., a_N$ which represent all possible input cases
2. all possible test predicates for all programs
3. variables for input data and test predicates
4. logical operators like AND, OR and NOT
5. existential and universal quantifiers

We are dealing with (possibly very large but) finite sets. Therefore existential and universal quantifiers can be defined in analogy to

$$OR_{y \in Y} Q(y) = Q(y_1) \text{ OR } Q(y_2) \text{ OR } ... \text{ OR } Q(y_m)$$

and

$$AND_{y \in Y} Q(y) = Q(y_1) \text{ AND } Q(y_m) \text{ AND } ... \text{ AND } Q(y_m),$$

where $Y = \{y_1, y_2, ..., y_m\}$.

An example of propositions (predicates) formulated within this language is "$OR_{x \in Y} NOT\ Q_1(x)\ AND\ NOT\ Q_2(x)$". In natural language it reads "There exists an input x in the input space Y for which neither program 1 nor program 2 is guaranteed to operate correctly." The test predicates Q_i are considered variable and one can ask for the case descriptions D for which the stated proposition A is valid. All these case descriptions form the *logical scope* LS(A) of the proposition:
LS(A) = {D | A} = set of all D for which A gets the value TRUE.

Theorems on programs and systems can be formulated elegantly by the use of the following

Convention on Quantifiers: The propositions $AND_{v \mid B} E$ and $OR_{v \mid B} E$ are called universal and existential quantifier respectivley. Furthermore we write $\Pi_{v \mid B} E$ and $\Sigma_{v \mid B} E$ for sums and products. In theses expressions v is an m-tupel of index variables: $v = (v_1, v_2, ..., v_m)$. The *range* of a quantifier is the set of all possible combinations of constants (values of theses variables) for which condition B is TRUE. For each element of the range there is one operand of the quantifier. This operand is obtained by replacing in E all index variables by the respective constants. The replacement procedure is restricted to the free variables.

Example:

$OR_{(i, j) \mid (1 < i < 4)\ AND\ (0 < j < 3)}$ (i+j=3)
= (2+1=3) OR (2+2=3) OR (3+1=3) OR (3+2=3)
= TRUE OR FALSE OR FALSE OR FALSE = TRUE

The universal quantifier "$AND_{v \mid B} E$" reads "for all combinations of indices v in the range B expression (predicate) E is true". The existential quantifier "$OR_{v \mid B} E$" has the interpretation "there exists a combination of indices v in the range B for which E is true".

The following equivalences hold:

$AND_{v \mid B} E = AND_v (B \leq E)$

$OR_{v \mid B} E = OR_v (B \text{ AND } E)$

$\Sigma_{v \mid B} E = \Sigma_v (ord(B) \cdot E)$

$\Pi_{v \mid B} E = \Pi_v (ord(B) \cdot E + 1 - ord(B))$

where "≤" reads "implies" (like in Pascal). "AND_v" is an abbreviation of "$AND_{v \mid TRUE}$", i.e. the range of the quantifier extends over all possible combinations of indices. The same holds for the other quantifiers. The function ord is defined as follows: ord(FALSE) = 0 and ord(TRUE) = 1.

For all these quantifiers (OR, AND, Σ, Π) all possible variables of arithmetic or logical type as well as case descriptions (predicates) may serve as indices. The proposed formal notation is sufficiently powerful to talk about program dependability.

Definitions and Rules

In the following theory the case descriptions play the same role as the elementary events known from the usual treatment of probability theory.

Logical probabilities are introduced by defining a measure μ(D) for case descriptions which has the two elementary properties: $0 \leq \mu(D) \leq 1$ and $\Sigma_D \mu(D) = 1$. The *logical probability of a proposition* A is defined to be the sum of all measures of the case descriptions belonging to its logical scope:

$p(A) = \Sigma_{D \in LS(A)} \mu(D) = \Sigma_{D \mid A} \mu(D)$

In both cases the sum extends over all case descriptions for which the condition A is valid.

From this definition Kolmogoroff's axioms can be derived as rules or theorems respectively (Papoulis, 1965). The most important rules are given below.

1st rule: $0 \leq p(A) \leq 1$ for all propositions A

2nd rule: p(TRUE) = 1 (The probability of tautologies equals one)

3rd rule: From A AND B = FALSE follows p(A OR B) = p(A) + p(B)

Corollary: p(A OR B) ≤ p(A) + p(B) for all propositions A and B

For brevity we shall write "AB" instead of "A AND B" and "¬" instead of "NOT". From rule 2 and 3 follows

4th rule: $p(\neg A) = 1 - p(A)$

5th rule: If A implies B, then $p(A) \leq p(B)$.

This follows from the fact that LS(A) forms a subset of LS(B).

The *conditional probability of A assuming B* is defined by p(A | B) = p(AB) / p(B). The 6th rule is concerned with relations between probabilities:

6th rule: p(A | B) / p(A) = p(B | A) / p(B)

But how can we calculate these probabilities? What is the value of e.g. $p(OR_{x \in Y} \text{ NOT } Q_1(x) \text{ AND NOT } Q_2(x))$? The difficulty lies in the fact that we usually do not know the values μ(Q). We do not try to get estimates on the measure μ. Nevertheless, far reaching theorems on programs and on program dependability can be derived without this knowledge.

FOUNDATIONS OF SOFTWARE DEPENDABILITY

Failure Probabilities

The system under consideration consists of only one program. The case description D is therefore given by the test predicate Q of this program.

Let r(x) denote the *data dependent failure probability:*

$r(x) = p(\neg Q(x)) = \Sigma_{Q \mid \neg Q(x)} \mu(Q)$

r(x) equals the probability that a program will fail on input x. The sum extends over the logical scope of the proposition "The program fails on input x".

Let q(x) denote the probability of input x. The fact that the program is characterized by Q (test predicate) and that the input of the program is given by x can be represented by the ordered pair (Q, x). We define a measure on these ordered pairs by the *postulation of independence* of case descriptions and input data:

$\mu(Q, x) = \mu(Q) \cdot q(x)$

The data independent failure probability p is defined as the mean value of the proposition "¬ Q(x)" over all case descriptions and all input data. The definition of probabilities has to be extended to product spaces in a straightforward manner. By calculating

$p = \Sigma_{(Q, x) \mid \neg Q(x)} \mu(Q, x) = \Sigma_{x \in X} \Sigma_{Q \mid \neg Q(x)} \mu(Q, x)$
$= \Sigma_{x \in X} \Sigma_{Q \mid \neg Q(x)} \mu(Q) \cdot q(x) = \Sigma_{x \in X} r(x) \cdot q(x)$

we get the result of

Theorem 1: $p = E[r(x)] = \Sigma_{x \in X} r(x) \cdot q(x)$.

E[Z] denotes the expected (or: mean) value of the random variable Z. Therefore we will call p the *mean failure probability* of the program from now on.

Probability of Correctness

The complement 1-p of the mean failure probability gives the probability of a correct result for a randomly chosen input. The measure 1-p is proportional to the *degree of correctness* of a program. Now we want to calculate a completely different measure on program dependability: the *probability of correctness k* of a program.

To be realistic we restrict the input space X to a subset Y. Y may contain all relevant input cases. The probability of correctness k is defined to be the probability of the proposition "On all relevant input cases the program yields correct results". More formalistic the proposition reads "$AND_{x \in Y} Q(x)$".

We take the complement of this measure, the probability of incorrectness

$u = 1 - k = 1 - p(AND_{x \in Y} Q(x)) = 1 - p(\neg OR_{x \in Y} \neg Q(x))$
$= p(OR_{x \in Y} \neg Q(x))$

Because we do not know the measure μ we are not able to calculate u (or k respectively). But during the following paragraphs it will be shown, that for more complex systems it is possible to express the probability of correctness of the whole system in terms of the correctness of its components.

SOFTWARE REDUNDANCY

Two Diverse Programs

Now let the system contain two programs according to one common specification. The *case description of the whole system* can be introduced as the ordered pair of the case descriptions (test predicates) Q_1 and Q_2 of program 1 and program 2:
$D = (Q_1, Q_2)$.

Let us assume the two programs have been developed by different programming teams both following the same specification. The two teams may possess the same or a different background of knowledge and they may have followed the same design method or they may have been forced to use different ones (forced diversity).

In general we have to distinguish between the case descriptions of program 1 and program 2. We shall write M_1 and M_2 instead of M, μ_1 and μ_2 instead of μ etc. If the programming teams possess the same background of knowledge and if the teams have freely chosen the methods we get $M_1 = M_2$ und $\mu_1 = \mu_2$. This yields then truly independent versions.

We define a measure on the ordered pairs (Q_1, Q_2) according to the *postulation of diversity*:

$$\mu(D) = \mu(Q_1, Q_2) = \mu_1(Q_1)\cdot\mu_2(Q_2)$$

The programs may be designed such that they are able to monitor the results of their counterparts and to issue an alarm if their results are different. We define the whole system to work correctly if the results of both programs are correct or if an error is detected and reported. This means that the system performs well on given input x if either program 1 or program 2 or both show the correct result. The correctness of the system with respect to the input data x can be described by the proposition "$Q_1(x)$ OR $Q_2(x)$".

m-Out-of-n Systems

The *m-out-of-n systems* consist of n diverse programs, monitoring each other according to a certain scheme. The system performs correctly in all cases where m of the n programs yield the correct results.

The systems with two diverse programs described above are 1-out-of-2 systems. One further very interesting class of m-out-of-n systems are the *t-fault diagnosable systems* (Preparata, Metze, Chien, 1967): "A system of n units is one-step t-fault diagnosable if all faulty units within the system can be identified without replacement provided the number of faulty units present does not exceed t."

On the condition that the number of programs yielding a faulty result is less than or equal to t the correct result can be found. Such systems must contain more then $2\cdot t$ programs: $n > 2\cdot t$. On the other hand there always exist t-fault diagnosable systems with exactly $2\cdot t+1$ programs. Obviously the relation $m + t = n$ holds in these cases. A 1-fault diagnosable system contains at least 3 programs. The smallest 1-fault tolerant system fitting this scheme is the 2-out-of-3 system where the three programs monitor each other.

FAILURE PROBABILITY OF 1-OUT-OF-2 SYSTEMS

"K(x)" be a formal representation of the proposition "At least one of the programs 1 or 2 yields the correct result on the input x": $K(x) = Q_1(x)$ OR $Q_2(x)$. The proposition K is called *predicate of correctness* of the system. To express system failure we write "$\neg K(x)$". Obviously $\neg K(x) = \neg Q_1(x)$ AND $\neg Q_2(x)$. The data dependent failure probability of the system $r_{div}(x)$ is defined by $r_{div}(x) = p(\neg K(x))$.

Theorem 2: $r_{div}(x) = r_1(x)\cdot r_2(x)$

Proof:
$$\begin{aligned}r_{div}(x) &= p(\neg K(x)) = \Sigma_{D\,|\,\neg K(x)}\,\mu(D)\\&= \Sigma_{(Q1,\,Q2)\,|\,\neg Q1(x)\,\text{AND}\,\neg Q2(x)}\,\mu_1(Q_1)\cdot\mu_2(Q_2)\\&= \Sigma_{Q1\,|\,\neg Q1(x)}\Sigma_{Q2\,|\,\neg Q2(x)}\,\mu_1(Q_1)\cdot\mu_2(Q_2)\\&= \Sigma_{Q1\,|\,\neg Q1(x)}\,\mu_1(Q_1)\cdot\Sigma_{Q2\,|\,\neg Q2(x)}\,\mu_2(Q_2)\\&= p(\neg Q_1(x))\cdot p(\neg Q_2(x)) = r_1(x)\cdot r_2(x).\end{aligned}$$

In analogy to the derivation of theorem 1 it can be shown that $p_{div} = E[r_{div}(x)] = \Sigma_{x\in X}\,r_{div}(x)\cdot q(x)$. Now the notion of covariance can be introduced by the definition:

$$Cov(r_1(x), r_2(x)) = E[(r_1(x)-p_1)\cdot(r_2(x)-p_2)].$$

The result of the calculations

$$\begin{aligned}Cov(r_1(x), r_2(x)) &= \Sigma_{x\in X}\,(r_1(x)-p_1)\cdot(r_2(x)-p_2)\cdot q(x)\\&= \Sigma_{x\in X}\,r_1(x)\cdot r_2(x)\cdot q(x) - p_1\cdot p_2\\&= \Sigma_{x\in X}\,r_{div}(x)\cdot q(x) - p_1\cdot p_2 = p_{div} - p_1\cdot p_2\end{aligned}$$

is presented in

Theorem 3 (Littlewood, Miller 1989): The formula $p_{div} = p_1\cdot p_2 + Cov(r_1(x), r_2(x))$ holds for diverse 1-out-of-2 systems.

If the programming teams possess the same background of knowledge and if the teams can freely choose the methods the failure probabilities of the program can be denoted by p ($p = p_1 = p_2$). And the covariance can be replaced by the variance σ^2. Theorem 3 now takes the form of

Theorem 4 (Eckhardt, Lee, 1985): In the case of truly independently developed program versions the formula $p_{div} = p^2 + \sigma^2$ holds for 1-out-of-2 systems.

In the quoted papers the two above theorems are based on urn models. To get the results of theorems 3 and 4 on the basis of small populations of diverse programs the *urn model with replacement* has to be chosen. This model is inadequate, because it is possible to choose identical program versions which therefore will show the same failure behaviour. The *urn model without replacement* fits the situation of software diversity much better. Unfortunately some difficulties are emerging by the use of this model: Within the derivations of the above theorems some limiting processes are necessary.

It seems to be more satisfying to derive the theorems 3 and 4 in full generality on the basis of logical probabilities. Within the area of logical probabilities the axioms (postulations) and premises of the theory of software dependability can be stated clearly and unequivocally.

General results for m-out-of-n systems can be derived following Eckhardt/Lee (1985).

PROBABILITY OF CORRECTNESS: REDUNDANT SYSTEMS

The 1-Out-of-2 System

From the formulas for the probability of incorrectness u of the simple system we can derive the probability of incorrectness u_{div} for the diverse system by substituting "$Q_1(x)$ OR $Q_2(x)$" for "Q":

$$u_{div} = p(OR_{x \in Y}(\neg Q_1(x) \text{ AND } \neg Q_2(x)))$$

The probabilities of incorrectness of the two programs are denoted by u_1 and u_2 respectively.

The proposition "$OR_{x \in Y}(\neg Q_1(x) \text{ AND } \neg Q_2(x))$" implies proposition "$(OR_{x \in Y} \neg Q_1(x)) \text{ AND } (OR_{x \in Y} \neg Q_2(x))$": The former gets the value TRUE if there exists an input x on which both programs may fail. And the second proposition takes the value TRUE if there exists an input x_1 on which program 1 may fail and an input x_2 (possibly different from x_1) on which program 2 may fail. From this follows that the logical scope of the first proposition forms a subset of the logical scope of the second one.

By using the 5th rule for logical probabilities we get

$$u_{div} = p(OR_{x \in Y}(\neg Q_1(x) \text{ AND } \neg Q_2(x)))$$
$$\leq p((OR_{x \in Y} \neg Q_1(x)) \text{ AND } (OR_{x \in Y} \neg Q_2(x)))$$

The further deriviations follow the lines of the proof of theorem 2. This yields

Theorem 5: $u_{div} \leq u_1 \cdot u_2$

It is worthwhile to compare theorems 4 and 5: In the case of truly independently chosen program versions theorem 4 implies $p_{div} \geq p^2$ whereas theorem 5 yields $u_{div} \leq u^2$.

Theorem 4 says that diversity may be useless in such cases where it is used to mask errors of poor programs. On the other hand theorem 5 is encouraging: If the design process aims at correct programs and if the quality of the software is very high such that low figures for u can be achieved, then diversity may pay off.

m-Out-of-n Systems

The programs of the m-out-of-n system are numbered: $S_1, S_2, ..., S_n$. The set of indices is denoted by I: I = {1, 2, ..., n}. A case description D of the whole system is given by the n-tupel of the test predicates of the programs: $D = (Q_1, Q_2, ..., Q_n)$.

The m-out-of-n system performs correctly on input x if at least m programs yield the correct result on x. The predicate of correctness K(x) of an m-out-of-n system is given by

$$K(x) = OR_{J \mid c(J) \geq m} \text{ AND}_{j \in J} Q_j(x) \text{ AND}_{i \in I \setminus J} \neg Q_i(x)$$

where J denotes a subset of I and c(J) denotes the cardinality of J. The existential quantifier "$OR_{J \mid c(J) \geq m} E(J)$" reads: "There exists a subset J of I containing at least m elements for which proposition E(J) holds." The set I \ J contains all elements of I not contained in J. Obviously relation $c(I \setminus J) = n - c(J)$ holds.

The predicate of correctness K(x) reads: "There exists a subset J of I containing at least m elements such that all programs S_j, j∈J, yield the correct result on input x. All the other programs S_j, j∈I \ J, are not always working correctly on x."

Let the negation of the predicate of correctness "$\neg K(x)$" assume value TRUE. We take J to be a subset of I thus that the proposition "$\text{AND}_{j \in J} Q_j(x) \text{ AND}_{i \in I \setminus J} \neg Q_i(x)$" holds. This set of indices consists of less than m elements: c(J)<m. Conversely "K(x)" would hold contrary to the assumption. Therefore proposition "$\neg K(x)$" implies "$OR_{J \mid c(J)<m} (\text{AND}_{j \in J} Q_j(x) \text{ AND}_{i \in I \setminus J} \neg Q_i(x))$". The latter proposition implies "$OR_{J \mid c(J)<m} \text{AND}_{i \in I \setminus J} \neg Q_i(x)$" which is equivalent to "$OR_{J \mid c(J)>t} \text{AND}_{i \in J} \neg Q_i(x)$", where t=n-m.

Obviously it is sufficient to consider sets J containing exactly t+1 elements. Finally we get: Proposition "$\neg K(x)$" implies "$OR_{J \mid c(J)=t+1} \text{AND}_{i \in J} \neg Q_i(x)$".

Let $k_{m\text{-out-of-}n}$ denote the probability of correctness of the m-out-of-n system and let $u_{m\text{-out-of-}n}$ denote its probability of incorrectness. From the above mentioned implications and by using the 3rd and 5th rule for logical probabilities we derive

$$u_{m\text{-out-of-}n} = p(OR_{x \in Y} \neg K(x))$$
$$\leq p(OR_{x \in Y} OR_{J \mid c(J)=t+1} \text{AND}_{i \in J} \neg Q_i(x))$$
$$\leq p(OR_{J \mid c(J)=t+1} \text{AND}_{i \in J} OR_{x \in Y} \neg Q_i(x))$$
$$\leq p(OR_{J \mid c(J)=t+1} \text{AND}_{i \in J} U_i)$$

where U_i is defined by $U_i = OR_{x \in Y} \neg Q_i(x)$. The proposition "$U_i$" says that "program i is not correct" or that "there exists a relevant input x on which program i fails (sometimes)". Obviously we have $p(U_i) = u_i$, where u_i denotes the probability of incorrectness of Program S_i. The corollary following the 3rd rule yields

$$u_{m\text{-out-of-}n} \leq \Sigma_{J \mid c(J)=t+1} p(\text{AND}_{i \in J} U_i)$$

The probability of proposition "$\text{AND}_{i \in J} U_i$" is the sum of the measures of all case descriptions $(Q_1, Q_2, ..., Q_n)$ for which the proposition holds. Due to the postulation of diversity

$$\mu(Q_1, Q_2, ..., Q_n) = \mu_1(Q_1) \cdot \mu_2(Q_2) \cdots \mu_n(Q_n)$$

and due to the fact that the test predicates Q_i are contained in different conjunctive components U_i of the predicate, the summation over the Q_i can be done separately. From this follows

$$p(\text{AND}_{i \in J} U_i) = p(\text{AND}_{i \in J} OR_{x \in Y} \neg Q_i(x))$$
$$= \Pi_{i \in J} p(OR_{x \in Y} \neg Q_i(x)) = \Pi_{i \in J} p(U_i) = \Pi_{i \in J} u_i$$

This is summarized in

Theorem 6: $u_{m\text{-out-of-}n} \leq \Sigma_{J \mid c(J)=n-m+1} \Pi_{i \in J} u_i$.

Example 1: Obviously theorem 6 generalizes theorem 5:
$u_{1\text{-out-of-}2} \leq \Sigma_{J \mid c(J)=2} \Pi_{i \in J} u_i = \Pi_{i \in I} u_i = u_1 \cdot u_2$.

Example 2: $u_{2\text{-out-of-}3} \leq u_1 \cdot u_2 + u_1 \cdot u_3 + u_2 \cdot u_3$.

Theorem 6 can be strengthened. We do this for 2-out-of-3 systems using the fact, that from the 3rd rule for logical probabilities the following equality can be derived:

$$p(A \text{ OR } B) = p(A) + p(B) - p(AB)$$

Theorem 7:
$$u_{\text{2-out-of-3}} \leq u_1 \cdot u_2 + u_1 \cdot u_3 + u_2 \cdot u_3 - 2 \cdot u_1 \cdot u_2 \cdot u_3$$

Proof:
$$\begin{aligned}
u_{\text{2-out-of-3}} &\leq p(\text{OR}_{J \mid c(J)=2} \text{ AND}_{i \in J} U_i) \\
&= p(U_1 U_2 \text{ OR } U_1 U_3 \text{ OR } U_2 U_3) \\
&= p(U_1 U_2 \text{ OR } (U_1 U_3 \text{ OR } U_2 U_3)) \\
&= p(U_1 U_2) + p(U_1 U_3 \text{ OR } U_2 U_3) - p(U_1 U_2 U_3) \\
&= p(U_1 U_2) + p(U_1 U_3) + p(U_2 U_3) - 2 \cdot p(U_1 U_2 U_3) \\
&= u_1 \cdot u_2 + u_1 \cdot u_3 + u_2 \cdot u_3 - 2 \cdot u_1 \cdot u_2 \cdot u_3
\end{aligned}$$

PROGRAMMING EXPERIMENTS AND URN MODELS

May a population of K truly independently developed program versions on the basis of one common specification be given, let N be the number of representatively chosen test cases and let n_i denote the number of test cases where i of the K versions fail, then the probability that a randomly chosen version fails on a randomly chosen test case equals p^*:

$$p^* = \sum_{i=1}^{K} (i/K) \cdot (n_i/N) = 1/(K \cdot N) \sum_{i=1}^{K} i \cdot n_i$$

The value p^* may serve as an "estimate" of the failure probability p of one single program. An "estimate" p_{div}^* of the system failure probability p_{div} of the diverse 1-out-of-2 system can be calculated by means of the *urn model without replacement*

$$\begin{aligned}
p_{\text{div}}^* &= \sum_{i=2}^{K} i/K \cdot (i-1)/(K-1) \cdot n_i / N \\
&= 1/(K(K-1)N) \cdot \sum_{i=2}^{K} (i-1) \cdot i \cdot n_i
\end{aligned}$$

Let F denote the number of incorrect versions, i. e. K-F versions are always working correctly. The "estimates" of the probabilities of incorrectness are $u^* = F/K$ and $u_{\text{div}}^* = F/K \cdot (F-1)/(K-1)$ where u_{div}^* is defined under the hypothesis that a 2-version system fails if both programs contain faults.

Example: A group of 16 programmers have solved the following problem independently: A search algorithm shall be designed to find the position where a certain word (sequence of characters) appears for the first time in a given text file.

Finally there existed 16 versions of the program: K = 16. A 17th program - the gold version - had been designed following a correctness proof. Error detection was made by comparing the output of each of the 16 versions with that of the gold version.

Test results: There were 1000 randomly chosen test cases. All versions worked correctly on 814 test cases. Two versions failed on 163 test cases. These two versions and a third version failed on 23 test cases. Faults were detected in these three versions only.

This yields the following "estimates" for p, p_{div}, u and u_{div} respectively:

$$\begin{aligned}
p^* &= 0.025 \\
p_{\text{div}}^* &= 0.002 \\
u^* &= 3/16 \\
u_{\text{div}}^* &= 0.025
\end{aligned}$$

The basis of a population of only 16 versions is much too small for drawing conclusions about the efficiency of software diversity. Moreover, we had set up a rather artificial experimental situation (class room assignment): The above problem is extremely small and the "best practice" development (based on a correctness proof) had been reserved for the gold version only.

In real world situations random drawings of programs are not available and there is no gold version at hand: All programs (perhaps one silver and one bronze version) are the objects of our probabilistic reasoning and cannot be used to get estimates of the unknown probabilities p, u, p_{div} and u_{div}.

REFERENCES

Baber, R. L.: *The Spine of Software*. John Wiley, 1987

Carnap, R.: *Logical Foundations of Probability*. Chicago 1950

Carnap, R.; Stegmüller, W.: *Induktive Logik und Wahrscheinlichkeit*. Springer, Wien 1959

Eckhardt, D. E.; Lee, L. D.: A Theoretical Basis for the Analysis of Multiversion Software Subject to Coincident Errors. IEEE Trans. on Software Engineering, SE-11(1985)12, 1511-1517

Grams, T.: *Denkfallen und Programmierfehler*. Springer, Heidelberg 1990

Gries, D.: *The Science of Programming*. Springer, Heidelberg 1981

Littlewood, B.; Miller, D. R.: Conceptual Modeling of Coincident Failures in Multiversion Software. IEEE Trans. SE-15(1989)12, 1596-1614

Papoulis, A.: *Pobability, Random Variables and Stochastic Processes*. McGraw-Hill, New York 1965

Preparata, F. P.; Metze, G.; Chien, R. T.: On the Connection Assignment Problem of Diagnosable Systems. IEEE Trans. EC-16(1967)6, 848-854

ISSUES OF SECURITY

METHODS OF PROTECTION AGAINST COMPUTER VIRUSES

K. Gaj*, K. Górski*, R. Kossowski* and J. Sobczyk**

*Warsaw University of Technology, Institute of Telecommunications, Nowowiejska 15/19, 00–665 Warsaw, Poland
**Warsaw University of Technology, Institute of Automatic Control, Nowowiejska 15/19, 00-665 Warsaw, Poland

Abstract. The paper presents an examination of the threat to computer systems posed by computer viruses and an assessment of counter-measures which reduce the risk of an infection. In the first half existing viruses are analysed and classified, and available protection methods are evaluated. The authors' experiences in dealing with viruses under PC-DOS are reported and brief information is given about an anti-virus program developed by one of the authors. In a more general approach the second half of the paper describes mechanisms which can be built into operating systems to reduce or eliminate the risk of virus infection. The proposals deal with assuring a high level of virus resistance, regardless of actual virus code, without a substantial downgrade in performance. General security measures are described and two ideas suggested for universal virus detection mechanisms.

Keywords. safety; system integrity; system failure and recovery; computer viruses; operating systems

1 INTRODUCTION

Computer viruses have become an increasing threat to a majority of computer systems over the last years. Personal computer systems are especially exposed to the danger of a virus epidemic, as they do not have almost any built-in mechanisms of protection against viruses. Nevertheless, it appears that none of the existing computer operating systems are satisfactorily resistant to the wide variety of existing viruses.

In the first part of this paper after giving a formal definition of a computer virus, we describe a typical structure of the virus code. Then, we propose a detailed classification of computer viruses known so far. The presented classification is important and useful because viruses belonging to the same group, according to this classification, can be found out and removed using the same or similar method. More precisely we deal with viruses on IBM PC and compatibles, under PC DOS. Then, we discuss known to us methods of protection against computer viruses. The proposed classification of such methods is useful as it facilitates a credible assessment of preventive capabilities of existing operating systems. It also enables to evaluate a large number of protection programs available under those operating systems. An original program developed under PC DOS, using some of the presented methods of protection, is described. Our experiences in preventing a virus epidemic, under PC DOS, in the Department of Electronics Engineering at the Warsaw University of Technology are reported.

In the second part of the paper we present a more generalised approach to virus protection and discuss the need for and the possibility of making operating systems more resistant against viruses. We describe several security mechanisms which help in limiting the spread of viruses. Towards the end two ideas are suggested for detecting the presence of viruses in the system. The emphasis in this part of the paper is on measures which work against viruses irrespective of their code or their characteristics (such as rate of replication).

2 DEFINITION OF A COMPUTER VIRUS

A computer virus is a piece of a program that exists within an otherwise normal program. When this program is executed, the virus code seeks out other programs and replicates itself. Several more or less formal definitions of a computer virus were reported. The most appropriate according to us seems to be Burger's definition (Burger 1989) which describes a computer virus as a program that is characterized by the following features:

1. modifies software that does not belong to this program;
2. can make modifications in a group of programs;
3. can detect modifications made by itself;
4. on the basis of 3., avoids repeated modifications of programs.

Additionally

5. Modified software gets features enumerated in 1.-4.

According to this definition a program that lacks one or more of above mentioned features should not be named a virus.

3 BASIC FUNCTIONAL BLOCKS OF A COMPUTER VIRUS

In the code of a majority of viruses the following functional blocks can be distinguished:

1. Virus token (identifier, marker) i.e. part of the code which enables a virus to recognize that a given program is infected. Running this part of the computer code usually does not cause any real effect. Virus identifier usually consists of instructions like nop (no operation), jump to the next instruction etc. Functions of the token may also be fulfilled by the file parameters. For example, a virus can change the time and date of file creation to specific values, which are later recognized as a mark of infection.

2. Virus kernel i.e. the fundamental part of a virus, which according to the definition given above, must always be implemented. It performs all functions associated with virus replication. Its basic tasks are:

   (a) to find a program uninfected so far,
   (b) to assure the possibility of its modification,
   (c) to alter the program in such a way that it assumes the structure of the infected software.

3. Manipulation (executive) unit. It appears optionally. If it is used to "let the user know" that the virus is in the system. It can be as harmless as displaying a moving ball or falling letters on the screen, or as vicious as deleting files, formatting a hard disk or overwriting randomly chosen sectors on a disk. Activation of a manipulation unit usually depends on fulfillment of some special conditions (number of times the program was run, free disk space reduced below fixed level etc.).

4 CLASSIFICATION OF COMPUTER VIRUSES

A fundamental feature of computer viruses is their ability for recurrent multiplication. Thus the most important criterion of virus classification seems to be the time and mechanism of virus multiplication. According to this criterion we can distinguish the following types of viruses:

1. Viruses which infect other programs while running the infected program (carrier program).

   When the carrier program is run then actually the virus code is run at first. The virus kernel looks for an appropriate executable file without a virus token. When such a file is found the virus replicates, writing its code to this file. Then, if the virus has got a manipulation unit, a condition of its triggering is checked. If this condition is fulfilled the manipulation unit is activated. Then, the carrier program or its indestructed part is run.

2. Viruses which infect other programs during system service (request, interrupt, extra code, trap) procedures.

These viruses install themselves in the operational memory and change system behaviour in such a way that, during one or more system services, the virus code is invoked at first. We can further classify these viruses according to the time of the virus code installation into those which install themselves during:

A system boot;

B start of the computer program.

The additional criterion of classification of viruses is the type of disk space occupied by the virus code. Most viruses append or overwrite executable files, the others occupy system areas of the disk.

According to the additional effects of multiplication viruses can be divided into two groups which cause respectively:

1. irreversible changes on the disk by overwriting existing files or system areas of the disk,

2. reversible changes assured by appending files or rewriting those areas of the disk or file that are to be overwritten into free areas on the disk.

According to the function of the manipulation unit we can divide viruses into those:

1. that do not possess a manipulation unit,

2. whose manipulation unit causes only temporary effects,

3. whose manipulation unit causes permanent damages of hardware or software stored on the disk.

For a particular operating system - PC DOS - we can give the presented above classification in greater detail. For example we can further classify viruses from the group 2, i.e. those which install themselves in the operational memory. The following moments of installation can be distinguished:

A1 reset from an infected floppy diskette or fixed disk;

   A virus changes the contents of a boot record on the floppy diskette or fixed disk. After reset, the program from the boot sector is activated. This is actually the virus code. It can read into operational memory the remaining parts of the virus program from other sectors of the disk. Then, it installs itself as a resident program and changes vectors of selected interrupts so that they point to the virus code.

A2 loading infected system;

   A virus code replaces standard interrupt service procedures or internal system commands in system files (IBMDOS.COM, IBMBIO.COM, COMMAND.COM). Thus the loading of the infected system is equivalent to installation of the resident virus code.

B running an infected program;

   Running a carrier program causes installation of the virus code and appropriate changes of selected interrupt vectors.

The additional criterion of classification of viruses is the type of disk space occupied by a virus code. Under PC DOS, viruses from groups 1 and 2B overwrite or append executable files i.e. COM, EXE and overlay files. We can associate every virus from these groups with types of executable files it can infect. The most numerous are viruses which infect only COM files and those which are able to infect all mentioned above types of executable files. Viruses from group 2A1 write themselves to boot sectors of floppy diskettes and hard disks. The most numerous are viruses which infect only floppy diskettes and those which infect floppy diskette boot and one of the fixed disk boots (standard or master boot).

Viruses from group 2A2 overwrite or append one of the system files like COMMAND.COM, IBMBIO.COM, IBMDOS.COM, driver SYS files etc.

5 METHODS OF PROTECTION AGAINST VIRUSES

Fig. 1 illustrates methods of protection against viruses known to the authors and their most successful implementations under PC DOS. The basic task in the struggle against viruses is detection and localization of infection. Although it does not solve all problems, it allows at least to delete infected files and to stop the spread of a virus. After that we can proceed with programs that are able to cure infected disks of detected types of viruses.

The best solution is to assure detection of the virus code before the carrier program is run. This usually relies on finding out a virus identifier or a characteristic fragment of code, so this method can not be used, when we do not know a given virus. Additionally, a virus can defend itself against such methods by changing its code during replication. An unknown virus can be looked for on the basis of specific instructions included. Such method is usually not very efficient because of difficulties in pointing out suspicious instructions, and a virus can defend against it using self-modification or self-encryption.

Resident anti-virus monitoring procedures allow for detecting a virus while running the viral code and usually protect against effects of a virus operation. Before executing every suspicious system function, these procedures pause execution of a program and ask the user, whether to continue or halt the program. The most efficient are those resident procedures that monitor functions characteristic for a virus kernel. Unfortunately, a sophisticated virus may avoid or deactivate monitoring procedures.

One of the most efficient seems to be detection of a virus on the basis of effects of its operation, after a virus has spread in a computer system. One of the most simple is the method of bait in which an especially prepared file is stored in an easily accessible place on a disk, and periodically compared with its copy hidden under a meaningless name. The other more general method is based on preparing a file containing information about chosen files and directories on the disk. This information includes, for example, for each file: its length, attributes, date and time of its creation, and two checksums computed using different algorithms. After some time, a second file containing the same information is created. Contents of both files are compared and differences are reported. By analysing these differences, the spread of viruses can be simply detected.

Quite effective, but rather troublesome is the method based on the use of harmless viruses that spread in the computer system and detect other viruses or protect against effects of their operation.

In the avoidance of computer viruses very helpful are those programs that enrich defensive capabilities of an operating system. For DOS such programs allow, for instance, to create unconditionally write-protected and protected by a password disk partitions.

6 FIGHTING AGAINST VIRUSES UNDER PC DOS

6.1 Strategy of fighting against viruses

In Fig. 1 the most successful, known to us, implementations of methods of protection against viruses under PC DOS are presented. All of them are system-independent programs, as PC DOS has not got any preventive mechanisms built in. Among presented programs the basic, and at the same time the most effective ones, are those that enrich operating system security functions. For example programs for disk management, which allow to define several users of a single PC DOS system with different rights of access to created by a super user

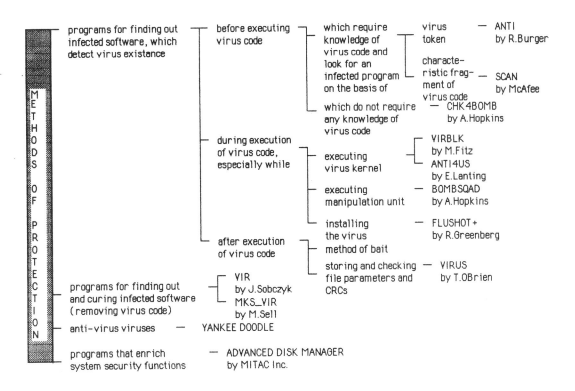

Figure 1: Methods of protection against computer viruses.

disk partitions. Every user is identified using a password. Similar and even more powerful mechanisms of protection are offered by programs for network management (for instance in Novell NetWare). Those mechanisms include, for example:

- password protected accounts,
- two level user hierarchy (supervisor, user),
- file ownership,
- file level write access protection,
- extended set of file attributes including *right to modify file attributes*, and *execute only* attribute.

Our experiences show that the mentioned above group of methods can significantly reduce the risk of virus infection. They are especially recommended to be used in student computer laboratories. Unfortunately presented methods have also got significant drawbacks. They allow viruses to spread freely in a limited area of disk space accessible to a particular user and additionally they are entirely irresistant to faults of a super user.

Thus, other methods seem to be worth considering. Methods that do not need the knowledge of the virus code are more general, but not always effective. Those that are based on the knowledge of a virus are less general, but have an advantage that they allow not only to detect a virus but also, in some cases, to remove effects of its operation.

Using public-domain software, it is good to check it at first using programs for finding out the virus code during the execution of a virus. Unfortunately, available programs (like VIRBLK, BOMB-SQAD, FLUSHOT+) are not resistant against all possible types of viruses, and may be rather troublesome as they slow down system operation and invoke a user.

If an unknown virus gets through this first barrier, it can be caught on the basis of effects of its operation. Method based on calculating, storing, and periodically checking CRC of disk files is easily implementable and very effective. Its drawback lies in the fact that because of the amount of time needed for calculating CRC's it cannot be repeated very often, and because of this a virus can cause significant damages in the system before being detected.

Among the methods which make use of the knowledge of a virus the most effective are those which identify a virus on the basis of characteristic fragments of code. For example, program SCAN has, included in a special ASCII file, characteristic fragments for about one hundred viruses. Because this program has only got the option of searching for viruses and not removing them, the necessity for writing an effective program for curing infected software appeared.

6.2 Description of the Virus Detector and Remover under PC DOS - VIR

The program VIR has been written by one of the authors of this paper - Jerzy Sobczyk. The first version of this program appeared in January 1988. This description concerns the version 3.01, which dates from May 1990.

VIR searches through disks, specified by the user, and additionally through the operational memory, looking for all viruses familiar to the author. If such viruses are found in the memory, partition table, boot sector or in any file on the disk they can be removed without affecting system operation. Original partition table, boot sector or file contents are reconstructed whenever possible.

Virus detection is done by comparing the program code surrounding the start address of the program with the virus code. Memory resident viruses are disarmed by modifying their code to do nothing but just pass control to the previous service routine. This ensures consistency of the interrupt service chain.

Some viruses can cause different kinds of damages by overwriting contents of disk files or sectors. This program tries to handle this by erasing damaged sectors and files (if requested).

Additionally, VIR has also capability of calculating, storing and checking CRC of most important parts of tested files (or whole files on request) and thus reporting all changes on the checked disk.

Main goals considered during design and implementation of VIR were: reliability and speed. The program is designed to be as safe as possible. Direct disk access (logical or physical sector addressing) is used only when necessary (e.g. for erasing boot sector viruses, recovering clusters used to hide a virus code and marked as bad ones, etc.). All files related tests are performed via DOS function calls. This feature enables usage of the program on logical drives, and on network drives as well.

7 VIRUS RESISTANCE AT THE OPERATING SYSTEM LEVEL

Extending our interest from PC-DOS to other operating systems we can distinguish two types of replication of viruses in the file system:

1. replication to programs on the basis of permissions possessed by already infected programs,

2. breaking of system security in order to gain permissions to replicate to programs so far inaccessible.

The breaking of system security in 2 usually has the purpose of obtaining the privileges of other users, possibly also the system operator (Super-user). With these privileges the virus can then access parts of the system otherwise out of bounds. Typical example of type 2 is breaking of passwords. Another is to force the overflow of a buffer allocated on the stack by some program in such a way that return addresses are overwritten and result in a jump to the virus, which then executes as if it was part of that program.

From the point of view of protection we can divide our countermeasures into two categories:

- passive protection which aims to make it impossible to gain undeserved privileges ie. it protects against the second type of replication while at the same time it limits how far viruses can spread without appropriate permissions,

- active protection which is directed solely against virus replication irrespective of how it occurs (whether it is of type 1 or type 2).

This can be extended to classify viruses according to whether their replication involves the breaking of system security or whether they are 'well behaved'.

Passive protection is made up of mechanisms which enhance general system security and do not rely on viruses to prove their worth. We must however assess the need for virus-specific active protection. After all it is clear that damage caused by viruses can also be caused by non-virus programs eg. Trojan Horse programs. Also, anti-virus algorithms need to be on all the time to guard against a potential infection which may take place once in a decade or not at all. Nevertheless in some systems the additional security, not dependent on user cooperation, may be desired. If developed, active protection may offer the detection of viruses, help in removing them from the system, identification of their authors. The early detection of an infection is important because many viruses are written to carry out their manipulative task when some conditions are met and the task often involves inflicting damage (eg. corrupting a data file) on the virus's closest sorroundings. Both factors mean that the potential damage increases with the number of copies of the virus. As far as system performance is concerned we need to make the algorithms event-driven so that they will be activated only in certain circumstances (such as modification of executable files) which occur infrequently.

The conclusion is that the main aim should be to improve system security in the sense of limiting the user's (and hence the virus's) ability to inflict damage and in the sense of preventing illegal spread of viruses to other users. This does not imply that we should not develop virus-specific methods. In some cases they may be more attractive to implement and use. Currently widely-used operating systems lack both adequate passive protection and any forms of active protection. It is often attempted to make up this lack of security by non-system programs such as those described in the first part of this paper. The problem with such an approach is twofold:

- the fact that these programs lie outside the operating system may make them vulnerable, the virus may circumvent them,

- they usually require active user involvement which may make them inconvenient, in other words the responsibility for protecting against viruses lies with the user and not the operating system.

We can expand the above issues into a list of recommendations for active anti-virus mechanisms in operating systems:

1. base the mechanism on features which best differentiate between viruses and 'legal' programs, avoid false alarms,

2. design it to be independent of virus code and characteristics, build only upon those features which all viruses (both present and future) are certain to have or else try to predict how viruses will evolve and make it possible to adapt the mechanism to those changes,

3. make it transparent to the user, independent of other security schemes in the system,

4. minimise its impact on system performance.

The measures presented in the next section in addition to being effective against viruses on their own, also serve to prevent the virus from tampering with active protection mechanisms.

8 PASSIVE PROTECTION

Many anti-virus programs for PC-DOS monitor accesses to disks. Whenever write access is attempted to an executable file the anti-virus program steps in and asks the user for permission to carry out the request, listing the name of the file which is to be accessed. Thus the user is made aware of all such accesses and can decide whether they are justified or suspect. The monitoring is carried out by intercepting calls to the operating system (software interrupts) by entering the address of the monitoring routine into the interrupt vector table. This has two drawbacks: the virus code can just as easily restore the address of the original routine into the table thus circumventing the anti-virus program or it could come with its own disk access routines and never use the system function. The reason for this lies in the lack of resource protection in PC-DOS. What is needed is the **channeling of access to resources (such as disks) through the operating system**. This is now widely implemented in the form of two (or sometimes more) modes of operation with a clear distinction between user-level and system-level programs in terms of their ability to access resources. Resource protection is a prerequisite for other mechanisms introduced below.

Virtual memory is also missing from PC-DOS. The consequence of virtual memory is the separation of processes running in the system. Without this one can imagine virus code in one process influencing the actions of another process (for example substituting its own values for original parameters to system calls) and in effect gaining the other process's privileges. With virtual memory communication between processes can be handled by the operating system with messages passing.

Also concerned with memory management and protection is **segmentation**. Here we can divide our programs into smaller entities (segments) and to each one assign an appropriate level of protection. For example we can place all our code in a read-only segment. Now it is the processor which automatically prevents unforeseen situations from occurring and possibly causing damage - for example an erroneous address calculation in our program will be detected before that address is mistakenly used to modify part of our code. Similarly we can prevent the processor from interpreting words fetched from data or stack segments as instructions. Such effects could be exploited by viruses eg. for entering a higher privilege level. Segmentation could also be used to prevent the self-encryption of viruses which makes detecting them more difficult.

Security in secondary storage is customarily handled by maintaining information about the rights of each subject (active component of the system, process acting on behalf of a user) to access a given object (collection of data, system service, resource). This information can be kept in one of two ways:

- associated with each object is a list of subjects who have rights to access that object,

- associated with each subject is the list of objects it can access.

The first method is called **access lists**, the second - **capability lists**. They are equivalent in the sense that if fully implemented they both carry the same amount of information. However operating systems used to limit the amount of information maintained for performance reasons. Thus PC-DOS does not differentiate subjects, while the UNIX file system classifies subjects with respect to a given object as owner, member of the same group as the owner, or any other user. Within each class it is possible to grant or deny access separately for reading, writing or executing a file. Currently capability lists are gaining popularity because they can be combined with segmentation and extended into capability-based addressing which covers in a uniform way access to all system components, not just files.

The importance of employing capabilities in a file system endangered by viruses stems from two facts. One is that they provide a high degree of protection against the manipulation unit of a virus, that is we limit the possible damage inflicted by the virus. The second is that we make its propagation restricted to certain paths and because permissions to modify executable files are rare we make its propagation between files of different users unlikely. The default situation should of course be the denial of access To protect more effectively we should also enforce the rule that any modification of a program causes its capabilities to become invalid. The effect of capabilities is to force the virus to attempt to break system security more often to achieve the same extent or rate of spread.

Once a virus infection does take place it may be important to trace the origin of the offending code, perhaps for the purposes of legal action. However: *"Among the various novel and potentially worrisome features of these spreading attacks is the anonymity they can confer upon the attacker ; once the Trojan Horse instructions have spread to a large number of programs or systems it may be difficult or impossible to determine where the spread started or to identify the original attacker."*(Chess). We feel that there is a need for a strong **authentication mechanism**, possibly based on public-key encryption algorithms (for example the RSA algorithm) and like those in force in other information networks. Together with the methods described in the next section this would allow the identification of the program from which the infection started in one system, the identification of its origin in the network and eventually the naming of the person responsible.

A lot has been written about password authentication but the Internet worm (Spafford 1989) exposed a weakness of the password scheme used on machines on that network, where passwords had been kept encrypted but in user-accessible space. Therefore any number of attempts could have been made at breaking them. To avoid this the password table should have been kept in protected memory and access to it channeled through the operating system. With this arrangement we could for example limit the number of attempts made by a process at 'guessing' the password or introduce a delay between subsequent attempts increased after each unsuccessful one.

9 ACTIVE PROTECTION

We will now move on to describing two ideas for methods of detecting the activity of viruses in the system. First of all we should note that we cannot determine if a section of code is a virus by analysing it. Viruses are not really different from other programs in the system - it is their intention which sets them apart. To overcome this we need to look at the behaviour of the system - looking at a single program is not enough (unless we know what to look for).

So far viruses have been known for their rapid spread in file systems. This could suggest a mechanism based on the monitoring of activity in the system (file creations and modifications) and raising the alarm when this activity rises above a specified level. However the introduction of such mechanisms would quickly motivate authors of virus programs to limit the rate at which their viruses spread. The effect would still be to infect all the programs in the file system but the slower rate of infection would make it undetectable. In other words we must design our mechanisms to deal with viruses regardless of how fast they spread.

9.1 Code Propagation Monitoring

One idea is to monitor the propagation of modifications between files. By propagation we mean that if file A modifies file B which subsequently modifies file C then we can talk of (potential) indirect modification of file C by file A. In a typical uninfected system we can divide executable files into three groups:

1. the majority which do not modify any other executable files,
2. those which create new executable files but do not modify any existing ones (eg. linkers),
3. the minority which do purposefully modify other executable files.

If we imagine a graph with files in its nodes and arcs representing the act of one file writing to (creating or modifying) another file then we can make the following observations:

- in an uninfected system there will be arcs coming out from nodes representing files from groups 2 and 3. For each such file we will be able to extract a tree representing the maximum extent of propagation of modifications from that file. We will expect that the chances of encountering paths encompassing more than one node (+root) in this tree will be small;

- in an infected system the activity of the virus will appear superimposed over the graph described above. The virus will steadily develop a tree connecting new programs to those already infected, with the root of the tree pointing to the program from which the infection originated. This activity will be observed regardless of the rate at which the virus spreads. The characteristic feature of the infection tree will be the ordering of arcs in each path by time. We consider trees and not sub-graphs because from the point of view of virus protection we are not interested in a multiple infection of the same file.

The question is when to decide that we have a case of virus infection. We can develop rules which will base the decision on the size (longest path, breadth) of the maximum tree extracted from the graph. We can also take into account the type of programs in that tree and the potential for further growth. In the event of an alarm being raised we can use the infection tree to trace back the possible propagation of the virus to provide two pieces of information of great help in the process of removing the virus from the system:

- the program suspected of starting the infection,
- all programs suspected of being infected.

There is no difficulty in using this method in distributed file systems since we only need to operate on program identifiers regardless of where the programs actually are.

9.2 Modification Structure Monitoring

A different approach is to detect the virus whenever it attempts to copy itself to a new executable file. On every modification of such a file we compare it (the modification) with the record of any changes made to the program currently executing the write access. If we detect that the two programs have been changed in a 'similar' way then we will draw the conclusion that the code which lies 'behind' those modifications is trying to reproduce itself and we will regard it as a virus. The key problems of this method are mentioned in the preceding sentence:

- how to judge 'similarity',
- how to extract modifications and how to associate them with code fragments,
- how to represent and store both modifications and the full structure of a program.

Discussed below are some aspects of the use of this method for virus detection. To start off with we must define a way of representing the structure of a program. We have tried to investigate whether the static execution tree will be an appropriate representation for our task.

9.2.1 Comparing modifications.

The static execution tree (SET) is constructed as follows: starting from the entry point of the program we scan the code for procedure calls. Every procedure call results in a branch being added to the tree and the code of that procedure being scanned recursively. System function calls result in a branch being added but no further scanning. Indirect function calls (when we cannot statically determine the target address eg. when the address is to be found during execution in a processor register) also cause a terminating branch to appear. The leaves of the tree will consist of system function calls or indirect function calls or calls to procedures which do not call anything. Unfortunately because we wish to have a representation of the whole program and because some parts of the code may be reached only by indirect calls we may have to construct several trees in such a way that every fragment of the program is covered by one of the trees.

One of the problems which we will have to overcome to successfully implement this method is to judge whether the code of the virus will result in a large enough sub-tree so as to provide a basis for meaningful comparison. While we cannot force virus programmers to program in a structural and readable way we can ensure that they need to execute at least the three basic tasks of the virus (scanning the file system for a victim, opening the target file, writing to that file) as separate system calls and any destructive activities through still other. We will then have more information at our disposal.

9.2.2 Extracting modifications.

This seems to be the most difficult aspect of this method due to the variety of viruses which we may encounter in the future. We have tried to predict some possible innovations and present the most important features below. We must note however that viruses which we have seen to date belong to only a small subset of these classes.

1. overwriting/non-overwriting: an overwriting virus may overwrite some function calls from the original program which means that some branches will disappear from the SET, worse still we may face the situation where those overwritten calls are replaced by calls from the virus code, the problem is how to detect this situation. To further confuse us the virus may make random changes in the whole file;

2. dispersed/continuous: this refers to whether the virus is written as one fragment of code or as several fragments which communicate and synchronise their execution through global variables. For dispersed viruses we will extract several modification sub-trees and we need to decide how to compare sets of sub-trees or how to combine each set into one sub-tree;

3. self-modifying: here we can distinguish three types of self-modification

    - modifications of some fragments of code which do not change the structure of the virus code,
    - modifications which do cause a change of structure eg. addition of procedure calls or system function calls,
    - modifications which cause a change of class eg. cause a continuous virus to become dispersed in the next generation.

9.2.3 Storing modifications.

The problems here are following:

- what to do with consecutive modifications, storing them separately causes problems if we think of a virus which transfers itself in several stages;

- when extracting modifications after a write access what should we compare the new version against, after a few modifications the original SET may be meaningless, perhaps we should at the same time save the modifications separately and combine them into the full SET;

- how to deal with modifications which use the following sequence of actions: read code from old file into memory, erase old file, create a new file under the same name, write modified code into this file.

Despite all the problems the idea of comparing modifications has two very attractive features. One is the speed with which the virus would be detected - on its first act of replication. The second is that a successful detection leaves us with the structure of the suspect code fragment. This can be used to search through a library of such structures for known viruses. The entries in the library would also contain information - perhaps in the form of automated procedures or proper programs - on how to remove the virus from the system. In the case that the virus was found in the library those steps would automatically be executed. If it was not found then the system operators would need to check that the alarm was indeed caused by a virus, develop steps to remove it and then update the library.

10 CONCLUSIONS

A thorough strategy designed to fight off viruses needs to address the following four issues:

1. avoidance, that is preventing viruses from entering the system in the first place,
2. limiting the spread of viruses,
3. detection of viruses,
4. identification and removal of those viruses which have been detected.

In this paper we have discussed the middle two issues - passive protection treats the issue of limiting the spread of viruses, while active protection aims to detect virus activity in the system. Avoidance of viruses seems to be impossible to attain by technical means since they can be introduced into the system by a properly authorised user. The removal aspect also seems at this time to be impossible to treat in a general, uniform way for all viruses. What is needed is some form of an automatic virus remover generator.

We have said earlier that we feel more effort should be devoted to general security measures than to virus-specific measures. Experience with Novell NetWare shows that passive protection mechanisms can drastically limit spread of viruses. We must above all consider the cost-effectiveness of implemented measures. The mechanism which we regard the most effective against viruses is the use of **capabilities for controlling access to files.** In some systems though introduction of capabilities may be undesirable or too complicated (eg. existing systems). There we may find the demand for pure anti-virus mechanisms.

We have tried to focus on methods which promise to work against a large number of viruses irrespective of their actual coding and characteristics and which distinguish as well as possible between an infected and a clean system. Although they are based on signalling only a probable presence of viruses they have features which help in removing the virus.

We have also tried to predict possible new classes of viruses. Nevertheless someone who is willing to undertake a lot of effort will almost always find a way of avoiding protection mechanisms. This brings us to the question of who writes viruses and for what purpose. The development of viruses and anti-virus mechanisms resembles an 'arms race' situation. Each new protection mechanism should increase the effort needed by the other side to overcome it by more than was needed to develop the mechanism. One more point we would like to mention here is that if writing and spreading viruses will be treated as a crime then the focus may switch from ways of stopping the infection to ways of identifying the person responsible. Such a mechanism and the threat of legal action may be more effective than other technical measures described above in preventing virus infections.

Acknowledgements

We would like to thank Mr Artur Tomaszewski for his valuable contribution.

References

Burger R. (1989). *Das große Computer-Viren-Buch* Data Becker

Chess D.M.(1989). *Computer Viruses and Related Threats to Computer and Network Integrity* Computer Networks and ISDN Systems, July 10, 1989 p.141

Spafford E.H. (1989) *The Internet Worm: Crisis and Aftermath* Communications of the ACM, June 1989, p.678

THE NEED FOR A STANDARD METHOD FOR COMPUTER SECURITY REQUIREMENTS ANALYSIS

R. J. Talbot

Admiral Management Services Limited, 15 Victoria Avenue, Camberley, Surrey, GU15 3JH, UK

Abstract. The definition of requirements for computer systems is known to be difficult; it is argued that this is especially so for those requirements which are security related. It is suggested, by analogy with the ADP industry, that there is a need for a standard security requirements analysis method, employing standard techniques and standard notations, and for a class of individuals called security requirements analysts. Further, characteristics for such a method are proposed; in particular, it should be developed in a modular fashion such as to be embeddable in systems development methods. Finally, the rôle of the analyst is contrasted with other rôles for individuals involved in secure systems procurement.

Keywords. Security; requirements analysis; requirements analysts; standards; development methods; covert channels.

INTRODUCTION

Many procurement contracts for systems processing sensitive information continue to be let with inadequately defined computer security requirements. In common with other IT systems they often suffer problems arising from incomplete, inconsistent and ambiguous specifications. However, there are aspects specific to computer security which make the capture and expression of its requirements particularly difficult. This means that the scope and granularity of specifications is often inappropriate to define the functionality of systems that address the real security risks. Further problems arise when architectural and implementation constraints are introduced, and when varying degrees of trust are required in different security features.

In recent years, national standards applicable to the evaluation and certification of secure computer systems have been produced. These standards require further interpretation for each of the parties involved in the procurement, development and certification of secure systems but they do provide a framework which, once sufficient "case law" has been built up, should help to improve the specification of computer security requirements.

This paper proposes that a discipline of standard concepts and terminology should be imposed upon the secure systems procurement community, through the development, training in and use of a standard method for performing the task of security requirements analysis. It argues for such a method on efficiency grounds and by analogy with the ADP industry. Suitable characteristics for such a method are presented. It is proposed that it should be embeddable within a general requirements analysis method, and possible approaches are discussed. It requires a tool-kit of appropriate modelling techniques to be assembled and shown to be adequate, along with supporting guidance on the appropriate employment of the tools.

The role of a specialist security requirements analyst is also discussed, and related to those of security officer, procurer, developer, evaluator and certifier. This is considered to be a key role requiring considerable in-depth and broad-based skills and experience.

The various sections of this paper expand upon this Introduction.

PROBLEMS WITH REQUIREMENTS DEFINITION

The General Problem

The production of complex computer systems is well-known to be an *essentially* difficult task; see, for example, Brooks (1987). This arises from the fact that a computer system is used to model, and sometimes to control, some part of its environment, which is itself structured in a complex manner. As a consequence it is difficult even to state the way in which the computer system should behave in its intended environment - its **requirements** - let alone to decide *how* it might be built to have such a behaviour - its **design**. Notice that making any statement, be it about requirements or design, necessitates the taking of *decisions*.

A commonly accepted view is that the two activities of **requirements definition** and of **design** should be separated, basically because both activities are individually sufficiently difficult that mixing them will inevitably cause bad decisions to be made; a development which imposes such a discipline will be termed **top-down** in this paper. If a top-down discipline is not employed, then a successful system development becomes unlikely; even if it does still happen to succeed, then it is likely to have been wasteful of resources. (Real life analogies abound. Generals making the wrong *strategic* decisions can nullify all the correct *tactical* decisions of their subordinates; setting off walking from A to B without first deciding upon the direction of B from A is never a good idea).

A strict temporal separation of requirements definition from design, which expects all requirements to be defined and baselined before any design work is undertaken, is the **Waterfall Model** for systems development. In this paper there is no intended implication that this is considered to be the only valid model, for, taken strictly (or narrow-mindedly), it denies the usefulness of prototyping, for example. Certainly, other systems development models have their merits; e.g the **Spiral Model** of Boehm (1988). What is really required is a development about which one can make justifiable claims. This means that the development will have to be structured in some way, making use of abstraction techniques, and that there will have to be traceability between design levels. In other words, it is considered that a top-down description *must* be possible at some stage.

The major aim of the developers of a computer system must always be to produce the most cost-effective solution to the users' computing problems. In doing so it is to be remembered that many users do not understand the computer systems which are being proposed as solutions. However, they do understand their problems and should therefore be able to understand a statement of their requirements, if suitably expressed and explained. In many industries it is accepted that a specialised group of individuals is needed for eliciting, analysing, reporting upon and defining requirements for users. They are termed **requirements analysts**, and they are often provided with a **requirements analysis method**.

Specific Problems With Security Requirements Definition

There are particular difficulties in stating security related requirements. They have to do with stating prohibitions, or, at least, restrictions on **covert channels** in the implemented system. The problem arises from the notion of security, the idea of a covert channel and the nature of computer systems.

The **notion of security** predates the computer; indeed, it is ancient. It always had (and has) to do with controlling information flows between various parties; preventing certain information from reaching certain persons, preventing the introduction of disinformation, or ensuring the timely arrival of certain information with certain persons. A security conscious organisation, or individual, decides on precisely which flows to allow and to disallow in a **security policy**.

The **idea of covert channels** has to do with the fact that humans are endowed with reasoning powers; they can infer facts from other facts. Hence, even if an organisation is successful in controlling the direct flow of items of information referred to in the security policy, it remains possible that a malicious person can infer facts they wish to have, but should not be allowed to have, from those which they are allowed to have, (or which they cannot be prevented from having). If that someone does so, then they are said to have found and exploited a covert channel. (There is no generally agreed definition of the term "covert channel" in the literature. This is a rather all-embracing one which it is convenient to adopt in this paper).

The **nature of computer systems** is that they are models of some part of the real world which interests their users; they are used to store, manipulate and transfer such information. They are also artefacts; machines made by people and understandable to other people, who can observe their behaviours and, even, take them apart and examine them. However, whilst understandable *in principle* computer systems can still be enormously complex affairs *in practice*, to the extent that even the developer can become unaware of features of the developed system. These characteristics imply that complex computer systems provide a fertile breeding ground for covert channels.

Clearly, in stating security requirements for a system, an organisation always has to say which control flows are to be permitted and which are to be prevented: the **direct security requirements**, as we shall call them. Doing so may be relatively easy since the entities involved are all known at the time of writing the requirements. Even so, great care is necessary in framing requirements. For example, requiring that something should be stored in encrypted form is not the same as requiring that it should not be stored in unencrypted form.

It is the statement of those requirements which restrict the possibilities for covert channels which causes greater problems. The entities involved in covert channels are *not* necessarily, or even usually, known at the time of writing the requirements. Intelligent attackers may gain their desired information indirectly (i.e. by inference from facts not prescribed for in the direct security requirements) through an understanding of detail of a sort not mentioned in, nor usually appropriate to, the statement of requirements. Timing channels provide the obvious example. Apparently, one is in a "Catch-22" situation regarding the scope and granularity of the security requirements specification; requirements should not pre-empt design decisions, but, in that case, they cannot address all the security requirements. McDermid (1989) refers to this as a problem of **abstraction failure**. The accepted way out of this dilemma is to state some security requirements which apply to all levels of design and implementation. They are of the form:

- There shall be no covert channels

- No covert channel shall have a bandwidth greater than ...

- Any allowed covert channel shall be audited ... .

To take away from this discussion is the fact that the framing of security requirements is an enterprise which requires considerable skill and expertise, for it is harder than the job which traditional requirements analysts do. This being so, given the accepted role for the latter individuals, there is surely a need for appropriately trained individuals, for **security requirements analysts**.

The Impact Upon Development

The major impact of the presence of security requirements upon a development is to make it take longer. In part, this is simply because those systems which organisations decide to place security requirements upon are more often than not the systems which they least wish to have fail; they are valuable systems. Once one is concerned about how a system could *possibly* fail to work as expected, then one has far more work on one's hands than in the more usual case where one judges, often generously, that it ought to be OK. (It is the difference between asking someone "Do you think it'll work?" and asking them "Is there any way in which you can conceive of it going wrong?").

There are several specific ways in which it is readily apparent that secure systems, as opposed to ones which are otherwise important, will require more effort. One is in the continual need to assess the system for covert channels. This may be seen as a manifestation of the "could it conceivably go wrong?" problem. Is there any way at all, in this open-world in which one can look at the same phenomenon from myriad different perspectives, that an attacker can learn, or do, something that he should not? Another is the accepted necessity to use formal modelling and proof techniques in order to justify one's confidence in the implementation of the security requirements. Not only does one then require highly trained and intelligent personnel to develop and manipulate the models, but one needs, arguably, even better people to decide what it is that should be modelled, and to appreciate the scope and limitations of such modelling activities. In particular, one runs across the abstraction failure and open-world problems together; you can only prove assertions about things you have modelled, and you cannot model everything.

STANDARDS AND "CASE LAW"

In the last section it was argued that requirements definition is both a crucially important and essentially difficult activity, *security* requirements definition being particularly so. Often, in situations where important and difficult decisions recur, it is useful to save time and effort by providing standard solutions to the questions arising. In the computer industry this principle is applied to the development process by demanding the provision and use of appropriate **standard techniques**, preferably integrated into **standard development methods**. For example, in the ADP industry, in the UK, say, one has methods such as SSADM employing techniques such as data modelling and entity life histories. Of course, the adoption of standards should never be taken as an excuse for not thinking individually about each case to which they are applied, but, by and large, they do impose a discipline which is advantageous.

Some years ago US standards were introduced applicable to the development and evaluation of secure computer systems; since then other nations have worked on their own standards, and a harmonisation of a number of European standards is imminent.

An important additional benefit of such standards is the introduction of greater **objectivity** to the process of evaluating and certifying secure systems. It is clearly insufficient to simply have "gut feelings" about the security features of a system; confidence must be justifiable, and one way to do so is by judging against standards. However, because of the enormous variety of computer systems which are built, it is not possible to give detailed prescriptive standards, and this has not been attempted in the standards referred to. Instead, the standards have been written in a way which continues to call for interpretation on a case by case basis. Evaluations and certifications which have been and are being pursued provide something analogous to "case law" in the legal system. Whilst a history of such case law should be helpful, it is to be remarked that computing is not as static as most situations with which the law courts have to deal - witness the problems which the lawyers have with computer crimes.

These standards clearly influence the appropriateness of standards elsewhere; principally, on the development methods to be used, since developers will be keen, or may even be contracted, to pass evaluation and receive certification. For example, current and all expected future standards consider that higher levels of confidence in a system satisfying its security requirements can only be justified if those requirements have been expressed in some mathematically formal language and the system's high level design has been *proven* to satisfy them. This has obvious knock-on effects for the developer in terms of choice of techniques and method.

A METHOD FOR SECURITY REQUIREMENTS ANALYSIS

The Need

The preceeding sections have set up a context for justifying the major proposal of this paper; that a standard method should be developed and adopted for the definition of security requirements, a **security requirements analysis method**. In summary, that context comprises of a recognition of various necessities:

• To (logically) separate requirements definition from design

• To enhance and supplement usual requirements analysis techniques with ones for dealing with specific issues relating to security

• To understand the impact of these issues on the implementation of systems

• To be able to give objective and appropriate justification of one's confidence in the security of systems.

If the ADP industry, which is probably the computing sub-culture that has most practical experience of producing large systems, has found itself in need of standards, then how much more so is there a need for a security requirements analysis method. A symptom of this is the amount of time lost at present because of misunderstandings about words and their meanings; a discipline of the sort which we all accept in learning to speak our own, or a foreign, language is desperately required. Words like "function" are used by different people with different meanings; this is very dangerous as it means that they draw differing interpretations of statements and conversational partners can talk past each other without realising it. One way in which the understanding of the whole community could be advanced would be by a concerted effort to learn the same formal languages, to read the same books, such as Hoare's (1985) book on CSP, Jones' (1986) on VDM, Spivey's (1989), say, on Z, and Milner's (1989) on his various process calculi. The community must find the time to understand, and then adopt, standard languages. They can be used in standard techniques in standard methods.

It has been indicated above that another effect of the use of standard methods for ADP requirements analysis, which one would like to exploit in the security field as well, is to buffer the non-technical user from the details of the computer system. The way this works in practice is worth noting; the ADP requirements analyst interviews users, maybe several times, and produces models of the required system; he or she then generates costed options for systems solutions and presents them to the users for their selection. Now, the models which are shown to the users are not to the same level of detail as the analyst has developed them; the analyst presents the tip of an iceberg of supportive information: he or she draws upon this additional information in generating the costed options and in answering the users' questions. Similarly, there should be no need for the security officer of a secure system to understand more than the top-level features of a mountain of more detailed material.

As already mentioned, models in formal languages will be expected at some points if sufficient confidence is to be achievable in the security features of the system which are considered most sensitive. One would expect the buffering "process" to include making it unnecessary for security officers to understand such notations. Indeed, ADP analysts are themselves buffered from the necessity of learning truly formal, mathematical techniques by standard methods such as SSADM. (There is a formal mathematical theory to some techniques, such as relational data analysis, but the technique is taught and applied as a prescription which does not require the analyst to know anything at all about the underlying theory of relational algebras). This may also be possible in the security field, although it is less likely that a serious analysis (or analyst) can avoid thorough understanding of formal techniques.

The possibility arises that some parts of the analysis will require formal languages, and others, for less sensitive features or those with no security relevance whatsoever, will not. This implies the need to be able to embed formal language descriptions into less formal descriptions of the system at large.

The Method And Its Development

In summary, there are certain characteristics of a putative security requirements method which would seem forced upon us:

- An integrated set of languages, tools and techniques

- A programme for assembling them for use which is tailorable to make it appropriate to the great diversity of secure systems and devices which one wishes to produce: e.g. to application domain, and to size of project

- It should be embeddable within general requirements definition methods.

A natural approach to developing such a method would be to follow the "modular approach" to methods espoused by the CCTA in the production of Version 4 of SSADM, (as was explained by CCTA representatives to the audience at the Spring Meeting of the SSADM Users Group on 10th May, 1990). This approach is an exploitation of the metaphor that methods are like software and should therefore be developed in a modular fashion. (The metaphor has been accepted in discussions towards standard European methods for 1992). Each method should then be *specified* as a black box, i.e.

- the generic inputs and products

- the pre- and post-conditions to the use of the method.

making different methods for doing the same job "plug compatible". Embedding one method inside another one is then as natural a process as calling one software module from another. For example, SSADM, an ADP systems development method, can be embedded within a project management method (such as PRINCE) and "called" whenever a strategic study method throws out the need for a new system. In the present case, for a security requirements analysis method, carefully specifying the method will similarly allow embedding it into larger methods. Much of the work required in justifying a development method for secure systems could then be transferred to development methods purveyors; they would be presented with a well-specified interface to the embedded standard security requirements method and required to show how their method uses its products.

THE NEED FOR SECURITY REQUIREMENTS ANALYSTS

The ADP industry does not just have analysis methods, it has analysts. They are the buffer between users and the technical detail of systems of which those users require no knowledge. They are specialists in eliciting, analysing and specifying requirements (for ADP systems) in such a way that the users and the implementers have a common knowledge of what is to be built. The industry recognises that this job is both difficult and critically important.

In place of "users", the security field has "security officers", but there remains the same core problem: not everyone can be a specialist in every area – knowledge and expertise are distributed. The security requirements analyst should have an understanding of methods and techniques for eliciting, analysing and defining security requirements, but does not need to understand the wider issues involved; these remain the concern of the security officers. The role is also to be contrasted with that of the evaluators or certifiers, who must assess the confidence accruing from all aspects of the system construction and operation; however, the skills required are largely common to both roles, and one could conceive of training and using evaluators or certifiers as analysts. Indeed, this option is attractive inasmuch as such individuals invariably have a high clearance status, which is going to necessary for security requirements analysts.

SUMMARY AND CONCLUSIONS

Systems with imposed security requirements are of a wide variety of types. They include ADP systems, but extend to real-time embedded ones, and hybrids between these two types; they may be vast networks, or hardware only devices that you can pick up and carry around. Given this enormous variety, and the inherent difficulties in specifying security requirements, there is even more need for a dedicated class of specialist **security requirements analysts**, and for a **method** to support them than there is for requirements analysts and requirements analysis methods in the rest of the computer industry.

REFERENCES

Boehm, B.W. (1988). A spiral model of software development and enhancement. *Computer*, May 1988, 61-72.

Brooks, F.P. (1987). No silver bullet: essence and accidents of software engineering. *Information Processing '86*, ISBN No. 0-444-70077-3, H.-J. Kugler ed., Elsevier Science Publishers B.V. (North Holland) (c) IFIP 1986.

Hoare, C.A.R. (1985). *Communicating Sequential Processes.* Prentice Hall International, ISBN 0-13-153289-8.

Jones, C.B. (1986). *Systematic Software development Using VDM.* Prentice Hall International, ISBN 0-13-880725-6.

McDermid, J. (1989). Assurance in high-integrity software. In *High-Integrity Software*, C.T. Sennett ed., Pitman Publishing, ISBN 0 273 03000 0, p251.

Milner, R. (1989). *Communication and Concurrency.* Prentice Hall International, ISBN 0-13-115007-3.

Spivey, J.M. (1989). *The Z Notation: A Reference Manual.* Prentice Hall International, ISBN 0-13-983768-X.

THE TESTING OF REAL-TIME EMBEDDED SOFTWARE BY DYNAMIC ANALYSIS TECHNIQUES

D. Hedley

Liverpool Data Research Associates Ltd., 131 Mount Pleasant, Liverpool, UK

Abstract. Dynamic Analysis techniques which test programs thoroughly have been available for many years. The application of these techniques to real-time embedded software systems must overcome some particular problems associated with embedded systems. The paper discusses some solutions to these problems and describes a tool currently being used to test real-time embedded software written in a variety of languages.

Keywords. Program testing, Program validation, Dynamic Analysis, Real-time embedded systems.

INTRODUCTION

Large real-time embedded systems are becoming more common. Very often, these systems require a very high reliability as, for example, in nuclear power stations, air traffic control and avionics software. However, major difficulties often occur when attempting to validate software which is used in these systems. In particular, the use of advanced testing tools has often been found to be troublesome. In many cases, these tools cannot cope with embedded software and often, only a subset of a language can be analysed.

Much present-day software validation (especially in the U.K.) consists of applying Static Analysis techniques to code and analysing the results for inconsistencies. However, independent surveys of validation methods have invariably shown that Dynamic Analysis is the single best means of discovering errors in software, and is, in particular, much more useful than Static Analysis.

To release software into the world without at least ensuring that it has been fully executed by its test data is asking for trouble; in the case of safety-critical software arguments could be made that this practice is criminally negligent.

There are three main problem areas which must be overcome with Dynamic Analysis of embedded systems: the information generated by Dynamic Analysis must be transmitted from the target machine to the host machine for analysis; the program under test may no longer fit into the target machine if the code is expanded to monitor execution; and monitoring the execution may change the run-time characteristics of the code.

The above problem areas obviously make Dynamic Analysis more difficult on embedded systems, but the problems are by no means insoluble. This paper deals with methods of solving these problems and describes a Dynamic Analysis testing tool which is currently being successfully used for the testing of real-time embedded software systems written in a variety of languages.

DYNAMIC ANALYSIS

Methods of software validation can be subdivided into two classes: Static Analysis techniques and Dynamic Analysis techniques. Static Analysis does not require the program to be executed: the Static techniques analyse the program code looking for anomalies and inconsistencies. Dynamic Analysis techniques are those which rely on the program being executed and analyse information gathered while the program is running.

For the past 20 years, it has been possible to test ordinary software by means of Dynamic Analysis techniques to ensure at the very least that each line has been executed before release by some test data. Metrics for measuring the effectiveness of Dynamic

Analysis, the Test Effectiveness Ratios (TERs) were first proposed in 1962 [Brown] and later extended to form a hierarchy of measures [Woodward]. The most basic metric (TER1) gives the ratio of lines of code executed by test data compared with the total executable code. TER2 gives the proportion of branches executed; TER3 gives the proportion of a particular class of important subpaths (LCSAJs) executed. The class of TER metrics extend to TER∞, the proportion of complete program paths executed.

To maximise TER1 should be the minimum level of testing necessary for most programs. It is somewhat harder to achieve TER2 than TER1, and much greater effort is required to achieve the others. For example, to maximise TER3 each line needs to be executed an average of 8 times along different local subpaths.

The authors recommend testing to the TER3 level for critical systems. The achievement of TER3 implies that the software has been rigorously tested and the errors remaining in the code after such a test are minimal. In many areas, the potential errors which will be discovered by testing to TER3 are known; for example, each loop must be executed zero (if possible), once, twice and three or more times, thus eliminating all common looping errors. The level of Dynamic testing required by DEF-STAN 0055 normally requires testing to the TER3 level.

Achieving TER4 is much harder again. Some work has been reported using the TER4 measure [Duran], but TER3 is the highest level reached by any commercially available tool. For critical software, testing to the TER3 level is considered essential.

The values of the TER metrics are arguably the most important figures that can be given to a customer or standards authority regarding the testing of a program.

In general, it is easier to extract Static-based metrics than Dynamic-based ones from any software, especially that of an embedded system, and easier to construct automated systems to extract these metrics.

However, the gathering of Static Analysis based metrics gives information about the quality and consistency of the code rather than the correctness. Static Analysis is also beset by the 'maximal output' problem, whereby the analysis of any large sized piece of code produces a volume of output which requires several trees of paper to print. This output then needs to be visually examined to find the important messages. Many real-time systems are large: millions of lines of code are quite common. Only the most critical parts can be processed in this way. A further problem is that after modification, the same Static Analysis validation needs to be carried out on the modified system.

Surveys of error detection capability using a variety of methods undertaken by authors with no vested commercial interests [e.g. Howden, Girgis, Lauterbach] are uncommon in the literature. However, such surveys have one common conclusion: that Dynamic Analysis is better than any other single method for finding errors. To emphasise the benefits of the different methods of testing, give yourself the choice of travelling by two aircraft containing embedded avionics systems: one validated purely by Static Analysis or one validated only by Dynamic Analysis. Would you prefer your software to have been checked statically for inconsistencies or dynamically to examine the behaviour of the system whilst running?

Obviously, the use of Static and Dynamic Analyses together are better than either on its own. For critical software, the more techniques applied to the software, the safer the software will be. Any tool or person analysing software can trigger thought processes and come up with some useful information.

What no tool can do is the hardest part of testing - ensuring that all requirements are present. Nothing can verify software against missing requirements that have not yet been thought of. These requirements have the property of not being obvious to the requirements analysers. They often concern unusual situations, but ones which are not necessarily rare in use. No amount of testing or validation can guarantee that these faults can be found. They will only show up in practice or through chance.

EMBEDDED SYSTEMS: PROBLEM AREAS AND SOLUTIONS

Let us consider the particular problems of real-time embedded software: that is, time-critical software running under embedded constraints. Examples of such software are found in avionics systems, nuclear reactor control systems and railway signalling systems.

Development systems for real-time embedded software usually contain at least two computers: the host machine and the target machine. The host is generally a commercially available computer on which the program is developed and on which any tools or other software development aids are kept. The target is used for program execution. Often the target has only rudimentary facilities compared to the host. Programs are often cross-compiled on the host and executable code is sent to the target to be executed.

Given that a program was written in the correct language and dialect, tools have been available in the marketplace to aid the Dynamic Analysis of ordinary software for some years. There are fewer automated tools available to the serious tester of embedded software. It is only recently that tools have become generally available for the testing of embedded systems.

Traditional Dynamic Analysis techniques require the program being analysed to be instrumented, i.e. probes are inserted into the program to analyse its run-time behaviour. These probes are normally in the form of procedure calls at specific points in the program selected by an automated tool. When executed, these procedures produce information regarding the route taken through the program along with the possibility of producing information on any pertinent program states or variable values. This contrasts with a user inserting executable assertions into his own program [Hennell] which can enable other information of interest to be produced.

This type of instrumentation is called intrusive instrumentation, as the program being analysed has now been changed to incorporate some extra code. It is possible to carry out non-intrusive Dynamic Analysis of a program by interrogating other elements of a system. For example, it may be interesting to know exactly what data is passing along a particular route, or to keep track of the values held by particular registers. This can normally be done by non-intrusive means, i.e. the original program is not modified. However, the information produced by such methods is rather limited compared to that available by intrusive instrumentation.

One argument against intrusive instrumentation is that the program being tested is not the same as the original uninstrumented program. The insertion of probes into the code could in general change the functionality of this code, as well as changing the more obvious characteristics such as size and run time speed. Obviously, if the instrumentation consists purely of extra 'print' routines, the functionality of the original program should remain unchanged. The test data produced by testing the instrumented code will traverse the same logical paths in the source code program whether or not it has been instrumented.

However, it is also possible to use the instrumented (and hence tested) version of the source code in real applications. This would normally require that the probes have their output shut off, by a logical variable for instance. If this is done, the penalty is that the code in the final application runs a little slower than it would have done otherwise.

The philosophical problem which arises here is that there are times when one has to accept the results of a demonstration of particular functionality. In practice this happens extensively. There is a continuous process of translation in many software activities which requires that the originality functionality of software is preserved whilst adding housekeeping functions, for example during compilation.

Applying Dynamic Analysis to real-time embedded systems is complicated by three major problems:

(1) the program being tested is run on the embedded system (the target) and the information required by Dynamic Analysis must be made available on the host.

(2) instrumenting a program increases the amount of code to be compiled.

(3) instrumenting a program causes the instrumented program to do more work and hence run more slowly.

The action of each of the probes results in an item of information being generated which must be transmitted back to the host. Communication is obviously easiest if there is a physical line tying the two machines together. In this case, the Dynamic Analysis information can be returned along the line, possibly sharing it with other information. The probe, therefore, must pass information to a special driver which collects the data and transmits it to the host. The nature of this driver will vary from implementation to implementation depending on hardware facilities.

Often, however, the host/target link is only one-way, i.e. there is no return link from the target to the host. In fact, if the target is operating in a moving environment, a fixed link is not possible. However, there is no reason why this information cannot be returned to the host by other means - the authors have experience of remotely produced paper tape being used, and of information produced on magnetic tape being flown back to the host.

Code expansion can be a problem when the target machine or the language code segment size is limited. It is impossible to instrument code without increasing its size. Similarly, it is impossible to instrument code to make it do more work while keeping the real-time characteristics of the code the same.

Again, some parts of the code may be very time critical and the instrumentation of these may be difficult. However, it is not necessary to instrument the entire program; only a subset of the program need be instrumented at any one time, thus achieving the analysis of the entire program gradually.

Liverpool Data Research Associates (LDRA) were a major partner in a recent Esprit project (TRUST) funded by the EEC to investigate and improve the testing of real-time embedded systems. During the course of this project, up-to-date Dynamic Analysis techniques were successfully used in various real-time embedded systems. As part of the project, new software probe instrumentation techniques were developed which take less space and less extra execution time than the basic strategy used for non-embedded software.

USEFUL TOOLS

The LDRA TESTBED is the only tool widely available for obtaining Dynamic Analysis metrics for real-time embedded systems. It is currently available for, among others, the common embedded high-level languages Ada, C, Coral 66, Fortran, Pascal and PL/M 86 (and their compiler-dependent dialects and features) and Intel 8086-based and Motorola 68000-based Assemblers.

As well as producing Static Analysis metrics on code quality and complexity, TESTBED gives the TER1, TER2 and TER3 metrics for Dynamic Analysis - the only commercially available tool to do so.

The use of Dynamic Analysis as a basic element in validation activities is rapidly becoming essential. When an independent licencing or assessment body requires an objective measurement of the extent of the testing activity then the TER metrics (or equivalent) are the only serious contenders. They clearly indicate the commitment to testing although their interpretation in terms of in-use reliability is still not well understood. Therefore proposed standards such as DEF-STAN 0055 (for military systems), the ISO standard for the functional safety of programmable electronic systems and DO178b (for avionics systems) expect the software producer to show the results of Dynamic Analysis and some types of Static Analysis.

CONCLUSION

The full Dynamic Analysis of real-time embedded software is possible for most such systems, although there are problem areas which may need attention in many cases. For any system with critical properties or which must be validated to a particular international standard, such analysis to a high level is essential.

Tools have been available commercially for several years to carry out such analysis on non-embedded software, and these are now being used successfully on many embedded systems

REFERENCES

Brown, J.R., "Practical Applications of Automated Software Tools", Report TRW-SS-72-05, TRW, Redondo Beach, California, 1972.

Girgis, M.R. and M.R.Woodward, "An Experimental Comparison of the Error Exposing Ability of Program Testing Criteria", Proceedings of Workshop on Software Testing, Banff, Canada, IEEE Computer Society, 1986.

Hennell, M.A. and E.Fergus, "A Comparison of Static and Dynamic Conformance Analyses", Proceedings of SAFECOMP'90 (elsewhere in this publication), IFAC Publications, Oxford, U.K., 1990.

Howden, W.E., "Reliability of the Path Analysis Testing Strategy", IEEE Transactions on Software Engineering, Vol.2, No.3, pp.208-215, 1976.

Lauterbach, L. and W.Randell, "Six Test Techniques Compared: The Test Process and Product", Proceedings of NSIA 5th Annual Joint Conference, Washington DC, 1989.

Duran, J.W. and S.C.Ntafos, "An Evaluation of Random Testing", IEEE Transactions on Software Engineering, Vol.10, No.4, pp.438-444, 1984.

Woodward, M.R., D.Hedley and M.A.Hennell, "Experience with Path Analysis in the Testing of Programs", IEEE Transactions on Software Engineering, Vol.6, No.3, pp.278-286, 1980.

Copyright © IFAC SAFECOMP'90,
London, UK, 1990

THE ELEKTRA TESTBED: ARCHITECTURE OF A REAL-TIME TEST ENVIRONMENT FOR HIGH SAFETY AND RELIABILITY REQUIREMENTS

E. Schoitsch*, E. Dittrich*, S. Grasegger*, D. Kropfitsch*, A. Erb**, P. Fritz** and H. Kopp**

*Australian Research Centre Seibersdorf, A-2444 Seibersdorf, Austria
**ALCATEL, A-1210, Vienna, Austria

ABSTRACT. ELEKTRA is the electronic railway interlocking system developed by Alcatel Austria in cooperation with the Austrian Research Centre Seibersdorf. High safety and reliability requirements are met by means of a certain system architecture combining redundancy and diversity principles additional to an extensive software quality assurance programm. Since testing still is a very important activity during all phases of a systems life cycle, the architecture of the test environment, the procedures and tools need a careful design too, for ELEKTRA, an application dependent approach has been chosen for efficiency and effectiveness (as approved to a generic approach).

The tools ACHILLES (module test in a simulated process environment of concurrant processes and signal flow) and PAMIR (system test on a real-time basis with host-target support) are described in more detail.

Both tools are designed for (automated) regression testing and analyses and for efficient support of system assessment. Test cases are derived from the user requirements specification and by simulating normal operation of the station.

Experiences of the test team will be reported.

Keywords: Railway interlocking, redundancy, regression testing, software diversity, system architecture, test tools, verification and validation.

THE SYSTEM ARCHITECTURE OF "ELEKTRA"

Alcatel Austria has developed, in cooperation with its research institute Alcatel-Elin RC and the Austrian Reserach Centre Seibersdorf, the electronic interlocking system ELEKTRA, the first installation of which has started operation December 1989. (Erb, 1989; Schoitsch, 1989).

High safety and reliability requirements are met by means of a system architecture combining redundancy and diversity principles for reliability and safety respectively. This architecture has to be taken into account during testing and maintenance too. In order of importance, the principles to be proven are (as far as the railway as a fail-safe system is concerned):

- safety (the system has to be free of safety relevant design faults)
- reliability (MTBF for a total system failure has to be 10 years)

Figure 1 shows the architecture of the ELEKTRA interlocking system. The high standard of safety and reliability which is needed to accomplish the requirements of a public railway system is gained by combining two separate concepts, one for SAFETY, the other one for RELIABILITY (Theuretzbacher, 1986):

SAFETY is acquired by using appropiate software engineering techniques during the development phase (fault avoidance), including diverse testing on two separate sites, and by applying DIVERSITY to a very high degree in the systems architecture (fault tolerance). The high degree of diversity in software is achieved by implementing two totally different software channels, each one on another computer. One channel is designed by techniques of conventional, well structured high level programming, the second one ("Safety Bag") checks all safety relevant conditions and actions by applying the rules via a real-time expert system . So even diverse ways of thinking are applied ! (The user requirements specification of the railway interlock system is proven by a hundred years of experience !).

RELIABILITY is achieved by applying the diverse system three times in an identical replication and by majority voting (2 out of 3) (Triple modular redundancy, software voting (VOTRICS (Wirthumer, 1989)).

TESTING IN THE CONTEXT OF "DEPENDABILITY"

Dependability is an umbrella-term covering all principal characteristics known in connection with safety related software and is widely used in that context (Redmill, 1990 a,b).

It is the property of a Programmable Electronic System (PES) that allows reliance to be justifiably placed on the service it delivers (Laprie, 1990).

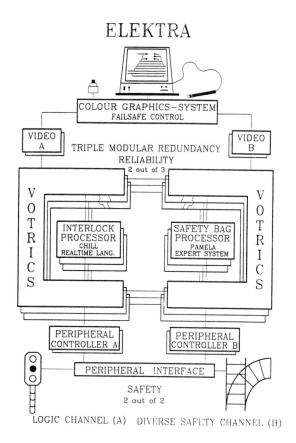

Fig. 1: System Structure of ELEKTRA

The most important result of earlier work within EWICS TC7 and of the last SAFECOMP '89 was that "dependability" of a system is not just a single system property or attribute and, therefore, cannot be covered by just one means. The attributes of dependability (see Fig. 2a) are now included in definitions of the latest draft proposal for the IEC standards of SC65A, WG9 and WG10 (IEC, 1990a):

"Dependability is the collective term (umbrella term) used to describe the general performance quality of a system and its influence factors: Reliability, Availability, Maintainability, Safety and Security" (see Fig. 2a).

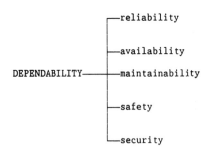

Fig. 2a: Dependability and its constituent properties

Dependability may be improved by means of procurement type (fault avoidance, fault tolerance) or validation type (fault removal, fault forecasting) (Fig 2b). Methods and tools may be classified according to this scheme.

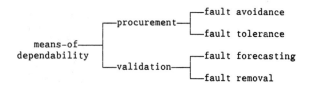

Fig. 2b: Means of Dependability

Some of the dependability attributes are contradictionary, therefore, to obtain sufficient dependability of a system, according to the goals of the system considered, a global strategy and a selected bundle of means has to be applied. This includes in case of ELEKTRA:

Fault Avoidance:

"The use of design techniques and implementation methods which aim to prevent, by CONSTRUCTION, fault occurence or introduction".

Fault avoidance is achieved by an extensive quality assurance programme including strict design and implementation guidelines (procedures), reviews and audits.

Fault Forecasting:

"The use of techniques to estimate, by EVALUATION, the presence, creation and consequences of faults".

For the "Proof of Safety" failure modes and effect analysis (FMEA) for all hardware elements of the system, a functional description of the system and a description of all selfchecking mechanisms had to be delivered (Erb, 1989).

Fault Tolerance:

The built-in capability of a system to provide continued correct service despite of a limited number of faults (hardware or software).

This is achieved by the system architecture of ELEKTRA: The two-channel approach using diverse techniques for each channel for safety of the system and identical triplication of each channel (triple modular redundancy) with 2 out of 3 voting.

Fault Removal:

"The use of techniques to minimize the presence of faults by VERIFICATION throughout the development phases".

In case of ELEKTRA, fault removal includes testing (procedures, tools, hardware- and software environment) on several system levels at different design stages and is included in the global system design strategy. This includes:

- Module Tests
- Functional (subsystem/system) Tests
- Safety Tests

Software and system tests have to be embedded into the global dependability considerations of the system. These systems attributes (redundancy, diversity) have to be proven to be guaranteed during the whole life cycle of the product, which includes all test steps and the ability to guarantee these features by an appropriate test strategy too.

Software tests in general are used during the Design and Development Phase, the Manufacturing and Installation Phase, and the Operations and Maintenance Phase to assess system safety, reliability and maintainability and to identify software defects for removal. To achieve efficiency in testing, data requirements of all interested parties must be carefully considered, and tests designed to minimize overlap and repetition. Data required for assessment of safety, reliability and maintainability must usually be obtained from tests being performed to generate data for several other purposes (IEC, 1990c).

THE ELEKTRA TEST STRATEGY AND ENVIRONMENT

The complexity of such systems makes it difficult to test the system in all conceivable conditions. Faults introduced during the development phase remain in a system despite of careful testing or can be introduced during the maintenance work. It is therefore important that testing is carried out in a reproducible manner using carefully planned and regulated procedures.

There are two aspects where special considerations need to be given to testing in connex with design, development and, later on, maintenance:

a) control of error reporting, change authorisation and software/system configuration.

b) regression testing, verfication, validation and assessment.

Therefore, a structured test strategy was implemented for ELEKTRA, in addition to and complementing the design- and implementation strategy (Erb, 1989).

As basis for the strategy served the IEEE and ESA Software Engineering Standards (IEEE, 1987; ESA, 1987).

"Test Strategy is the selected combination of functional tests (external, "Black Box Test") and structural (internal, "White Box test")" (IEC, 1990c).

The mix of test types chosen consists of:

- module test (internal test)
- functional test (external test)
  - subsystem tests
  - system tests
  - final test of field installation
    (not considered here).
- safety tests (functional tests with
  respect to safety)

For the overall test strategy as well as for each level of testing, a document was designed, describing the test environment and defining rules and guidelines, procedures and documents to be used and to be produced in each step. These testplans correlate to the system structure shown in fig. 1 and the test bed in fig. 4.

Module Test

Module tests are performed by the designers themselves, before the modules are integrated into the software of the ELEKTRA system. A strictly controlled release procedure defines, when, how and with what additional information (including results of module tests) a new module may be released. The files are archivated after release on a common directory, which is not accessible to the designer to prohibit uncontrolled change of code or documents. The module test documentation, which is produced by testing with the tool ACHILLES on the VAX-host environment as simulator, is basis for the software module review.

Functional Tests

Functional tests are performed on the subsystem level and on the system level.

The subsystems considered are (see fig. 1, 4):

- Peripheral Controllers
- Central Controllers (CC-level:interlock processor, safety bag)
- Video Input/Output Controllers (VIO)

In each case, the test environment consists of the respective target processor, a simulation of the environment on a test processor (16 + type), and the host-based software tools MultiTasking Monitor MTM and PAMIR (in case of the CC-tests) on the VAX host computer. Functional testing is done by sending CHILL-signals to stimulate the required function.

The RTM-testprocessor shown in fig. 4 serves mainly to support work on the peripheral controllers, which are of 8-bit type.

The diagnosis processor (DGP) which is shown in fig. 4 allows to monitor and collect all data concerning the operation of an installation and regularily sending them to the service team via public networks, but because of safety it is prohibited that any interaction with the interlocking system may take place.

For system tests, in the Research Centre Seibersdorf as well as in Vienna at Alcatel, a complete test lab has been installed containing all the equipment for an interlock system as it has to be installed at the field site. Additional to that, test processors and simulators as described here and later, including interactive devices like a push button board simulating signalling devices, track relays, axel counters, switches and level crossings, are provided. This allows to test independently at both sites and improved very much the efficiency of the tests during critical phases of the project.

Safety tests

Safety tests concern the peripheral input patterns and the behaviour of the diverse channels and are described later with PAMIR.

THE TEST TOOL "ACHILLES"

ACHILLES (Advanced CHILL and Environment Simulator) is a tool for debugging CHILL-286 programs (16 + processor) running under VAX/VMS. ACHILLES simulates both the CHILL code and the process environment, which is normally contained in the runtime executive. ACHILLES is independent from the target hardware and is able to support any target operating system by allowing different kinds of ENVIRONMENT SIMULATION. This makes ACHILLES open for simulating any combination of CHILL-environments with different operating and database systems.

ACHILLES provides white box testing (verify values of variables), black box testing (behaviour of a module to its environment, which signals are sent and received) and multi process testing. The structure of ACHILLES is shown in fig. 3.

Functional Description

ACHILLES is an interpreter of a special intermediate code ("O-Code") which is included in a file ("SIMFILE") generated by the Alcatel - ELIN RC CHILL compiler with usage of a special qualifier(/SIM). This file contains also the original symbol table of the compiler. So there is no loss of debug information from compilation to execution of a CHILL program by ACHILLES. Usually an application consists of more than one module, so ACHILLES has to "link" all modules by resolving references to external variables, procedures and processes. During this process of linking ACHILLES also performs consistency checks e.g. on mode (=type) declarations.

Simulation of a multi tasking language like CHILL does not mean interpreting one program statement after the other but means also supporting process management and communication. This part of an application usually is embedded in a multi tasking real time operating system. ACHILLES is able to support various operating systems by giving the user a lot of possibilities for configuring his special ENVIRONMENT SIMULATION. ENVIRONMENT SIMULATION means in this context the simulation of the functionality which is realized by a real time OS on the target system. During testing with ACHILLES, these functions have to be substituted by the simulator.

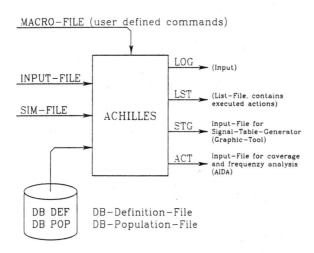

Fig. 3: Structure of the ACHILLES test tool.

After "linking" of all CHILL modules to the testbed, ACHILLES waits for input from the user to start the simulation. There are three possibilities how ACHILLES may get input commands:

- interactive testing: the user may type in the commands,

- input file: the input is read from a file and

- "mixed" mode: e.g. starting with an inputfile, switching to interactive mode, stepping through a procedure, and then reading the next command from the inputfile.

ACHILLES is also useful for performing regression tests because every test session is reported to a logfile which can be used as a command input file in another session.

Two very powerful runtime checks (which usually cannot be performed with traditional, objectcode oriented debuggers) shall be mentioned also:

- ACHILLES detects the usage of variables or parts of variables which have not yet been initialized (or are explicitely set undefined, e.g. loop counters at the end of a loop) up to granularity of single bits e.g. within a bitstring.

- ACHILLES controls "lifetime" of objects (as defined in CHILL) and raises an exception if a pointer references an object whose lifetime has already expired.

- ACHILLES provides graphic output visualizing the CHILL-signal flow between processes in form of "szenarios" (signal table generator).

ACHILLES collects statistics about coverage and frequency of CHILL statements wich are written to a special output file. One ore several of these files serve as input for AIDA (ACHILLES INTERACTIVE DYNAMIC ANALYSER), another tool, which analyses these coverage and frequency statistics.

THE TEST TOOL PAMIR

PAMIR (Processs Video and MultiTasking Monitor Integration Regression Tool) is a software-tool for automated subsystem and system tests for boards based on the 16+ processor familiy of Alcatel Austria. This includes the subsystem tests of the central controller level (Interlock- and Safety Bag Processor). In case of system tests (functional test), this affects the control level (failsafe process video system) and the central controller level (both channels) (see fig. 4). This is achieved by recording, journalling, simulating and analyzing the communication between the respective processors on both levels.

PAMIR resides on a VAX-computer integrated in a VAX-Cluster and is written in VAX-PASCAL under VMS, such utilizing the power of an advanced computer environment. The interaction between the tester or the test system and the application processors, which are Intel processors (16 bit) on boards designed and manufactured by Alcatel Austria, is done via IEEE-bus and/or asynchronous lines (configuration see fig. 4).

PAMIR Functionality

The software and hardware configuration used allows real-time tests to be performed manually (interactively) (by doing the same actions as a signal man or station master at the video terminal via light pen) or automated via command files executed at the VAX host computer. Command files may be edited and such test cases added or extended in functionality, and even timing constraints or conditions may be altered or added. All manual input may be recorded, all message exchange between processors be monitored or interfered with. All actions and results of are documented for later use, especially for regression tests to prove correctness of the system after changes (new versions, extended tests). The overall time behaviour of the system could be dynamically analyzed, and systematic fault elimination was done in an efficient manner.

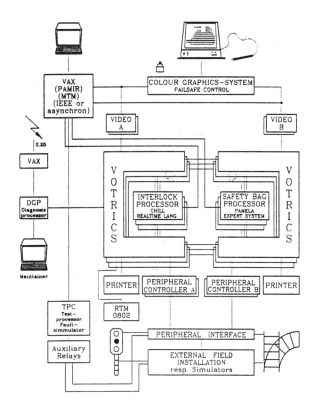

Fig. 4: Configuration of the ELEKTRA Test Bed.

The architecture of a real time system depends, of course, on the dimension of time. Therefore, it is not only necessary to be able to perform tests in true real time but also to interpret the results with respect to timing differences and differences in sequence of events, which might be legal or illegal. The automation of the interpretation of test results made it necessary to add some sophisticated programs to the tool which extract relevant information, do some intelligent filtering and compress and format the results.

Input files for testcases are generated from manual input via the video system or from the requirements specification of the Austrian federal railways, such providing some diversity in producing test input.

The architecture of ELEKTRA (two diverse channels) makes it necessary to prove, that illegal requests from one channel are detected by the other one and requested actions are refused. Another method is to generate a faulty status in one processor by error-seeding into the local data base. The test tools described allow such interactions and interventions via the interfaces to the central processors.

For the level of peripheral controllers (PC-A, PC-B), another tool TESTGEN allows the automatic generation of testcases for subsystem tests, since the input is strictly binary (power set). Since the railway periphery is connected to the interlocking pripheral controllers via two electrically distinct interfaces, hardware faults are bpossible only in one channel. To allow to consider only one failure at a time in the FMEA, detection of all those faults has to be guaranteed. To perform such tests, all possible bit patterns have to be generated, and executed using an environment simulator.

PAMIR allows in an implementation dependent manner and with moderate effort to simulate all situations defined in the reqirements specification of the Austrian Federal Railways, to repeat these test cases and to compare the results. The hardware interface to the railway periphery is also part of these tests, train movement and certain malfunction of railway periphery may be simulated by test hardware (test processor) and auxiliary relays (status and fault simulation). Regression testing is an indispensable feature for the "proof of safety", which has to be signed by an independent assessor (consultant), which is necessary according to Austrian Law.

Some outstanding features of PAMIR are:

- user friendlyness (simple command syntax, on-line help, definable function keys, command stack, show-function for parameter set up, statusline to show status of all processors).

- mixed mode: on-line change from interactive mode to command file input and vice versa, restart of at certain points of the input file or at point of interrupt)

- asynchronous execution of output processing, real-time behaviour

- configuration of tool to all possible hardware test configurations of ELEKTRA

- automatic generation of regression test files no difference in performance and functionality between interactive or automated test runs.

- test szenarios: combination of test runs to more complex test szenarios and definition of variants of such szenarios.

- evaluation of results by a set of off-line tools (data reduction, filtering, support of evaluation).

The test strategy includes an extensive reporting system. Input and output files generated by the various test tools are a very helpful support to testers, designers and maintainers.

CONCLUSIONS

Only a bundle of means and tools allow sufficient testing of a complex application to "allow reliance to be justifiably placed on the service it delivers" (Laprie, 1990). This includes a combination of hardware and software tools.

Since the system properties, the architecture of hardware and software and application specific attributes play a most important role in safety related systems and have to be considered primarily for assessment and the "proof of safety", we do not believe that a "generic" approach to V&V is possible with commercially justifiable means (if at all!), or that static tools for analysis are all that is needed. Therefore the application dependent approach was choosen.

Separation of tests according to vertical or horizontal levels of the system, regression testing and high level automated support of testing and interpretation of results proved to be absolutely necessary. The tester has to have the same experience and knowledge as the designer and user, just testing against specifications is not enough. Combinations of tests, stress tests and additional tests relying on users experience are necessary. Fortunately, there are several in railway signal engineering not only at the main contractors site but also at the Research Centre Seibersdorf.

The attributes of the system architecture have to be taken into account, the preservation of system properties such as diversity and redundancy, self checking features etc. has to be controlled throughout the software life cycle. This guarantees that there will be no compromise concerning safety, since it is a well accepted fact that the production of a reasonable large and complex fault free real time software is not possible nor can be guaranteed in practice.

REFERENCES

Alcerra, D., and Galivel, Ch. (1989). Validating the SACEM Railway Control System using the IDAS Software Test and Debugging Tool. Safety of Computer Control System 1989. pp. 59-64

Erb, A. (1989). Safety Measures of the Electronic Interlocking System "ELEKTRA". Safety of Computer Control Systems 1989, Pergamon Press, London, pp. 49-52

ESA (1987). Software Engineering Standards. ESA PSS-05-0 Issue 1, Paris-CEDEX.

Grümm, K. (1989). An effective strategy and automation concepts for systematic testing of safety related software. Safety of Computer Control System 1989. pp. 71-89

IEC (1990a) Software for Computers in the application of industrial safety related systems. Draft 65 (Sec) 94.

IEC (1990b) Functional Safety of Programmable Electronic Systems. Generic Aspects. Draft 65A (Sec) 96.

IEC (1990c) Software Test Methods. Committee Draft TC56/WG10, Jan. 1990, 56 (Sec) 307.

IEEE (1987). Software Engineering Standards. Willy&Sons Inc., New York.

Krebs, H. (1989). Programm diversity - an attempt to a quantitative description. SAFECOMP '89.

Laprie, J.C. (1990). Dependability. Basic Concepts and Associated Terminology. Report LAAS No. 90.055, March 1990.

Redmill, F.J. (Ed.) (1988). Dependability of Critical Computer Systems 1. (EWICS TC7). Elsevier Applied Science, London, New York.

Redmill, F.J. (Ed.) (1988). Dependability of Critical Computer Systems 2. (EWICS TC7). Elsevier Applied Science, London, New York.

Schoitsch, E. (1989). The Interaction between Practical Experience, Standardization and the Application of Standards. Safety of Computer Control Systems 1989, Pergamon Press, London, pp. 17-24.

Sethy, A. (1989). Works on the verification of software safety. SAFECOMP '89.

Smith, D.J. and Wood, K.B. (1987). Engineering Quality Software. Elsevier Applied Science, London, New York.

Theuretzbacher, N. (1986). Using AI-Methods to improve Software Safety. IFAC SAFECOMP '86, Sarlat, France.

Wirthumer, G. (1989). VOTRICS-Fault Tolerance realized in Software. Safety of Computer Control Systems 1989. pp. 135-140.

Example for PAMIR Input File:

```
INP: DEFAULT.PAMIR
LOG: PB_3_7-M.PAMIR_LIST

%%%%%%% Signalfreistellung

-------> 3.7.2 A siehe 3.6
------->       B siehe 3.8.3 D, 3.8.1.1 A
------->       D.3.3, D.3.5 siehe 5.4
------->       D.4 siehe 5.9
------->       E siehe 5.3.6.5 C
------->       G.1, G.3 siehe 3.2.1 C
------->       H siehe 3.6.2 A
-------> 3.7.3.1 siehe 5.2.2.3 A
-------> 3.7.4.1, 3.7.4.2 A,B

-------> 3.7.5 A Haltlampenausfall

MTM_DGP: S:LS5
WAIT   : DELAY_RELAY

LP_ACT :     1   1829AC  ; SCH1
WAIT   : DELAY_LP
LP_ACT :     1   1812AK  ; H1
WAIT   : DELAY_LP
LP_ACT :     1   3745AD  ; /Z
WAIT   : DELAY_Z_LONG

-------> Signal geht auf Frei

LP_ACT :     1   1812AK  ; H1
WAIT   : DELAY_LP
LP_ACT :     1   3882A@  ; /FH
WAIT   : DELAY_LP
MTM_DGP: I:AT
WAIT   : DELAY_AT

MTM_DGP: R:ALL
WAIT   : DELAY_RELAY

-------> Haltlampe Signal A ausfallen lassen

WAIT   : DELAY_PAUSE

LP_ACT :     1    424@C  ; A
WAIT   : DELAY_LP
LP_ACT :     1   1974AJ  ; R2
WAIT   : DELAY_LP
LP_ACT :     1   3742AC  ; /ZO
WAIT   : DELAY_Z_LONG

-------> Signal geht auf Frei

LP_ACT :     1   1974AJ  ; R2
WAIT   : DELAY_LP
LP_ACT :     1   3882A@  ; /FH
WAIT   : DELAY_LP
MTM_DGP: I:AT
WAIT   : DELAY_AT
-------> Haltlampe A wieder in Ordnung

WAIT   : DELAY_PAUSE

-------> 3.7.5 C siehe 10.1.1.3

-------> 3.7.6.1 Signalbegriff bei Ausfahrt auf St
------->         weiteres siehe Anlagenpruefung

   ;              ;
   ;              ;
   ;              ;

CLO: LOG
```

The ELEKTRA Testbed

EXAMPLE for PAMIR REGRESSIONSTEST and EVALUATION:

Trace 1,2,3,4,5: Extraction of "show tracebuffer" and "get record"

Example "show tracebuffer"

```
MTM DGP: sh tb
%dgp-> : Tracebuffer entry 1 is a SIGNAL
%dgp-> :    Destination:    [TVC_A_CMD_VOTER;0]
%dgp-> :    Signal:         S_VA_EA_9287_INDICATION_SIGN
%dgp-> :    Source:         [VA_COMMAND_INPUT;0]
%dgp-> :    Timestamp:      36:24:32.00
%dgp-> :    Priority:       36
%dgp-> :    Process State:  TVC_A_CMD :TVC_A_CMD_VOTER\ Line 144 + 0005H
%dgp-> :    Queue:          INVALID NAME STRING
%dgp-> :    S_VA_EA_9287_INDICATION_SIGN
%dgp-> :       .F_NAME = E_TRACK
%dgp-> :       .F_NUMBER = 7
%dgp-> :       . = E_SET_RSA
%dgp-> :       . = E_ROTTE
%dgp-> :       . = E_SIGNAL_MAN_1
%dgp-> : Tracebuffer entry 2 is a SUSPENSION
%dgp-> :    Instance:       [TVC_A_CMD_VOTER;0]
%dgp-> :    Timestamp:      :25:15.35
%dgp-> :    Process State:  TVC_A_CMD :TVC_A_CMD_VOTER\ Line 144 + 0005H
%dgp-> :       E_SIGNAL_MAN_1)
%dgp-> : Tracebuffer entry 3 is a SIGNAL

results in:

%dgptb1: S_VA_EA_9287_INDICATION_SIGN (VA_COMMAND_INPUT, TVC_A_CMD_VOTER)
%dgptb2: S_VA_EA_9287_INDICATION_SIGN (E_TRACK, 7, E_SET_RSA, E_ROTTE,
         E_SIGNAL_MAN_1)

or (include timestamps)

%dgptb1: S_VA_EA_9287_INDICATION_SIGN (VA_COMMAND_INPUT, TVC_A_CMD_VOTER,
         :25:15.35)
%dgptb2: S_VA_EA_9287_INDICATION_SIGN (E_TRACK, 7, E_SET_RSA, E_ROTTE,
         E_SIGNAL_MAN_1)
```

Example "get record"

```
MTM CCA: get rec,rla_main_signal,0,rea_masi_speed
%cca-> : Record 0 of Relation RLA_MAIN_SIGNAL in UNPACKED format
%cca-> :
%cca-> : RLA_MAIN_SIGNAL(0)
%cca-> :    .REA_MASI_SPEED
%cca-> :    .F_SPEED_DISPL = E_IL_MEDIUM_SA
%cca-> :    .F_SPEED_STAT  = E_SPEED_OK_SS
%cca-> :    .F_SPEED_CHANGE = FALSE

results in

%ccarec: RLA_MAIN_SIGNAL (14, REA_MASI_SPEED = (E_IL_STOP_SA, E_SPEED_OK_SS,
         FALSE))
```

EXAMPLE FOR SIGNAL FLOW GENERATION (ACHILLES) (printed on a text printer):

The stg-file (stg=signal table generator) contains the signal flow (".stg") and the database accesses. The identifiers are named according to the naming conventions.

FLANKENSCHUTZÜBERPRÜFUNGEN VOM SBP

```
         MASIG  ADSIG  PRSIG  SWI_S  LCSI_R       SBP    INS_VO
MNGTRR  ADMTRR SIGREP SHSIG TRACK SWI_D SBL60A DIS_VO COMM_H

         S_SB_IL_5401_AUTOM_INT_INST(e_switch,22),
         (e_masig,5],e_flank_prot_superv_ai,e_line_up_route,
         e_commitment)

         S_SB_IL_5401_AUTOM_INT_INST(e_switch,24),
         (e_masig,5],e_flank_prot_superv_ai,e_line_up_route,
         e_commitment)

         S_SB_IL_5401_AUTOM_INT_INST(e_switch,25),
         (e_masig,5],e_flank_prot_superv_ai,e_line_up_route,
         e_commitment)

         S_SB_IL_5401_AUTOM_INT_INST(e_switch,27),
         (e_masig,5],e_flank_prot_superv_ai,e_line_up_route,
         e_commitment)

         S_SB_IL_5401_AUTOM_INT_INST(e_masig,5],
         (e_masig,5],e_global_locking_ai,e_line_up_route,
         e_commitment)

         S_IL_EA_4205_ELEM_DISPL([E_MASIG,5],0,
         e_route_locked_goal_d)
```

```
PUT_RECORD
   Relation (number/name):   23: RLA_MAIN_SIGNAL
   Record number: 5
   5..REA_MASI_FEST

GET_RECORD
   Relation (number/name):   27: RLA_PROTECTION_SIGNAL
   Record number: 0
   3..REA_PRSI_SI_LO

         S_IL_SB_4501_AUTOM_INT_COM([E_MASIG,5],[E_MASIG,5],
         E_SIG_CONN_ESTABL_AI,E_LINE_UP_ROUTE)
```

SOFTWARE COVERAGE METRICS AND OPERATIONAL RELIABILITY

A. Veevers

Department of Statistics and Computational Mathematics, University of Liverpool, P.O. Box 147, Liverpool, UK

Abstract. A relationship between operational reliability growth and coverage is described. A simple method of fitting the relation is illustrated using an example. The coverage metrics TER_1, TER_2 and TER_3 are used and their *strengths* as reliability predictors are compared.

Keywords. Reliability; coverage metrics; program testing; software engineering.

INTRODUCTION

Software quality assurance techniques which rely on formal methods of program proving and on subjective assessments of competing procedural strategies are aiming for perfect (up to specification) software. In anything other than trivial applications this ideal is virtually impossible to achieve with certainty. If it were otherwise, there would be no further need for the massive effort which continues to be directed towards the quality and reliability assurance of software.

Traditionally, the performance uncertainty in hardware products has been handled by overdesigning and by quantitatively assessing product reliability. Neither of these techniques carry over directly to software systems. Veevers, Petrova and Marshall (1987) concluded that the most relevant concept of software reliability is that of *operational reliability* defined as the probability that fault-precipitating conditions do not arise in a specified period of use in the appropriate operational environment. During the software development phase, reliability growth modelling (Bastani and Ramamoothy, 1988; Mazzuchi and Singpurwalla, 1988) using test cases drawn at random form a realistic operational profile (random testing) provides a way of quantifying the confidence currently placed in the software. Independently of this, software engineers have developed numerous metrics which quantify aspects of a program and testers deliberately select test cases in order to optimize one or more of these metrics. The relation between most of these metrics and operational reliability is obscure.

Veevers and Marshall (1990) used a mathematical argument to establish a functional relationship between coverage metrics and reliability. This paper provides empirical support for the relationship through an example.

RELATION BETWEEN COVERAGE METRICS AND RELIABILITY

It is assumed that any segment of code covered during the correct execution of a test case will also produce the desired result for future cases following exactly the same route. In a random testing phase each additional portion of fault-free coverage that is obtained must increase confidence in the software. Such additional coverage also reduces the size of the region of code conditions which remain unexplored. It is only from within this region that contributions to the unreliability can arise.

Based on this observation Veevers and Marshall (1990) suggested that under random testing an increase dc in coverage c produces, on average, a corresponding increase in reliability proportional to the current amount of unreliability. The proportionality factor, p, represents the *strength* of the relationship between a particular coverage metric and reliability. Taking $r(c)$ to represent reliability as a function of coverage the above can be expressed as

$$r(c + dc) = r(c) + p.dc.(1 - r(c)). \quad (1)$$

In the limit as $dc \to 0$, this can be written as the differential equation

$$r'(c) = p(1 - r(c)), \quad (2)$$

which has the solution

$$r(c) = 1 - ke^{-pc}, \quad (3)$$

where $k \geq 0$ is a constant. The value of k could reflect the amount of prior knowledge about the reliability of the software when coverage is zero. In particular, $k = 1$ corresponds to $r(0) = 0$, i.e. in the absence of any prior information and any successful use of the software a state of total unreliability is a rational belief to adopt.

The derivation of Eq. (3) is based on the dual growth of the coverage metric and the operational reliability in a random testing phase. As faults are discovered and perfectly corrected, reliability increases as does the coverage metric. If Eq. (3) correctly relates the two growth rates then the possibility of predicting reliability from coverage values obtained using systematic testing strategies arises.

FITTING THE MODEL

The parameters of the model given by Eq. (3) can be fitted from data which consists of pairs of values $(r(c), c)$. Noting that

$$-\ln(1 - r(c)) = -\ln k + pc, \quad (4)$$

a least-squares fit of the regression line of $-\ln(1 - r(c))$ on c may be made. This provides estimates of $\ln k$ and p from the intercept and slope, respectively, of the fitted line. Denoting the resulting estimates of k and p by $\hat{k}$ and $\hat{p}$, the reliability prediction equation for a coverage c is

$$\hat{r}(c) = 1 - \hat{k}e^{-\hat{p}c}. \qquad (5)$$

In order to obtain the data it is necessary to estimate operational reliability and to measure the corresponding coverage metric at various different times in the testing phase during which reliability growth is taking place. Suppose there are N bugs in the software at the beginning of the random testing phase. The failure rate at this point will be $\lambda_1 + \lambda_2 + \ldots + \lambda_N$, where λ_i denotes the failure rate of the ith bug when exposed to random testing. When a bug is discovered and perfectly removed the new failure rate will be the sum of the λ_i values corresponding to the remaining undiscovered bugs. An implicit assumption in models of this kind is that bugs do not mask each other. The extent to which this is not (approximately) true in practice will manifest itself in the lack of fit of the model. If a bug has just been corrected the reliability of the software over the next interval of t units of time is $\exp(-\lambda t)$, where λ is the current total failure rate. This follows from the fact that the failure rate is constant during fault-free intervals of time under random testing from the operational profile. In an experimental situation the same piece of software can be debugged many times, starting from a common initial state, until all bugs (except perhaps the extremely unlikely ones) have been removed each time. The failure rates of the individual bugs can be estimated from this experiment and single repetitions of it can be used to illustrate reliability growth.

Employing a suitable software tool during the experimental runs enables the values of selected coverage metrics to be obtained at any times of interest. If all the bugs are known then, for any experimental run, the successive values of λ can be reconstructed by simply adding the estimated failure rates of the remaining bugs as each bug is found and removed. Thus, pairs of values $(r(c), c)$ at each of a series of times throughout a random testing debugging phase can be obtained.

AN EXAMPLE

Beattie (1989), in another context, conducted an experiment using a fairly complex program, written in C, to determine the fares on passenger flights between the Caribbean and the USA. From the point in its development where the program began to give sensible answers to the point of final release, thirteen bugs were found. A series of repeat debugging random testing runs starting with the thirteen bug version each time was performed. The LDRA testbed was employed to provide the execution history of the test runs. Part of the information so provided was a set of values of coverage metrics at the instances of bug detection. Adapting this example for the present purpose, the thirteen bug failure rates have been estimated from twenty replicate runs and are shown in Table 1. The order of the bugs in this table corresponds to the average of 20 runs for which the achieved values of the test effectiveness ratio (TER) coverage metrics TER_1, TER_2 and TER_3 are shown. These metrics are, respectively, the proportion of executable statements, branches and LCSAJs actually executed during the run. An LCSAJ is a linear code sequence and jump as defined in Woodward, Hedley and Hennell (1980) along with the TER metrics.

TABLE 1 Estimated Failure Rates and the TER Coverages Achieved

Bug	Failure rate $\times 10^5$	TER_1%	TER_2%	TER_3%
1	50.463	31	18	10
2	11.487	49	33	20
3	8.215	46	31	18
4	7.412	45	31	19
5	6.744	48	31	18
6	4.540	52	36	22
7	4.222	50	34	20
8	4.180	46	32	20
9	3.229	54	38	23
10	2.646	59	44	28
11	1.411	76	59	40
12	1.099	76	60	42
13	0.538	94	82	62

The unit of time on which the failure rates are based is the time taken to execute one statement. Hence, taking $t = 1000$, say, in the estimation of reliability is equivalent to obtaining the 'reliability per 1000 statements'. This is a sensible reliability interpretation for software in the same spirit as quoting the 'one hour reliability' for a piece of hardware.

Figures 1, 2 and 3 show the plots of $-\ln(1 - r(c))$ against c for $c = TER_1$, TER_2 and TER_3, respectively. The fitted lines, Eq. (4), are also shown. It is evident that a reasonable fit is obtained in each case.

This example provides support for the relation between coverage metrics and reliability given by Eq. (3). Figure 4 displays the fitted formulae for the three metrics.

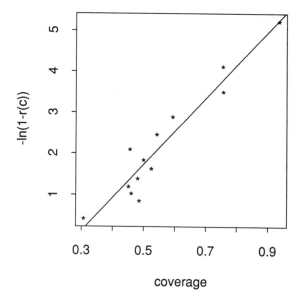

Fig. 1. Plot of $-\ln(1 - r(c))$ against $c = TER_1$, with the fitted line.

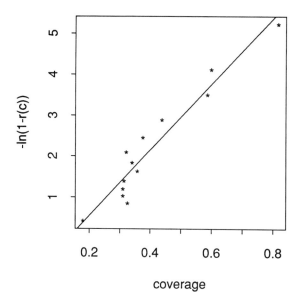

Fig. 2. Plot of $-\ln(1-r(c))$ against $c = \text{TER}_2$, with the fitted line.

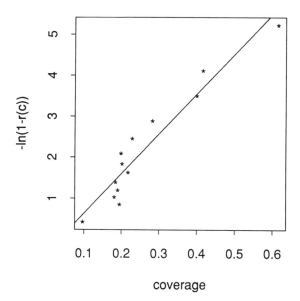

Fig. 3. Plot of $-\ln(1-r(c))$ against $c = \text{TER}_3$, with the fitted line.

Figure 4 provides a comparison between the three metrics based on their ability to predict the '1000 statement reliability' of the software. The estimated strength factors, p, for TER_1, TER_2 and TER_3 are, respectively, 8.0, 8.0 and 9.7 indicating that for the first two metrics the reliability rises towards the asymptote at the same rate whereas for TER_3 the rise is at a faster rate. Thus, the TER_3 metric is stronger than the other two in this sense.

CONCLUSION

Evidence has been provided to support a relationship between coverage and reliability of the form given by Eq. (3). At the present stage of development it is best used as a method of quantifying the relative performances of different coverage metrics. Further empirical support needs to be ob-

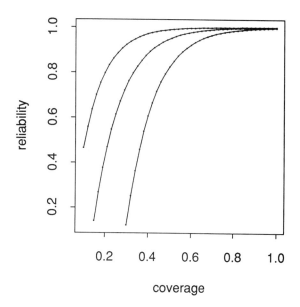

Fig. 4. Plot of the fitted formula relating reliability to TER_1, TER_2, and TER_3.

tained before confident predictions of reliability growth can be made from coverages obtained by systematic rather than random testing strategies.

ACKNOWLEDGEMENT

This article was prepared whilst the author was a visiting senior research scientist with the Division of Mathematics and Statistics, Commonwealth Scientific and Industrial Research Organization, Australia.

REFERENCES

Bastani, F.B., and C.V. Ramamoorthy (1988). Software reliability. In P.R. Krishnaiah and C.R. Rao (Eds.), *Handbook of Statistics*, Vol. 7. Elsevier, Netherlands. pp. 7-25.

Beattie, B. (1989). Structural units as alternatives to time in software reliability growth modelling. M.Sc. thesis, University of Liverpool, U.K.

Mazzuchi, T.A., and N.D. Singpurwalla (1988). Software reliability models. In P.R. Krishnaiah and C.R. Rao (Eds.), *Handbook of Statistics*, Vol. 7. Elsevier, Netherlands. pp. 73-98.

Veevers, A., and A.C. Marshall (1990). A relationship between software coverage metrics and reliability. Manuscript submitted for publication, SCM Department, University of Liverpool, U.K.

Veevers, A., E. Petrova, and A.C. Marshall (1987). Software reliability models, past, present and future. In B.K. Daniels (Ed.), *Achieving Safety and Reliability with Computer Systems*. Elsevier. Netherlands. pp. 131-140.

Woodward, M.R., D. Hedley, and M.A. Hennell (1980). Experience with path analysis and testing of programs. *IEEE Trans. on Software Engineering*, 6, 278-286.

QUALITY MEASUREMENT OF MISSION CRITICAL SYSTEMS

J. B. Wright*, F. Fichot*, C. Georges** and M. Romain**

OPL, 7, rue de Greffulhe, 75008 Paris, France
**Thomson-CSF (Division Systèmes Electroniques), Direction Assurance Qualité,
9, rue des Mathurins, 92223 Bagneux, France*

Abstract. The paper describes a complete and industrially-validated model for quantitative evaluation of the quality of a software-based system, and examines the deployment of such a model within the development, management and quality processes of software production. The installation and operation of the model is described within a Quality Control Department responsible for assuring the attainment of reliability and maintainability requirements for the on-board software components of missile guidance systems.

The model provides a framework for the quality analysis of both the software product and the development process, and supports measurements taken throughout the entire system lifecycle. The automation of various aspects of the model operation are discussed, and guidelines proposed for the choice and integration of tools for data collection and storage, heuristic calculation and report generation.

Experiences are reported of the phased installation of the model via a carefully managed technology transfer operation, and strategies outlined for the definition of didactic support required to educate both project and quality managers and software developers in the operation and analysis of the model, so as to ensure the successful adoption of the practices implied by its employment. In describing the observed benefits of use of the model, we discuss the future role in software certification of quality measurement technology, and examine the use of historical databases of quantitative project data in managing complex multi-disciplinary systems development projects.

Finally, generalisation of the existing model is discussed to demonstrate its widespread application within critical systems development environments controlled by industry-standard methodologies.

Keywords. Quality control; software engineering; software development; reliability; software tools

INTRODUCTION

The Electronic Systems Division (DSE) created within the Thomson group in 1967 is responsible for the development of ground to air and surface to air missile systems and command centres. In sub-contracting the majority of hardware development, the technical focus of the division remains the creation of software system components. Because of the critical role played by on-board software with regard to the structure and functioning of such weapons systems, the Quality Assurance Department of DSE is heavily oriented towards achievement of quality objectives for embedded software.

The Software Quality Service of the Quality Assurance Department has been in place since 1983 to manage quality assurance and control activities based upon software review, audit and inspection procedures. Due to the nature of the product under development and the extremely long life cycle of the systems concerned, particular emphasis is placed upon the attainment of software reliability and maintainability requirements.

OPL are an independent consultant group specialising in the transfer of software engineering technology. Since 1985, they have undertaken several major operations to define in-house methodologies for software development, project management and quality control, working particularly in domains where software forms the critical kernel of an industrial system (eg. aerospace, armaments, telecommunications satellites, nuclear plant administration etc).

Consultants of OPL have been working closely with the Electronic Systems Division of Thomson-CSF since 1988 to help define software quality management procedures, including the setting up of a programme aimed at measuring the quality of software produced within the division. The overall objectives of this programme are many fold, including :

- the development of internal procedures based around a quantifiable approach to quality assurance able to ease the negotiation of software quality requirements with both clients and sub-contractors.

- increased visibility of product quality during the whole of development, so as to focus management attention upon long term improvement of quality levels, thereby increasing the ability of the division to successfully tender and effectively operate within a highly competitive international marketplace.

- the estimation at the earliest possible date of the final product quality, in order to detect deviations between specified and achieved quality and to take corrective action as early as possible.

This paper describes a model for quality measurement which has been developed in response to these requirements, and reports some initial experiences in its practical application in the DSE software development environment.

NOTIONS OF SOFTWARE QUALITY

Attempts to manage the quality of computer software development must be conditioned by the unique character of the software product, as summarised by Pressman (1987):

- software is a logical rather than a physical system element; therefore success is measured by the quality of a single entity rather than many manufactured entities.

- software is not prone to wear out - if faults are encountered there is a high probability that each was inadvertently introduced during development and went undetected during testing.

- few, if any, spare parts (reusable software components) exist for the replacement of defective parts during software maintenance.

The response to date of the software development community to the so-called software crisis resulting from the lack of mature software development technology while applying software-based solutions to application domains of increasing complexity and criticality has been the promotion of a rigorous approach to software production, centered around a better understanding of the distinct processes of software development, management and quality. The resultant sets of methods, procedures and tools for software engineering have stressed the view of a software life cycle comprising a series of distinct phases.

International standardisation bodies have also been active in many areas to work to enforce adoption of such development and control methodologies. In the field of military software development, contractual obligations to comply with norms such as the US Department of Defense (1988) standard for mission critical systems development, have led DSE to organise quality assurance and control activities around certification of compliance with this standard. A strong emphasis of the resultant software life cycle is system documentation and the production of representations of the final software product based around high visibility of a set of *work products*.

Faced with such strong market-led pressures and the need to maintain software products throughout life cycles of typically 15 years duration, the quality measurement model described in this paper responds to the recognition that an effective software quality management programme can no longer rely upon a strictly qualitative vision of quality. The various life cycle work products are thus readily viewed as the sources of raw data able to be synthesised into a quantifiable evaluation of quality.

Initial responses to the need to measure software quality were proposed by Boehm (1978) and by McCall, Richards and Walters (1977) who developed a hierarchical vision of quality supported by a standardised terminology. This work has since been embodied in a draft standard published by the IEEE (1988), which includes notions such as reliability and maintainability within the definition of quality, and which resides upon the identification of:

- external (ie. user-oriented) quality characteristics, referred to as *factors*,

- internal (ie. developer-oriented) *criteria* necessary to ensure the realisation of the identified factors,

- a set of measurable attributes or *metrics* associated to each criterion, which act as probes to gather information relating to the quality of the software.

The precise way in which the higher level factor / criteria values of the DSE model are synthesised from the raw data collected as metric values is dictated, as subsequently explained, by the model structure and the associated heuristics.

DEFINITION OF A QUALITY MEASUREMENT FRAMEWORK

The quality measurement model developed by Thomson-CSF and OPL has been largely based upon the pioneering work mentioned above, with the following important extensions:

- enhanced coverage of test and specification phases. Quality measurements must be made throughout the whole development process, from product definition through to qualification, in order to support early and continuous quality assessment.

- coverage of production process quality in addition to product quality. This follows from the belief that the quality of any product (including software) depends both upon the intrinsic quality of the product (the rigour with which it has been defined, designed, produced and tested), and the quality of the processes by which it has been produced (the use made of advanced production and inspection techniques, the procedures by which the development has been managed etc).

- establishment of explicit relations between the factors, criteria and metrics and the phases of the development cycle during which measurements are taken. Each phase of the life cycle is treated as an indivisible series of activities resulting in a set of formally identified work products (equipment, service, code, documentation) from which quality measurements may be taken.

- use of a model memory to take account during calculations for a specific phase of results obtained during earlier development activities. This can be illustrated by consideration of the measurement of one of 18 criteria handled by the model, namely modularity. To calculate this value throughout software development, we use metrics applied successively to the software functional model, global architecture, low-level design and source code. The calculation of a value for this criterion may thus be successively developed, by work product inspection and metrics collection, at the end of the requirements specification, preliminary design, detailed design and coding phases. We observe that the effort consumed by software maintenance depends upon (amongst other things) the degree of modularity exhibited by the software. If we imagine that the modularity value has been evaluated very lowly during the initial specification and design phases but quite highly during the coding phase (perhaps due to the choice of a programming language which enforces modular structure at a syntactic level), we would clearly be seriously mistaken to estimate maintenance effort from a measurement of software modularity which did not take into account the low values calculated during the initial production phases.

Model Quantification

Amongst the 120 metrics supported by the model can be identified such wide-ranging examples as functional cohesion, number of faults detected by phase, degree of data structuring, readability / completeness of user guide etc.

Values for each metric can, according to the specific nature, either be taken directly from a work product or else calculated from a weighted checklist. The lists which have been used by the model refer to conformance of a work product with either a reference plan cited by the development methodology (eg. Software Quality Plan) or with the overall development methodology itself. This raw data is then processed in accordance with the following principles by which the objects of the model are quantified :

- thresholds are applied to measurements of terminal metrics (using three discretionary levels),

- values of non-terminal objects are obtained by calculating a normalised heuristic,

- object values can either be interpreted directly from the numeric values, or by threshold assignment.

The model definition includes identification of the following characteristics for each metric :

- development phase(s) during which the metric is calculated,

- source of the value (ie. identification of the work product from which the measurement will be made),

- data to be collected,

- calculation method and thresholds to be applied.

Reliability Measurement

Special mention is merited of the handling within the model of the Reliability factor. For phases upstream of testing, reliability is not calculated but rather predicted using models based essentially upon notions of robustness.

During the requirements specification and preliminary design phases, the reliability factor is evaluated from the Modularity and Test (Inspection) Coverage criteria, while at the detailed design and coding phases, account is additionally taken of measures of complexity and robustness.

Subsequently, during unit and integration testing, a reliability level can be evaluated either from the criteria used during the coding phase or, as is currently the case upon several DSE pilot projects, from estimation models aiming to measure the probability of failure-free functioning and test coverage rates.

Finally, during validation testing, the reliability level can be calculated from the Failure Rate criteria, which is itself evaluated from the three most commonly used operational reliability models (Musa 1987) :

- Littlewood-Verral

- Goel-Okumoto

- Musa.

These each calculate the Mean Time To Failure (MTTF) of the software as a function of evolution of failure rates. The reliability factor is evaluated from the predictive models used during the previous phases until such time as the operational models have attained a satisfactory confidence level (> 0.75). At this moment, the operational models are substituted to give a real value of the measurement of the software reliability.

General experience has been that the period of time necessary to reach such a confidence level has been long, sometimes even longer than the validation phase, in which case predictive results have been used throughout the operation of the quality measurement model. In other cases, inconclusive results have been provided by the operational models, depending upon the software failure rates.

INSTALLATION AND OPERATION OF THE MODEL

Acceptance and installation of the model within the DSE software development environment has been handled as a classical technology transfer action, supported heavily by an incremental approach and awareness raising programmes of strongly tailored training courses.

Following standard practices for technology transfer and in order to avoid anticipated problems in introducing a significant data collection overhead upon projects already in testing phases, an incremental approach has been defined for model introduction via a set of pilot projects.

Choice of Pilot Projects

The application of the model to an initial series of pilots was seen to serve a double purpose :

- to gain practical experience in the use of the model by those charged with its operation. The principal questions addressed here have concerned the overall perceptions of the model, identification of the practical difficulties encountered, sizing of the change in working practices implied etc.

- to validate the model. This will be performed primarily through regression analysis of metric thresholds and heuristic coefficients so as bring together, through the vehicle of the model, real results, *a posteriori* measurements and estimates made during development.

Within Thomson-CSF, a number of such pilot projects have been identified and the application of the model begun, with the objective being at the end of an 18-24 month trial period to have a model which can be truly considered to be in-place and accepted.

In addition, the highly modular nature of the model readily lends itself, within a given pilot project, to an incremental application considered essential to avoid environmental rejection :

- the initial model structure installed is that part, involving relatively few metrics, which concerns the calculation of production process quality.

- the next stage is to install the product quality axis of the model structure, except for the most costly and complex measures, ie. those concerned with reliability predictions and software ergonomy.

- the final step is to complete the model through implementation of the above-mentioned measures which were excluded at the previous stage.

Even given the strategy outlined above, it is clear that the density of information to be processed, both in quantitative terms (the number of elementary metrics) as well as qualitative terms (the nature and complexity of each elementary metric) requires a large effort and firm commitment on behalf of the organisation in order to achieve sucessful installation and operation of the model.

This is translated in practical terms as a need to ensure adequate handling of a number of essential tasks for the control of the volume of data concerned. The major terms of reference of such prerequisite work undertaken with DSE is considered below under the separate banners of educational support and model automation.

Training Support

The intrinsic complexity of the model necessitates the training of all parties affected by the day to day operation of the model. These users can be grouped under the following headings :

- software developer

- project or line manager

- quality manager.

In order to tailor the training provided as closely as possible to the specific needs of individuals, three principal levels of education were identified as being necessary to address the differing requirements of the above audiences :

- *general* : presentation of the underlying concepts of quality measurement, the static and dynamic structure of the model and the set of factors and criteria covered. The overall objective here was to enable the target audience to appreciate the technical and economic stakes, the basic principles and the general approach to installation. This training was particularly designed for line managers.

- *in-depth* : addressed to those responsible for the operational implementation of the model and the subsequent exploitation of the results which it provides, this training comprises a detailed presentation of each metric and heuristic of the model, and emphasises the existing links with the software life cycle, the collection of elementary data and the maintenance of the model. The principal audiences of such training have been quality managers and project managers.

- *selective* : a compromise between the above two courses, this training is particularly adapted to members of the development team. It adds to the general training a detailed presentation of a coherent subset of the metrics generated or manipulated by the trainee, eg. those metrics calculated during the coding phase, for a team programmer audience.

Toolset Support

The overall size and complexity of the model implies a heavy processing overload, centred upon the calculation of heuristics (mostly simple but numerous) and terminal metrics (numerous and very varied in nature, ranging from very simple to highly complex). In seeking to automate various aspects of the model operation, so as to reduce this overhead, it is clearly impossible to identify a single turnkey software engineering product able to support the whole of the model and the associated activities. Several classes of tool, however, can be identified which are at least capable of providing partial support in specific areas :

- *complexity analysers*, taking measurements of textual and logical complexity and of modularity from either source code or detailed design pseudo-code,

- *test coverage analysers*, able to provide indications of coverage afforded by a set of test programs, often in terms related to the control structure of the object under test,

- *configuration management* tools, able to track defect corrections through version control procedures,

- various *mathematical and statistical* tools.

For many metrics, such tool support may be considered a luxury; for certain such as the calculation of structural or conjunctoral excess ratios from progress tracking diagrams, it is even impractical. There remain others, however, such as the calculation of cyclomatic number for which it constitutes an absolute necessity. An incremental strategy to introducing tool support can be defined according to the relative priorities of these aspects, working from minimal to total support via a series of successive stages :

  - calculation of complex metrics

  - assistance in calculation of complete metrics set

  - calculation of heuristics

  - management and storage of quantifiable data

  - automatic generation of summary reports

  - automatic collection of elementary data.

Current experience with the model in DSE and elsewhere extends to use of dynamic and static analysers of the Logiscope[1] and Testbed[2] tools, as well as internally developed coverage and pseudo-code analysers.

The most obvious obstacle to achievement of the later stages of the above strategy lies in the increasing need to integrate a disparate set of heterogeneous tools. In a subsequent installation of the model undertaken by OPL within an airplane production environment, the Oracle[3] data base is being used as the mechanism for a common data management system accessible by a set of proprietary tools.

Data Synthesis and Presentation

For effective analysis of the vast volumes of data handled by the model, and the efficient implementation of management actions responding to the results provided, the data must be synthesised for presentation to disparate parties. The medium chosen for the required distribution of information throughout the development, management and executive hierarchies was a set of summary reports designed to exhibit the following characteristics :

  - graphics-based, highly readable layout,

  - presentation of information at multiple abstract levels, ranging from the general to the specific,

  - clear traceability between different presentational levels,

  - support for project and quality monitoring via differentiation of established objectives and achieved results.

Three classes of summary report were designed to meet these requirements :

  - Model Status Report, showing the quality levels of the factors and sub-factors of the model,

  - Factor Status Reports, illustrating the status through time (ie. by development phase) of individual factors in terms of the component objects (sub-factors or criteria). In this and the above report, colour coding conventions were used to increase the readability of the reports by associating different colours to value intervals.

  - Specific Study Reports which synthesise a defined subset of the total model data. Examples include studies to investigate the relative efficiency of inspection techniques (used from requirements specification through to coding phase) and test techniques (used during unit, integration and validation testing). This could be shown by plotting the number of faults, as a percentage of total number detected by fault or inspection techniques, against the cost associated with the use of the corresponding technique. Such a report could then be used to deduce :

    - detected failure rate and corrected failure rate during various development phases,

    - fault follow-up (detection / correction) by number, type and cost of repair,

    - the cost ratio production cost / fault repair (or correction).

CONCLUSIONS

The quality measurement model described in this paper is currently undergoing pilot test evaluation within the Electronic Systems Division of Thomson-CSF. OPL meanwhile are undertaking a number of other projects to apply the principles of the model to software development environments in the engineering and aerospace fields.

Early feedback from model usage suggests the need for complementary direct measures of cost of non-quality to be collected during fault correction activities undertaken during development and throughout the operational lifetime of the software. Such measures could

---

1 Logiscope is a registered trade mark of Verilog S.A.
2 Testbed is a registered trade mark of Liverpool Data Research Associates Ltd.
3 Oracle is a registered trade mark of Oracle Corporation.

also serve to validate the model, in observing that those projects obtaining the best quantitative indications of quality are also less susceptible to costly bug correction.

An important requirement has also been identified for consolidation of the model results :

- *horizontally*, across multiple projects within a department or division. The relative weight contributed by any particular project to the final global value for departmental quality will be a function of the criticality of the project, as modelled by the required level of reliability, intrinsic complexity of development etc.

- *vertically*, across departments. This would be achieved firstly in consolidating all projects relating to a given system, then for all systems within a given sector of corporate activity, and finally across all sectors addressed by the enterprise.

Such consolidation could be extended to encompass other engineering disciplines, eg. electronic component and circuit design, if quality measurement models could be developed for these domains with a similar structure and based upon the same fundamanetal concepts as that described in this paper. Expectations are high that this can be achieved in such areas where production is guided by a similar methodological framework to that imposed for software development by standards such as DoD 2167A.

Both short- and long-term benefits of adoption of the quality measurement model are anticipated within Thomson-CSF in several distinct areas :

- Establishment of in-house statistical databases to be integrated with other historical data and used for comparisons and estimates for subsequent project tendering and management.

- Guiding of design decisions by quantifiable evaluation of options with respect to criteria such as degree of module coupling, testability etc. In a similar fashion, overly complex elements of a large design can be identified and targeted for subsequent decomposition and refinement.

- Benefits to quality auditing from integration into software requirements specifications of quantifiable quality levels. When supported by an adequate base of economic analyses of cost of use (or non-use) of the model, a more scientific approach will be enabled to perform cost / quality tradeoffs at the contract negotiation stage in accurately estimating the cost of providing a product to attain a specified quality level.

To summarise, we find software measurement to be an exciting and challenging area in which to work, and one which is only now, after accumulating more than 12 years of theoretical basis, beginning to achieve widespread credibility within the software development community. We hope that this paper can contribute to the ongoing debate in the areas covered, and encourage the continued investigation into quantifiable software quality management.

REFERENCES

ANSI / IEEE (1988). P1061 Draft Standard for a Software Quality Metrics Methodology, Draft 15

Boehm, B.W., J.R. Brown, M. Lipow, G.J. MacLeod and M.J. Merrit (1978). *"Characteristics of Software Quality"*, ed. North Holland.

DoD (1988). Military Standard - Defense System Software Development, DOD-STD-2167A.

McCall, J.A., P.K. Richards and G.F. Walters (1977). *"Factors in Software Quality"*, 3 vols., NTIS AD-A049-014, 015, 055.

Musa, J.D., A. Iannino and K. Okumoto (1987). *"Engineering and Managing Software with Reliability Measures"*, McGraw-Hill.

Pressman, R.S (1987). *"Software Engineering - A Practitioner's Approach"*, McGraw-Hill.

SOFTWARE RELIABILITY ASSESSMENT — THE NEED FOR PROCESS VISIBILITY

C. Dale

Cranfield IT Institute, Fairways, Pitfield, Kiln Farm, Milton Keynes MK11 3LG, UK

Abstract. When dealing with ultra-high levels of reliability or with safety-critical systems, it is especially important to assess software reliability and safety, and to view these assessments as confidence building activities. Confidence should be built by examining various kinds of information about a given system, and making judgments regarding the compatibility of the information gathered with the level of reliability or safety required. Software and its development process are in general very abstract, so that the necessary information is usually difficult to find, unless special care has been taken to ensure that the software and its development process are made sufficiently visible. The main focus of this paper is process visibility: a five stage strategy is outlined for ensuring availability of the information necessary for confidence building. The initial stages are concerned with process based assessment: building confidence by ensuring that a development strategy appropriate to a given achievable target is being properly carried out. The final stages relate to product based assessment, which builds confidence by examining the product itself and its behaviour when executed.

Keywords. Confidence building; reliability; safety; software engineering; software metrics.

INTRODUCTION

There is an ever increasing trend for computers to be used in systems for which safety is a concern. Any explicit or implicit safety requirement associated with these systems usually implies a reliability requirement on the software within the computer; the only exception is where it can be shown that the software cannot impact safety of the system. Given such a requirement, it is important to be able to carry out assessments which generate confidence in the software aspect of the system.

Among those who should be interested in these assessments are the customer or procurer of the software (or the system of which the software is a component part), the project manager responsible for delivering the software product, and the software developer's quality manager.

Superficially, the customer is interested in the safety and reliability of the product or system only at the point of delivery, and whether the delivered system meets his requirements. On closer analysis, however, the customer's interest often needs to go much further back in the lifecycle than the time of delivery. When placing a contract for a bespoke system with critical safety and/or reliability requirements, it is wise to have confidence that the chosen supplier will be able deliver the required reliability. It is equally important that this confidence is maintained as the development of the software and the system proceeds.

The project and quality managers' needs for confidence from the earliest stages of development are more apparent. The project manager must be able to establish reliability targets for the software, formulate a plan to meet those targets, and monitor progress against the plan. The quality manager will want to be able to inspect measurements which demonstrate that a development process of the appropriate quality is being applied, and that the process is leading to product quality achievement.

There is then a clear need to be able to establish confidence in the safety and reliability of a system not only when it is delivered, but during and even before its development. This will come as no surprise to those familiar with safety analysis of engineering systems in general; many techniques exist for examining various kinds of information drawn from the entirety of the systems development cycle, and making judgments regarding the compatibility of the information gathered with the level of reliability or safety required (O'Connor 1981).

A significant problem which has to be overcome is that software is abstract in the extreme, and its development process also tends to lack the tangibility provided by the physical nature of traditional engineering systems. These factors mean that unless special care is taken, information necessary to the confidence building activity either does not exist, or does not get captured at the appropriate time.

Another problem is that software reliability assessment studies have tended to focus on measurement of reliability of completed or almost completed products (Dale, 1986). The need for assessment techniques earlier in the lifecycle, to aid the confidence building discussed above, has received scant attention, despite the fact that the proper management of the software reliability (and hence system safety) of systems containing software, from the viewpoints of both supplier and customer, demands the ability to carry out reliability assessments from the earliest stages of software development.

This paper describes in outline a five stage approach to the problem of building confidence in the safety and reliability of software-based systems. The first three of these stages are concerned with process based assessment: building confidence by ensuring that a *development strategy* appropriate to a given *achievable target* is being *properly carried out*. The final two stages relate to product based assessment, building confidence by examining the *product* and its *behaviour* when executed. These five stages are shown in Fig. 1, and are described in the sections which follow.

ACHIEVABILITY OF TARGETS

The first of the five assessment stages is to consider the achievability of the software reliability targets which are implied by the system level requirements for safety and reliability. It is important to note that safety and reliability concerns at the system level may each imply their own reliability requirements on the software: for example, safety requirements may imply a high level of reliability for a particular subset of input data; whilst system reliability requirements demand a more moderate level of reliability over the entire input range. Thus, there may be a multiplicity of reliability targets which the software has to meet; sometimes these targets will themselves be in conflict.

Often, reliability targets will be expressed at the system level, possibly based on some specified safety requirement. In carrying out high level system design, the system will be decomposed into its component parts, and a reliability apportionment exercise will result in the assignment of reliability targets to the different aspects of the system. It is important to make an initial judgement of feasibility of each of these component reliability targets at this stage, so that any necessity for reapportionment or redesign at the system level can be identified at the earliest possible - and thus cheapest - time.

Ultimately, the top level system design and the reliability apportionment exercise will result in some statement of reliability requirements associated with the software component or components. It is important that this statement is quantitative, and that it is made in terms which can be measured by observation of software execution. Thus statements such as *the software shall be fault-free* or *the software shall have fewer than x faults per 1000 lines* must be avoided - the first cannot be measured, and the second can be measured only in ways which are not directly related to the user-perceived failure behaviour, ie the execution of the software in time.

A number of different metrics are available which can be used to express the software reliability requirements in quantitative terms, without resorting to the unmeasurable notions above. The formal definition of reliability as the probability of successful operation, either for a given period of time, or on demand, is one candidate. Failure rate is another. Whatever metric is chosen, it is vital to have a clear understanding and definition of what constitutes a failure; this can only be achieved in the particular context of a given system.

A valid alternative to these metrics is to allocate well defined integrity levels to software components, such as those proposed by the IEC (1989a). Here the software is implicitly allocated the same target as other software in the same category.

Quantitative (or categorical) statements of software reliability requirements are important as a means of establishing a clear understanding of the level of reliability required. Anything less

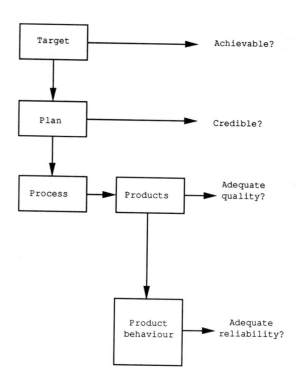

Fig.1 The five stages of confidence building

than such a statement leaves too much room for subjectivity and interpretation, for example between supplier and customer, or between system project manager and software project manager. Quantitative requirements also have motivational advantages, which flow from their objective nature: professional staff involved in the software development work know exactly what is required of them, and have something concrete to work towards. The other side of this issue is that if they do not believe the target is achievable, they may become very demotivated, so undermining further any chance of success.

The stated reliability requirement must reflect the various demands on individual functions of the software. Thus a protection system of some sort may be required to have a reliability greater than p for its *shut down when required* function, and a failure rate less than λ for its *keep running if safe* function. This example illustrates that a mixture of metrics can sometimes be necessary in the requirements statement.

It is also vital to remember, when specifying any required measure, that the conditions of measurement must also be specified, in order for the meaning of the quantitative reliability specification to be unambiguous.

Statements of reliability requirements should be for an appropriate level of reliability - neither too high nor too low. The reliability must certainly be sufficiently high for the application in question, but the requirement must at the same time be kept as low as permissible because of considerations of achievability and testability. Thus demands for perfection must be avoided - perfection is (at least) difficult to achieve and impossible to demonstrate. Similarly, high levels of reliability are difficult (and costly) to achieve and can be demonstrated only by extensive (and expensive) verification, validation and testing activities.

This does not mean that we should argue down the levels of safety demanded by society from engineering systems, rather that we ensure that demands for reliability of software are consistent with the societal need, and that excessive demands on software reliability are not made simply because the higher level system design is intrinsically inadequate to meet societal or customer needs.

The above arguments are intended to show that the software reliability assessment process should begin by looking at the achievability of the stated software reliability requirements, so as to establish initial confidence in the feasibility of the target. There is little point proceeding to subsequent phases of development if the software reliability target is judged to be unachievable at this stage; it is best to know early if the ultimate target is unlikely to be met.

The result of this stage of assessment will, if all is well, enable the software development to be planned in detail, in the confident expectation that the specified reliability and safety targets can be met. Planning - at least from the safety and reliability point of view - then becomes a case of identifying how the adjudged achievable target will actually be met. On the other hand, the first stage of the assessment process may identify a need for reconsideration of either the system targets, the high level design, or the reliability apportionment.

CREDIBILITY OF PLAN

Once an achievable target has been established, a plan can be drawn up to show how the product will be developed to meet its reliability (and other) requirements. This planned development strategy must now be evaluated, so as to generate further early confidence in successful achievement of the target. This avoids the danger of embarking on development with a high risk of failure to meet targets, as a result of defining a plan which is inconsistent with the specified requirements and the targets derived from them.

A particular problem in the safety-critical and ultra-high reliability areas is Miller's (1989) conclusion that, in critical applications where extremely reliable software is required, it may be impossible to assure the reliability in a statistically rigorous way; this means that through-development confidence building becomes even more important in these areas. The statistical methods to which Miller refers are concerned with the analysis of failure data from software testing, which by definition are available only after the software has been developed.

In theory, it is possible using statistical methods to verify any level of reliability, by testing a piece of software for long enough (with fixes being carried out in response to any failures encountered) to demonstrate that a given target has been achieved, with a given level of statistical confidence. Miller identifies three problems with this approach, in the ultra-reliability situation: the distribution of inputs used to exercise the software in testing may not perfectly fit the usage distribution which will be encountered in real operation; fixes may be imperfect; and the test time may be limited.

Even if the first two of these concerns are set on one side, Miller's analysis leads to a rule of thumb that a string of failure-free tests up to one order of magnitude longer than the acceptable mean time to failure is required, in order to assure a given target expressed in these terms. So, for example, three million test hours would be required to provide 95% confidence that a target mean time to fail of one thousand hours had been met.

If failures occur during the testing, the situation becomes worse: software reliability growth models (Dale, 1986) can be used to allow for the fact that fixes are being made in response to failures which occur, but for this situation Miller (1989) quotes Littlewood's observation that at least ten million hours of random testing and debugging are required to achieve an interfailure time of one million hours.

Even at more moderate levels, there remain the problems of mis-match between the environments used in testing and in real operation.

One of the implications of Miller's work is that a high level of *a priori* confidence is essential to any reliability or safety claims which are to be made later. It will not be practicable to demonstrate that a particular level of reliability has been achieved, simply by testing the software after it has been written. Any initial lack of confidence will, of course, have to be made up for by even more testing, on top of that implied by Miller's conclusions.

Returning to the second stage of confidence building, having looked forward to reinforce its importance, the main aim here is to establish that the planned development is likely to meet the target. A healthy degree of conservatism is necessary at this stage, because in many software-related areas, the current state of knowledge and understanding concerning which techniques and methodologies are likely to achieve which levels of reliability is meagre. In most cases, it is impossible to say with any confidence that method X is likely to meet target Y. Thus, extreme claims and targets have to be treated with equally extreme scepticism.

There are no well established techniques which can be used to make an assessment of likelihood of the planned development meeting the established target. Instead, an appropriate review or inspection process should be applied to the plan, and the experience of those involved in the review used to make the assessment as objective as possible. Any data on the way previous comparable projects were planned and carried out, and the levels of reliability achieved in practise, will be invaluable in this context. Such data is a very scarce commodity at the current time.

The second stage of confidence building may lead to recommended revisions to the development process, such as additional or more formal verification and validation activities. Once again, identifying the need for these at an early stage is much better than trying to put the problem right at a later stage. This is recognised in standards such as Draft Interim Def Stan 00-55 (Ministry of Defence, 1989), which emphasises organisational and planning aspects of development, as well as documentation and other procedural matters which fall within the scope of the next section.

QUALITY OF DEVELOPMENT

Having established an achievable reliability target, and produced a credible plan of how the target is to be achieved, the software development itself can commence with a healthy degree of confidence that the target will be met. The need now is to continue to build this confidence, as the product is built. To achieve this, the development process must be monitored to ensure that it is being carried out in a way consistent with the target and the plan.

This is the phase in which process visibility becomes a particularly apposite term: monitoring the development, producing evidence of what has been done, and arguing why this evidence enables an enhanced confidence in the reliability of the ultimate product is no more nor less than taking an essentially abstract process and making it visible.

This third stage of the assessment process is, then, concerned with determining whether the actual development is being carried out as planned, and whether it is of a standard consistent with the given reliability target.

In managing any kind of activity, the project manager needs to be able to monitor and control the critical success factors. In this case, it is the developing reliability characteristics which must be monitored and controlled; it is far from obvious how this can be done. The target reliability should have been expressed in a way which enables measurement of the final product reliability, such as probability of failure on demand, or failure rate; but how can the project manager hope to monitor this quantity at a point in the lifecycle when the software is not even executable?

The answer is of course that he cannot measure reliability in any direct sense until very much later in the development cycle, when it will almost certainly be too late to take the necessary remedial steps in an economical way. The challenge for the project manager is to identify factors which can be measured during development, to serve as indicators of the level of reliability likely to be achieved, based on the current stage of development.

In a well managed project, it should be possible for the project manager to report progress against a variety of objectives on a regular basis, and to have supporting evidence for his report. Problems are identified, and remedial action taken. In principle, the management of the reliability aspect should be treated no differently.

There are two kinds of evidence which the project manager can use to help him in managing the reliability: process information and product information; the latter will be dealt with in the next section. Process information is simply evidence that the project is being carried out in accordance with the plan, which was earlier adjudged to be capable of delivering the required reliability. This provides evidence that the job is being done properly, and thus gives some support to an argument that the target is likely to be met, as it provides indirect evidence about the developing product.

The key to providing this sort of evidence is to ensure that accurate records are maintained of the activities carried out, in at least as much detail as they were originally planned. The plan was the basis of an argument of how the target would be achieved; this needs now to be reinforced by evidence that the development really was carried out in the way that it was planned.

Inevitably, however, there will be some aspects of the development which do not happen exactly as planned. In these cases, the likely impact on reliability of the deviations from the plan must be identified, and if the impact is such as to undermine confidence in achievement by simply sticking to the plan, remedial action must be planned. In making these judgments, all aspects of the plan must be considered, not just the time and cost aspects which often preoccupy project managers - planned and actual levels of skill, tools, training, machine availability, and a host of others are all important, and the impact of deviations for all of these should be assessed. The guiding principle is that if some aspect of development was important enough to include in the plan, on which project credibility is based, then it is important enough for deviations to be assessed for their potential impact on the ultimate delivered reliability.

The records necessary to enable development process visibility are not difficult to produce and maintain, provided there is proper discipline in documentation, configuration control and quality assurance. These and accurate records of verification and validation activities are important supports to visibility. The collection, analysis and presentation of data concerning the development of the product can be utilised to help the manager and others to understand and communicate progress towards the stated reliability requirements.

Verification and validation activities especially need to be made visible. Reviews, inspections and so on are the times at which the intermediate products are subjected to maximum scrutiny, and their adequacy determined. The activities may be carried out by the programmer, within the team, within the organisation, or by an independent outside body: these are listed in increasing order of independence from the development of the object in question, which is usually associated with increasing visibility of the contribution towards reliability achievement.

One well-documented approach to process assessment appears in the HSE guidelines for programmable electronic systems in safety-related applications (Health and Safety Executive, 1987). These guidelines contain checklists applying to all stages of the development process, so recognising that reliable software results from actions carried out throughout development. Similar approaches underlie more recent work in international standardisation (IEC, 1989b).

PRODUCT QUALITY

In addition to ensuring that the development process is of the requisite quality, the project manager also needs information about the various intermediate products, so that there is some direct evidence of actual reliability and safety achievement; intermediate products may, of course, include specifications, documents and other non-executable objects.

This information is needed in order to show that the process is actually delivering the level of quality required: the process was planned in the way it was because it was believed that it would deliver the necessary level of quality, and process monitoring has been used to ensure that it has been carried out in that way; now confirmation is sought that the process is delivering product quality as anticipated.

The principle to be applied here is that measurement of various aspects of the software and other intermediate products of the development process enables assessment of the adequacy of the outputs. The details of the information used for this purpose will vary from project to project, depending mostly on the development methodology and techniques adopted, but it will often take the form of fault data, which can arise in a number of ways.

Formal review activities, such as Fagan inspections (Fagan, 1976), normally give rise to lists of faults identified. These lists may be categorised, showing the numbers of faults of each of a number of different levels of severity or potential consequence. In planning the achievement of reliability, targets may be set for these reviews, determining the intermediate acceptance criteria for the objects under review, and so helping the project manager monitor the achievement of quality. For example, a detailed design document may be releasable for coding to commence only when review indicates no major faults and less than three minor faults. There may also be criteria which determine whether a further full review is required following rework, or if review and approval of fault fixes will suffice.

There are many other ways in which software and the other products may be examined to ascertain their level of quality, one of which is static analysis - so called because it does not involve execution of software, only examination of the code (or other formal representation). This is normally a tool aided activity, aimed at analysing the syntax of the developed code to identify such things as unreachable code and various other non-preferred features. These do not always indicate the presence of specific faults, but do serve as indicators of potential problem areas; human expertise is still required to interpret the static analysis tool outputs, and identify whether the highlighted features are in fact related to faults.

Static analysis tools often provide in addition various measures of code complexity, which can be a further indication of problem areas. A related tool-aided area is that of symbolic execution, in which formulae are derived to show the relationships between code inputs and outputs; these can be inspected to determine whether they are in accordance with the specification, and so potentially indicate the presence (but not the precise location) of faults. Static analysis and symbolic execution are reviewed by Smith and Wood (1987).

Dynamic measures can also be made. These include test coverage measures (Hennell, Hedley and Ridell, 1983), which quantify the extent to which the software in question has been exercised by the testing to which it has been subjected. These measures are really measuring how well the testing process has been applied to the product in question, but are so intimately related to the product that they can almost be viewed as product measures.

Ensuring that both the process and the developing reliability-related characteristics of intermediate products are made visible in the ways described in this and the previous section above helps to substantiate claims that the actual development process was of the standard necessary to the achievement of the reliability targets, and that the quality of intermediate products is consistent with the levels of safety and reliability required in the delivered system.

PRODUCT RELIABILITY

The final stage of assessment is to measure the reliability of the final product, based on failure data collected during an appropriate regime of testing. This stage is carried out so as to provide further evidence to support the confidence already built during the previous four stages. It is almost never the case that a valid software reliability assessment can be carried out by addressing this stage alone, for reasons which are discussed below.

A host of statistical models exists for measuring and predicting the reliability of software, most of them based upon test data (Dale, 1986). All of these methods depend upon assuming that the environment in which the software is tested is representative of the operational environment of concern (or that a relationship can be defined between these two environments, enabling a mapping of the reliability in the test environment to the reliability in use). To date, very few organisations have been willing to make the investment necessary to understand the operational environment well enough to enable this assumption of representativeness to be validated - though some success has been reported from the IBM clean room development methodology (Currit, Dyer and Mills, 1986). Unless an attempt is made to replicate the real environment, any application of statistical models for software reliability prediction will result in information of an essentially qualitative value - numerical values do not always mean that reliability has been quantified.

Even when the real world environment has been replicated, an element of process assessment is still necessary to ascertain that the testing process is of the necessary level of quality, and that the original specification was properly validated. So even in these circumstances, statistical measurement of the completed product cannot stand in isolation as the only assessment which is applied.

When the highest levels of reliability are sought, as is often the case for safety-related systems, a more serious problem arises. The statistical models require amounts of data which cannot practically be obtained, in order to provide confidence in the measurement of the achieved level of reliability (Miller, 1989 and Dale, 1987). In these circumstances, the statistical models can provide the reassurance required only if there are strong *a priori* reasons for believing that the necessary level of reliability has been achieved.

Despite these limitations, statistical measurement of the achieved level of reliability should be carried out whenever reliability is an important requirement, either by applying a reliability growth model to data from testing, or by applying standard statistical techniques to a reliability test of the final product. In either case, the input data used should be statistically representative of data which will be seen in the environment of interest.

CONCLUSIONS

This paper has described a five stage software reliability assessment procedure, beginning with initial assessment of target feasibility, and culminating in measurement of the delivered level of reliability. In concept, the steps are simply those which need to be taken to manage the achievement of reliability, by enabling the responsible manager to monitor and control progress towards a target. They also provide the evidence necessary to support claims made to the customer or some independent assessor, concerning the reliability or safety of the product.

Ultimately, the degree of confidence in the safety of a system assessed using this procedure will depend on each of the five stages, but no formal method for combining the five stages is proposed, because of the extreme differences in the kinds of information available at the various stages. Rather, the five stages provide a framework within which an objective judgement of adequacy can be made, albeit of a qualitative nature.

The five stage procedure described is in essence an engineering approach to an engineering problem: confidence is built in parallel with the building of the software aspect of the delivered system.

REFERENCES

Currit, P.A., Dyer, M. and Mills, H.D. (1986). Certifying the reliability of software. *IEEE Transactions on Software Engineering*, SE-12, 3-11.

Dale, C.J. (1986). Software reliability models. In A. Bendell and P. Mellor (Ed.), *Software Reliability State of the Art Report*. Pergamon Infotech, Maidenhead. pp. 31-44.

Dale, C.J. (1987). Data requirements for software reliability prediction. In B. Littlewood (Ed.), *Software Reliability Achievement and Assessment*, Blackwell Scientific Publications, London. pp. 144-153.

Fagan, M.E. (1976). Design and code inspections to reduce errors in program development. *IBM System Journal*, 3, 182-211.

Health and Safety Executive (1987). *Programmable Electronic Systems in Safety Related Applications*. HMSO, London.

Hennell, M.A., Hedley, D. and Riddell, I.J. (1983). The LDRA testbeds: their roles and capabilities. In *Proc. IEEE Soft Fair '83 Conference*, Arlington, Virginia, July 1983.

IEC (1989a). *Software for Computers in the Application of Industrial Safety-related Systems*, Document 65A(Secretariat)94: Draft. IEC, Geneva.

IEC (1989b). *Functional Safety of Programmable Electronic Systems: Generic Aspects. Part 1: General Requirements*, Document 65A(Secretariat)96: Draft. IEC, Geneva.

Miller, D.R. (1989). The role of statistical modeling and inference in software quality assurance. In B. de Neumann (Ed.), *Software Certification*. Elsevier Applied Science, London. pp. 135-152.

Ministry of Defence (1989). *Requirements for the Procurement of Safety Critical Software in Defence Equipment*, Interim Defence Standard 00-55 (Draft). MoD, London.

O'Connor, P.D.T. (1981). *Reliability Engineering*. Heyden.

Smith, D.J. and Wood, K.D. (1987). *Engineering Quality Software*. Elsevier Applied Science.

ASSESSING SOFTWARE RELIABILITY IN A CHANGING ENVIRONMENT

T. Stålhane

ELAB-RUNIT, SINTEF, N-7034 Trondheim, Norway

Abstract. The reliability of a software system depends on its initial error contents and which errors are discovered during testing and use. Which errors are discovered and then corrected will depend on how the system is used. As a consequence of this, the reliability of a system will depend on its history (installation trail). Knowledge of this history can be used to predict the reliability at a new site and the test strategy necessary in order to obtain a specific reliability for a predefined site.

This paper shows how we can describe the history of a system by a function execution vector. The number of remaining errors will have a Poisson distribution with intensity depending on the execution vectors for all the previous sites and the initial error content. Once the cost of a failure in any system function is identified, it is also possible to find the optimal test strategy for test of the system at a specified site.

Keywords. Software engineering, program testing, reliability, computer software, prediction, coding errors.

THE BASIC MODEL

The MTBF

In this paper, we have chosen to focus on MTBF as an expression of software reliability.

If we let $N(t)$ be the number of times a software system fails in the interval $[0,t]$, then we have that (Finkelstein, 1983):

$$MTBF = \{\frac{\partial}{\partial t} E[N(t)]\}^{-1} \qquad (1)$$

In order to simplify our approach, we will work in the number-of-executions domain. Let us define

$X(t)$: the number of executions or transactions in the interval $[0,t]$. When no misunderstanding is possible, we will use X instead of $X(t)$

$w(t)$: $\frac{\partial}{\partial t} X(t)$

$N(x)$: the number of failures observed in X executions

With this in mind we can write

$$MTBF \cdot w(t) = \{\frac{\partial}{\partial x} E[N(x)]\}^{-1} \qquad (2)$$

We will use the notation MXBF (Mean Number of eXecutions Between Failures), defined as

$$MXBF = w(t) \cdot MTBF \qquad (3)$$

Since it is straight forward to move between MTBF and MXBF, we will use MXBF in the rest of this paper.

EXTENSIONS OF THE SIMPLE MODELS

If we assume that the failures are independent for each function i, then we have (Stålhane, 1988).

$$N_i \sim Bin(N_{oi}, F_i(x_i))$$

$$E(N_i) = E_x[E(N_i|X_i)] \qquad (4)$$

$$E(N_i) = N_{oi} E_x [F_i(X_i)]$$

It is not possible to procede without assuming a particular function $F_i(.)$. We can, however, use the following approximation.

$$E(N) \approx \sum N_{oi} F_i(p_i X) \qquad (5)$$

$$MXBF^{-1} \approx \sum N_{oi} p_i f_i(p_i X) \qquad (6)$$

Note that N_o and N_{oi} should be understood as $N_o(H)$ and $N_{oi}(H)$ respectively, where H is the system history.

INCLUSION OF THE SYSTEM HISTORY

The Number of Errors

Each time a programmer writes a program statement, there is a certain, small probability that he will make a mistake. Some data indicates that this probability varies across statement types. If we let the fault probability for statements of type i be π_i and l_i is the number of statements of type i, then we can write

$$N_o \sim Poisson(\sum l_i \pi_i) \qquad (7)$$

For the sake of simplicity, we will write

$$\theta = \sum l_i \pi_i \qquad (8)$$

By combining (4), (7) and (8) we get, for each function

$$N \sim \text{Poisson}(\theta F(x)) \quad (9)$$

If we assume that all discovered errors are corrected and let N_r denote the remaining number of errors, then it also follows that

$$N_r \sim \text{Poisson}(\theta \bar{F}(x)) \quad (10)$$

Assume that a system is installed at site 1, is used there for some time T_1, then installed at site 2 and used there for the time T_2 and so on. Then, if all errors are at risk at all sites, we have

$$N_o(i+1) = N_r(i) \quad (11)$$

Here $N_o(i+1)$ is the initial number of errors at site $i+1$ while $N_r(i)$ is the remaining number of errors at site i. The assumption that all errors are at risk at all sites is relaxed later. See (17) and (18).

It is now straight forward to show that

$$N_o(i+1) \sim \text{Poisson}(\theta \prod_{j=1}^{i} \bar{F}_j(x_j)) \quad (12)$$

It is possible to describe the history of a software system by an installation history diagram. Each line in the diagram describes the history of one instance of the software system, which is installed, used and maintained at a series of sites.

An example is shown in fig. 1. The upper line (S_1) in this example diagram shows that one instance of the system is installed first at site 1, then at site 2, then at site 3 and at last at site 4.

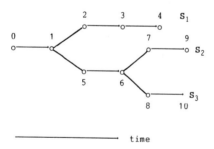

Fig. 1 Software system history diagram

The diagram in fig. 1 shows three different history paths, denoted S_1, S_2 and S_3 respectively. The expression (12) can now be rewritten as

$$N_o(i+1) \sim \text{Poisson}(\theta \prod_{j \in S_i} \bar{F}_j(x_j)) \quad (13)$$

We can thus state the following:

The number of errors in a system at the time of installation depends on

- how the system was designed, implemented and tested, through θ

- how it was used and maintained at the previous sites, through $F_j(.)$

- which sites it has been used at through S_i

In the same way, we see that

$$N \sim \text{Poisson}(F(x) \theta \prod_{j \in S_j} \bar{F}_j(X_j)) \quad (14)$$

$$E(N) = E[F(x)] \theta \prod_{j \in S_j} \bar{F}_j(X_j) \quad (15)$$

$$MXBF^{-1} = E[f(x)] \theta \prod_{j \in S_j} \bar{F}_j(X_j) \quad (16)$$

The Effect of Functions

We now need to consider our systems model. The model shown in fig. 2 is simple enough to be treated in a straightforward manner while still being applicable to a class of real systems.

Note that a more general systems model has been published by Cheung (1980). This model has, however, the drawback of using a constant reliability for each system module. In addition, the execution probabilities for each module are not connected to the use of the system in any simple, deductible way.

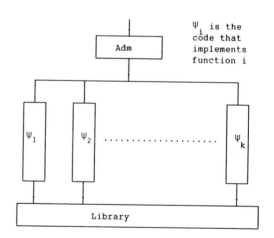

Fig. 2 A Simple Systems Model

We will assume that all errors reside in the functions code and that the administrative part and the library are error free. Experience has shown that this is a fairly realistic assumption.

Since each function is independent of all the others when the library is error free, we can define N_{ok} as the initial number of errors in the code implementing ψ_k. Let us further more introduce the following notation:

p_k : P(user executes ψ_k)

x_{kj} : the number of times ψ_k is executed at site j

$F_{kj}(x)$: P(no. of executions to failure at site j $<x|\psi_k$)

It follows that

$$N_{ok} \sim \text{Poisson}[\theta_k \prod_{j \varepsilon S} \bar{F}_{jk}(x_{kj})] \quad (17)$$

Only errors inside the functions with $x_j > 0$ are at risk and p_j expresses the probability that the function is used. It then follows from the theory of Poisson processes that

$$N_o \sim \text{Poisson}(\sum_{k=1}^{n} (\theta_n \prod_{j \varepsilon S} \bar{F}_{jk}(X_{jk})))\quad (18)$$

In the same way, we get

$$N \sim \text{Poisson}(\sum_{k=1}^{n} (F_k(X_k) \theta_k \cdot \prod_{j \varepsilon S} \bar{F}_{jk}(X_{jk}))) \quad (19)$$

Note that (18) and (19) are good approximations, even if each N_{ok} is only approximately Poisson distributed. This follows from the fact that a sum of stochastic processes can be approximated by a Poisson process as long as no process dominates the sum (Cox, 1980).

It is not possible to compute $E(N)$ in the general case, but the following approximation holds:

$$E(N) \approx \sum_{k=1}^{n} (F_k(p_k X) \theta_n \prod_{j \varepsilon S} \bar{F}_{jk}(p_k(j) X_j)) \quad (20)$$

$$MXBF^{-1} \approx \sum_{k=1}^{n} (p_k f_k(p_k X) \theta_k \prod_{j \varepsilon S} \bar{F}_{jk}(p_k(j) X_j)) \quad (21)$$

The results in (20) and (21) are the final result of this part of the paper. We have shown how the system reliability depends on the present usage (through P and $f(.)$), its history (through $F_j(.)$) and by its construction (through θ_k^j). We will now turn to the applications of these results.

HOW CAN WE USE THE RESULTS

Our Goals

The basic formulas will be used to attack three problems

- the optimal test pattern for a known site
- prediction of reliability at a new site
- reliability prediction in an incremental development model

Optimal Tests for a Known Site

Let us now use the index 0 for the in-house test. Thus, $F_0(.)$ describes the strategy for the system test and X_{ko} is the number of times the function Ψ_k has been executed during the test, X_o is the total amount of tests run and so on.

From (16) and (21) we can write

$$MXBF^{-1} \approx \sum_k MXBF(Xp_k)_k^{-1} p_k \quad (22)$$

At site 1 we have that (see (21))

$$MXBF_{k1}^{-1} \approx \theta_k \bar{F}_{ko}(x_{ko}) p_{k1} f_{k1}(p_{k1} X_1) \quad (23)$$

What we need is a testing strategy that gives us the largest possible MXBF. This is the same as minimizing

$$MXBF_{k1}^{-1} \approx \sum_k \theta_k \bar{F}_{ko}(x_{ko}) p_{k1} f_{k1}(p_{k1} X_1) \quad (24)$$

The only factors that we can influence during testing are $\{F_{io}(X_{io})\}$. We will consider a situation where we have a fixed number of tests (X_o) and uses the following cost function:

Total expected cost =

    cost of running the tests +
    cost of correcting discovered errors +
    cost of remaining errors to customer +
    cost of too low MXBF

We will use the following notation

  K : the cost of running one test

  C_o: the cost of correcting one error

  C_k: the cost of one error in function k when the error occurs at the customers site

  B : the cost of too low MXBF

Thus, we can write total expected cost, T as:

$$T = K \Sigma X_{ko} + C_o \Sigma \theta_k F_{ko}(x_{ko}) + \Sigma C_k \bar{F}_{ko}(x_{ko}) F_{k1}(x_{k1})$$
$$+ B \theta_k \bar{F}_{ko}(x_{ko}) p_{k1} f_{k1}(x_{k1}) + \lambda(\Sigma X_{ko} - X_o) \quad (25)$$

λ is the Lagrange multiplier.

Note that for a small MXBF, the last factor in (4.4) will dominate T and thus bring home the fact that nobody cares about error costs before the errors get rather rare. As long as they are frequent, the customer just counts the number of errors, high cost or not.

$$\frac{\partial T}{\partial X_{ko}} = K + C_o \theta_k f_{ko}(x_{ko}) - C_k f_{ko}(x_{ko}) F_{k1}(X_{k1})$$
$$- B \theta_k f_{ko}(X_{ko}) p_{k1} f_{k1}(X_{k1}) + \lambda \quad (26)$$

From experience, we now that K is small compared to C_o and C_k. We will thus look at the following equation:

$$\lambda + \theta_k f_{ko}(X_{ko})(C_o - C_k F_{k1}(x_{k1})$$
$$- B p_{k1} f_{k1}(x_{k1}) = 0 \quad (27)$$

In order to keep the notation simple, we will introduce the notation

$$h_k(X_1) = C_k F_{k1}(X_1 p_{k1})$$
$$+ B\, p_{k1} f_{k1}(X_1\, p_{k1}) - C_o \qquad (28)$$

The solution to (27) can then be written

$$x_{ko} = f_{ko}^{-1}\left\{\frac{\lambda}{\theta_k h_k(X_1)}\right\} \qquad (29)$$

$$X_o = \Sigma\, f_{ko}^{-1}\left\{\frac{\lambda}{\theta_k h_k(X_1)}\right\} \qquad (30)$$

We can now find λ from (30) and then optimum number of tests for function k from (29).

Let us consider the failure model of Jelinski-Moranda as a simple demonstration (Moranda, 1975):

$$f_{ko}(x_{ko}) = \varphi_k\, e^{-\varphi_k x_{ko}} \qquad (31)$$

By combining (29) and (31) we get

$$x_{ko} = -\frac{1}{\varphi_k}\ln\left[\frac{\lambda}{\varphi_k h_k(X_1)\theta_k}\right] \qquad (32)$$

From (32) and (30) we can now find λ from the equation

$$\ln(\lambda) = \frac{\Sigma\,\dfrac{\ln[\varphi_i h_i(X_1)\theta_i]}{\theta_i} - X_o}{\Sigma\,\dfrac{1}{\varphi_i}} \qquad (33)$$

From (32), at last, we find the optimum number of tests for function k as:

$$x_{ko} = \frac{1}{\varphi_k}\ln[\theta_k h_k(X_1)]$$

$$-\frac{1}{\varphi_k} \times \frac{\Sigma\,\dfrac{\ln[\varphi_i h_i(X_1)\theta_i]}{\theta_i}}{\Sigma\,\dfrac{1}{\varphi_i}} \qquad (34)$$

In the general case, $\{x_{ko}\}$ must be found by numerical methods.

Prediction of MXBF at a New Site

We have previously found (Stålhane, 1988) that $f(x)$, the error life time distribution, changed relatively little from site to site, even for different systems. As a first approximation we will thus assume $f_k(.)$ to be the same for all sites and that all site variability is included in $w(t)$, the usage intensity, and p, the function usage profile. From (22) this gives us

$$MXBF_2^{-1}(X_2) \approx \Sigma_k\, p_{k2}\, MXBF_{k2}^{-1}(p_{k2} X_2) \qquad (35)$$

If the system is in use at several sites, we can obtain a better estimate by using

$$MXBF_{k,n+1}^{-1}(x) = \frac{1}{n}\sum_{i=1}^{n} MXBF_{ki}^{-1}(x) \qquad (36)$$

This is equivalent to using the following estimator for $f_{k,n+1}$:

$$f_{k,n+1}(.) = \frac{1}{n}\sum_{i=1}^{n} f_{k,i}(.) \qquad (37)$$

This is the same as the non-parametric Bayesian estimator for f suggested by Colombo (1985):

$$f(.) = \frac{m}{m+n} f_{old}(.) + \frac{n}{m+n} f_{new}(.) \qquad (38)$$

Here m is the support for $f_{old}(.)$ and n is the number of observations used in the estimate of $f_{new}(.)$.

The MXBF for Incremental Development

Together with rapid prototyping, incremental development is one of the more promising methods for software development, (Graham, 1989). There are three reasons for this

- the method enables the producer to deliver part of the system to the user at an early stage and thus provides the opportunity for early user feedback

- contrary to rapid prototyping, the increments delivered are not meant to be thrown away. They are thus not a drain on the project resources

- the method can be combined with a risk-oriented way of managing software project, for instance through the Boehm (1989) spiral model.

The model suggested in (21) and fig. 2 is well adapted to incremental development. This can be shown in the following way:

Start by drawing the system history path as shown in fig. 1:

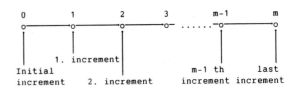

Fig. 3 Incremental development

For the sake of simplicity we will assume that each increment consists of a new function. The testing and usage of each function can be different. If we use our usual indices we have that

$$\text{MXBF } (X_1)_1^{-1} \approx f_1(x_1)\theta_1\bar{F}_0(x_0) \tag{39}$$

After installation of increment 2, we have that

$$\text{MXBF}(X)_2^{-1} \approx p_{12}f_{12}(x_2)\theta_1\bar{F}_0(X_0)\bar{F}_1(x_1) +$$

$$p_{22}f_{22}(x_2)\theta_2\bar{F}_2(x_1) \tag{40}$$

If we keep on including all m increments, we get

$$\text{MXBF}(X)_m^{-1} \approx \sum_{i=1}^{m=1} \{p_{im}f_{im}(p_{im}X_i) \cdot \theta_i \prod_{j \in S_i} \bar{F}_j(p_jX_j)\} \tag{41}$$

The vector **P** will change through the development and testing process. Thus, even if we always test all available functions with the same number of test cases, **P** will change. This will in turn cause MXBF to change. This type of change in system reliability is often observed in continuously growing systems, usually amplified by the users and testers shifting focus when a new function is introduced.

MANAGEMENT CONCERNS

One of the issues that should always be on top of the agenda for management is optimum use of resources. When it comes to testing, one of the reports that is important to management is a plot of number of errors found (and removed) per (1000) test cases or the daily/weekly development of MTTF/MXTF.

The majority of such plots that are published gives a rather chaotic impression. That this must be the case is readily seen from (19). When $\{S_i\} = \Phi$, we see that

$$\frac{E(N_k)}{\theta_k} \sim F_k(x_k) \tag{42}$$

The probability that $\Sigma E(N_k)/\theta_k$ in the general case will have a well known distribution is small indeed.

As shown by for instance Leach (1989), we can obtain a much better picture by logging each set $\{N_k, X_k\}$ separately. Fig. 4 shows the picture we get when we only report $\{N, X\}$, while fig. 5 shows the picture we get when we use the data set $\{N_k, X_k\}$.

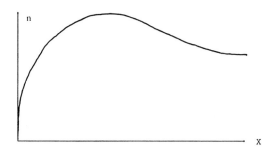

Fig. 4 All errors reported together

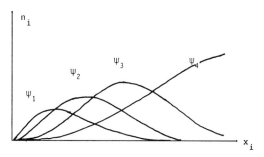

Fig. 5 Errors reported per function

If management only uses the plot shown in fig. 4, they would conclude that no reliability growth is taking place and that the whole project is in serious trouble. Some of the possible decisions could be:

- to scrap the system, since the reliability goal can never be reached

- to increase the total testing activity

By using the plot shown in fig. 5, however, they can easily see that everything is progressing in an orderly manner, except for ψ_4. Thus, testing should be stopped or reduced for all other functions, and all resources should be devoted to testing of ψ_4.

CONCLUSION

In this paper we have shown how we can use information on system usage to compute its reliability or design an optimal test set. The key information on system usage is a usage vector that describes the distribution of the number of executions over system functions. This knowledge can be used for several important purposes such as

- predicting the systems MTBF at a new site, once its installation path is known

- predicting the systems change in MTBF when a new increment is added to a system

- design an optimal test suite for a specific site.

Last, but not least, the model offers important insight for the manager when he uses failure data as a means for project monitoring both during development and after release.

REFERENCES

Boehm, B. (1989)
Software risk management
Tutorial at the 11th international
Conference on Software Engineering
May 5-18, 1989

Cheung, R.C. (1980)
A user-oriented software reliability model
IEEE Trans. Software Eng. SE-6, 118-125

Colombo, A.G. Constantini, D. and Jaarsma R.J.
(1988)
Bayesian, nonparametric estimation
of time-dependent failure rate
IEEE Trans. Reliab. R-34, 109-112

Cox, D.R. and Isham, V. (1980).
Point processes
Chapman and Hall, London 1980

Finkelstein, J.M. (1983).
A logarithmic reliability growth model for
single-mission systems.
IEEE Trans. Reliab. R-32, 508-511

Graham, D.R., (1989)
Incremental development: Review of
nonmonolithic life-cycle development models
Information and Software Technology
vol 31, no 1, Jan - Feb 1989, 7-20

Leach, L. (1989)
Large systems reliability
*Proc Sixth Annual Conference
on Large Software Systems*
Bristol Sept 26-29, 1989

Moranda, P.B. (1975)
Prediction of software reliability
during debugging
*Proc 1975 Annual Reliability and
Maintainability Symposium*, 327-332

Stålhane, T. (1988).
A load dependency model for software reliability
Ph.D. dissertation, 1988
Technical University of Norway

DEPENDABILITY EVALUATION OF WATCHDOG PROCESSORS

P. J. Gil, J. J. Serrano, R. Ors and V. Santonja

Dpto de Ingeniería de Sistemas, Computadores y Automática,
Universidad Politécnica de Valencia, Spain

Abstract: Many control systems need a good safety level, this can be done with the help of a Watchdog Processor, that not involve a high increase in the system's cost compared with other fault tolerant structures.

This paper presents Markov models for studying the safety, reliability and availability of a Watchdog Processor. To carry out a benefit analysis, the results of the models of the three systems: Simplex Systems, Reconfigurable Duplication Systems, and Stanby Sparing Systems, are compared.

The influence of parameters, such coverage, permanent faults, fault rate, etc, in all models is considered. For solving the models in the transient mode, a program based in the *randomization method* is used.

Keywords. Dependability evaluation; Reliability; Safety; Watchdog processors; Markov modeling.

INTRODUCTION

Usually, industrial control systems need a high dependability with a reasonable cost.

Following the approach (John,86) of analyze a system as an integral part of the design process and to use the results as an analysis to guide design decisions, this paper compares the dependability of the most simple structures used in the design of dependable computers.

Industrial control systems is based in replication. An approach based in a watchdog processor (Mammood,1988) is presented in this paper and we shall see that this is a viable alternative compared with duplication systems.

Other advantages of watchdog processor are high level in concurrent error detection and low complexity of the system.

The present paper proposes a model to analyze the dependability of a watchdog processor.

The proposed model is quite simple and has the same level of complexity as duplication systems used, in this paper, for comparison purposes. The differences between all the proposed systems are illustrated.

WATCHDOG PROCESSORS OVERVIEW

A Watchdog Processor (WP) (Mahmood,1988) *"is a small and simple coprocessor used to perform concurrent system-level error detection by monitoring the behaviour of a main processor."* Fig. 1. shows the general structure of a watchdog processor system.

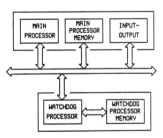

Fig. 1. Watchdog processor system.

Some techniques have been used for error detection like, structural integrity checking (Lu,1982), general rollback with multiple retries (Upadhayaya,1986), derived-signature control flow checking (Namjoo,1983). Generally the main processor transfers data to the WP, that monitors the main processor and collects the relevant information concurrently. This information is compared with data previously loaded in the WP memory, during the setup phase.

Some advantages of watchdog processor compared to replication systems is that the WP is cheaper, it can be added to any system without major changes, and the checking circuitry is totally independent of the checked circuitry.

A good survey of WP can be found in (Mahmood and McCluskey,1986).

SYSTEMS FOR COMPARISON

For comparison with the watchdog processor, a brief review on systems most often used to increase dependability in industrial control systems is presented. Two methods are considered: reconfigurable duplication and standby sparing. Also, a basic model of a simplex system is presented.

Simplex system model (S).

This system, with a single processor, is not able to detect or handle a fault. Let λ_m be the failure rate and μ the repair rate of the processor. To obtain the reliability function, a Markov model as shown in the figure is used. It contains two states:

Fig. 2. Simplex system model.

State (1) (initial state): the processor is fault free.
State (2) : the processor has failed.

The reliability R(t) of the system and its safety S(t) corresponds to the probability of being, at instant t, in state (1).

$$R(t) = S(t) = p_1(t) = e^{-\lambda_m t}$$

The availability model must take into account the repair rate μ. The model contains the same states as the previous one, but there is an additional transition from state (2) to state (1) with the repair rate.

The instantaneous availability of the system at instant t is, again, the probability of being, at that time, in state (1).

$$A(t) = p_1(t)$$

Reconfigurable Duplication System model (RD).

The system has two modules M_1 and M_2 that performs the same functions. Their outputs are compared in a comparator C that detects discrepancies between them. If the outputs of the two processors are different, the comparator will activate an error signal E_c. Each processor runs self-diagnostic tests that can reflect its own error through the output E_1 or E_2. These three error signals are the inputs to a switch circuit SW, that conects its output to the fault-free processor. If M_1 fails, the switch circuit will set $S = S2$, if M_2 fails then $S = S1$.

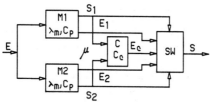

Fig. 3. Reconfigurable Duplication System.

The function of the comparator C is to increase the safety level of the system, in such a way that if one of the modules fails and the failure is not detected by the self-diagnostics the comparator will activate its output E_c. In this case, the switch is unable to change its output, as it can not detect the origin of the failure.

Let us assume that both modules have the same failure rate λ_m . The system will be repaired only when the two modules fail, with a repair rate μ (thus, we suppose that each module can not be repaired separately).

Let C_p be the coverage of the self-detection of failures in the two modules. Let C_c be the coverage of the detection of failures in the comparator.

To take into account the failures of the switch, its failure rate is considered in C_p and C_c. Thus the system is modeled as if the switch were fault-free.

The reliability Markov model of the system contains four states:

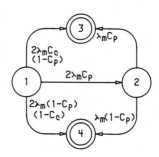

Fig. 4. Reconf. Duplic. Syst. Reliability model.

State (1) (initial state): Both processors are fault-free.
State (2) : One of the modules has failed and the failure has been properly detected and handled. The system does not fail.
State (3) : Safe fail of the system, that is, the system has failed but the failure has been detected and effectively handled.
State (4) : Unsafe fail of the system, the system has failed and the failure has not been detected.

The reliability is now the probability that the system is either in state (1) or state (2) (the two unfailed tates):

$$R(t) = p_1(t) + p_2(t)$$

The safety of reconfigurable duplication is different of reliability, as now there is a failed but safe state (3). So the safety of the system can be calculated as the probability of not being in state (4)

$$S(t) = 1 - p_4(t)$$

To obtain the availability model, two new transitions must be added to the previous model to reflect the repairs.

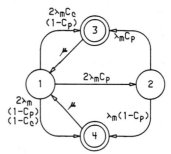

Fig. 5. Reconf. Dupl. Syst. Availability model.

The availability is the probability that the system is, at time t, in an operational state.

$$A(t) = p_1(t) + p_2(t)$$

Standby Spare System model (SS).

The standby spare system consists also of two modules. Each module runs self-diagnostic tests and can activate an error signal (E_1, E_2) if a wrong performance of the module is detected. Now the comparator is unnecesary as the two modules are not on-line simultaneously. The spare processor is brought on line, only when the operational processor detects a fault. As in the previous model, S_1 and S_2 are the modules outputs and S is the system output. The switch circuit inputs are S_1, S_2 and the error signals E_1 and E_2.

If we assume that initially M_1 is on-line and fault-free, the switch will connect its output S with S_1. In this situation, M_2 is just running self-diagnostics. If now, M_1 fails, the error signal E_1 will be activated and the system must reconfigure: M_2 must be brought on-line and the switch must connect its output to S_2.

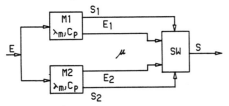

Fig. 6. Standby Spare System.

Let λ_m be the failure rate of each module, and let C_p be their self-detection coverage. As in the previous model, we will consider that the failures due to the switch are included in C_p. So C_p is really, the self-diagnostics and succesful reconfiguration coverage.

We will assume, again, that the system will be repaired only when the two modules fail, with a repair rate μ.

The reliability Markov model for this system contains five states, Fig. 7.

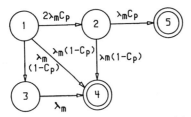

Fig. 7. Stand. Spar. Reliability model.

State (1) (initial state) : both modules are operating correctly.
State (2) : One of the modules has failed and the failure has been detected and handled in an effective manner. The system does not fail.
State (3) : The spare module fails and the failure is not detected. The system does not fail.
State (4) : Safe fail of the system.
State (5) : Unsafe fail of the system.

The reliability is the probability of the system being in state (1), (2) or (3).

$$R(t) = p_1(t) + p_2(t) + p_3(t)$$

The safety is the probability of being in a safe state.

$$S(t) = 1 - p_4(t)$$

The availability model can be obtained from the previous model by adding two new transitions, from the failed states to state (1).

The availability is the probability of being in states (1), (2) or (3):

$$A(t) = p_1(t) + p_2(t) + p_3(t)$$

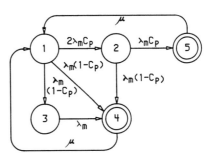

Fig. 8. Stand. Spar. Availability model.

WATCHDOG PROCESOR SYSTEM MODEL

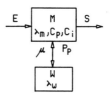

fig. 9. Watchdog processor system.

The model proposed assumes that the system (Fig 9.) is composed by two subsystems, the first, is the main processor and the other, the watchdog processor, trying to use the same modeling level than the previous models.

We assume two kinds of failures, permanent and intermitent-transient failures.

Obviously, the watchdog processor can not recover the main processor permanent failures, but it can do, in most cases, that the failures will be safe.

In the intermitent-transient failures case, the most frecuently kind of failures, the watchdog processor can recover the most of them.

Let λ_m be the main processor failure rate and λ_w the watchdog failure rate.

The watchdog processor is less complex than the main processor. So, we obtain the results for some λ_w values and with λ_m/λ_w varying from 1 to 10.

We assume that the permanent failure probability is P_p and then, $(1 - P_p)$ will be the intermitent-transient failure probability.

The permanent fault repair rate of the system will be μ.

We will assume, again, that the system will be repaired only when the system completely fails.

The repair rate of intermitent-transient failures is μ_i, it is independent of μ, and $\mu_i \gg \mu$.

Let C_p be the permanent failure coverage, and let C_i be the intermitent-transient coverage.

The safety and reliability Markov models.

The proposed model (Fig. 10) has five states:

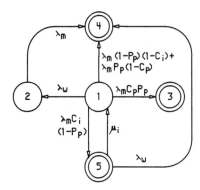

Fig. 10. WP reliability and safety model.

State (1) Both processors are fault-free.
State (2) Intermitent or transient main processor failure covered. Fail-safe
State (3) Permanent main processor failure covered. Fail-safe.
State (4) Unsafe fail.
State (5) Watchdog processor failure. The system does not fail.

The reliability, will be the probability that the system is either in state (1) or state (5):

$$R(t) = p_1(t) + p_5(t)$$

The safety will be the probability that the system is either in state (1), (2), (3), (5), or the probability that the system are not in state (4):

$$S(t) = p_1(t) + p_2(t) + p_3(t) + p_5(t) = 1 - p_4(t)$$

The transient failure repair rate is higher than the failure rate. So the system is in state (2) very few time and then we can reduce the states of the system (Fig. 11.).
The new states are:

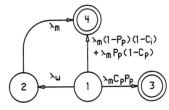

Fig. 11. WP reduced reliability and safety model.

State (1) Both processors are fault free.
State (2) Watchdog processor fails. The system does not fail.
State (3) Fail-safe.
State (4) Unsafe fail.

The new reliability will be the probability that the system is either in state (1) or in state (2).

$$R(t) = p_1(t) + p_2(t)$$

And the safety will be the probability that the system is in the states (1), (2) or (3), or the probability that the system is not in the state (4).

$$S(t) = p_1(t) + p_2(t) + p_3(t) = 1 - p_4(t)$$

To obtain the availability model, Fig. 12., two new transitions must be added to the previous model to reflect the repairs.

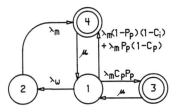

Fig. 12. WP availability model.

The availability will be the probability that the system is in the states (1) or (2)

$$A(t) = p_1(t) + p_2(t).$$

COMPARISONS AND RESULTS

In order to estimate the behaviour of a watchdog processor system, a comparative evaluation has been performed for the previous models. We used a program based in the "ramdomization method" for the transient analysis of reliability, safety and availability. Also, the influence of coverage in the watchdog processor system is studied.

Reliability and Availability

So as to compare and evaluate the reliability, the same failure rate of the main processor (λ_m) and coverage are assumed. Fig. 13. plots the reliability of differents models versus time. With the same failure rate ($\lambda_w = \lambda_m$) the reliability of WP is slightly better than the dual systems for $t > 750 hours$. We can observe an improvement of reliability when λ_m/λ_w increases. This improvement is less important as λ_m/λ_w approaches to 10.

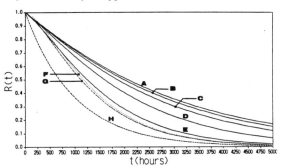

WP A $\lambda_m/\lambda_w = 10$ WP D $\lambda_m/\lambda_w = 2.5$ RD G
WP B $\lambda_m/\lambda_w = 7.5$ WP E $\lambda_m/\lambda_w = 1$ SS F
WP C $\lambda_m/\lambda_w = 5$ S H

Fig. 13. Reliability results.

Failure rate: $\lambda_m = 10^{-3} f/h$.
Coverages: $C_p = C_c = 0.9$, $C_i = 0.95$
Probability of permanent faults in watchdog processor: $P_p = 0.25$

The behaviour of availability is similar to reliability and the WP has the best avalilability as is shown in Fig.14.

Safety

Fig. 15 shows the results for safety with the same values for coverage and failure rates used in reliability analysis. The safest system is RD due to the use of a comparator. The safety of the S is also better than the WP, but an improvement in the safety of WP is observed as λ_m/λ_w increases. So, as far as safety is concerned the WP is not better than dual systems but is much better than simplex systems.

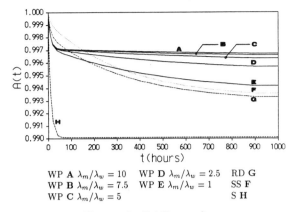

WP A $\lambda_m/\lambda_w = 10$ WP D $\lambda_m/\lambda_w = 2.5$ RD G
WP B $\lambda_m/\lambda_w = 7.5$ WP E $\lambda_m/\lambda_w = 1$ SS F
WP C $\lambda_m/\lambda_w = 5$ S H

Fig. 14. Availability results.

Failure rate: $\lambda_m = 10^{-3}$ f/h.
Coverages: $C_p = C_c = 0.9$, $C_i = 0.95$
Probability of permanent faults in watchdog processor: $P_p = 0.25$
Repair rate: $\mu = 0.1$

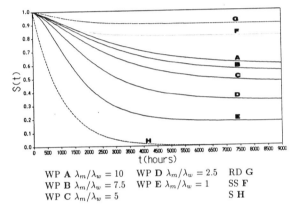

WP A $\lambda_m/\lambda_w = 10$ WP D $\lambda_m/\lambda_w = 2.5$ RD G
WP B $\lambda_m/\lambda_w = 7.5$ WP E $\lambda_m/\lambda_w = 1$ SS F
WP C $\lambda_m/\lambda_w = 5$ S H

Fig. 15. Safety results.

Reliability versus coverage

In this case, the influence of intermitent and transient failure coverage(C_i) is studied. Fig. 16 shows that this coverage has a great influence in the reliability of WP. As in the other systems we do not consider the coverage C_i, the reliability is a horizontal straight line. For values of C_i greater than 0.65 the WP is better than the other systems. The coverage of permanent failures do not have an influence in the reliability of WP as we have commented before.

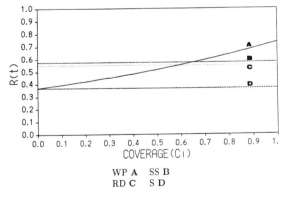

WP A   SS B
RD C   S D

Fig. 16. Reliability versus C_i

Failure rate: $\lambda_m = 10^{-3}$ f/h, $\lambda_m/\lambda_w = 5.0$
Coverages: $C_p = C_c = 0.9$
Probability permanent fault in watchdog processor $P_p = 0.25$

Safety versus coverage

Fig. 17 shows the coverage influence in safety. For high coverage the behaviour of the two replicating systems are better than WP.

As can be seen the WP system is the safest for low coverage (C_p) and high coverage (C_i).

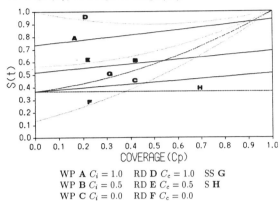

WP A $C_i = 1.0$ RD D $C_c = 1.0$ SS G
WP B $C_i = 0.5$ RD E $C_c = 0.5$ S H
WP C $C_i = 0.0$ RD F $C_c = 0.0$

Fig. 17. Safety versus C_i, C_c and C_p.

Failure rate: $\lambda_m = 10^{-3}$ f/h, $\lambda_m/\lambda_w = 5.0$
Probability permanent fault in watchdog processor $P_p = 0.25$

Permanent failures

The permanent failures have a great influence in the reliability. As can be seen in Fig. 18, if the P_p is lower than 0.25 ($1 - P_p > 0.75$), the reliability of WP is the highest. The most frequent failures are intermitent and transient, with a probability ($1 - P_p$) higher than 0.80, then, the WP always works with the best reliability.

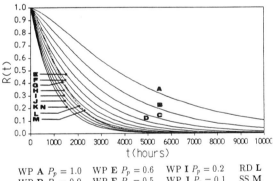

WP A $P_p = 1.0$ WP E $P_p = 0.6$ WP I $P_p = 0.2$ RD L
WP B $P_p = 0.9$ WP F $P_p = 0.5$ WP J $P_p = 0.1$ SS M
WP C $P_p = 0.8$ WP G $P_p = 0.4$ WP K $P_p = 0.0$ S N
WP D $P_p = 0.7$ WP H $P_p = 0.3$

Fig. 18. Reliability versus P_p.

CONCLUSIONS

In order to analyze the dependability of simple structures used in the design of dependable computers, a model for the watchdog processor scheme is proposed. With the results obtained and considering that:

- WP is cheaper than replication.

- The Watchdog can be added without major changes to a system.

We can conclude that the WP is a good choice for the design of reliable processors in industrial control systems.

For furher studies we are modeling the proposed scheme with stochastic Petri nets.

REFERENCES

Mahmood. A, and E.J. McCluskey (1988). Concurrent Error Detection using Watchdog Processors — A survey. *IEEE Transactions on Computers.* Vol.37, No.2, pp. 160-174.

Johnson B.W., and J.H. Aylor, (1986). Reliability and Safety Analysis of a Fault–Tolerant Controller. *IEEE Transactions on Reliability,* Vol.R-35, No.4, pp. 355-362.

Lu, D.L. (1982). Watchdog Processors and Structural Integrity Checking. *IEEE Transactions on Computers,* Vol.C-31, No.7, pp. 681-685.

Johnson B.W. (1989). *Design and Analysis of Fault Tolerant Digital Systems.* Ed. Addison-Wesley Pub.

Upadhyaya, J.S. and K.K. Saluja,(1986). A Watchdog Processor Based General Rollback Technique with Multiple Retries. *IEEE Transactions on Software Eng.,* Vol.SE-12, No.1, pp. 87-95.

M. Namjoo, M., (1983). CERBERUS 16: An Architecture for a General Watchdog Processor. *Int. Symp. Fault–Tolerant Computing. FCTS–13,* Milano, Italy, pp. 216-219.

Melamed, B. and M.Yadin,(1984). Randomization Procedures in the Computation of Cumulative-Time Distributions over Discrete State Markov Processes. *Operations Research,* pp.927-945.

PRACTICAL EXPERIENCE IN THE ASSESSMENT OF EXISTING SAFETY CRITICAL COMPUTER BASED SYSTEMS

B. W. Finnie and I. H. A. Johnston

The Centre for Software Engineering Ltd, Bellwin Drive, Flixborough, Scunthorpe, DN15 8SN, UK

Abstract. The principles underlying the proposed international standard for the safety of software have been applied to the assessment of a number of computer based systems. It has been found that the methodology of safety engineering has not yet transferred to computer systems and software. This results in hazards emerging late during the design or assessment process, with corresponding cost and delay. Company guidelines for good software engineering practice are lacking and individual engineers seek to apply good practice without the support of a consistent company framework. The assessment process itself can be made more cost effective by the adoption during the design process of methods which will make the analysis and understanding of the system less difficult. The same methods of working will also have a beneficial effect on the design process itself.

Keywords. Safety; Standards; System integrity; Software engineering; Computer evaluation; Coding errors; Failure detection.

INTRODUCTION

The Centre for Software Engineering Ltd has applied the principles of the proposed international standard for the safety of software (IEC, 1989) to a number of assessments of safety related computer systems. The intended applications of these systems have ranged from private cars to offshore oil platforms. The independent assessment of computer based systems is relatively new to the industries involved and a number of lessons emerge which will be of value in increasing the cost effectiveness of third party assessment. An awareness of the assessment process and some of the problems will help others to organise future safety related projects to take account of the assessors' needs. In practice steps which ease the assessment process will also be beneficial to the development programme.

ASSESSMENT OF SAFETY RELATED SYSTEMS

The Centre for Software Engineering Ltd uses an ordered approach to the assessment of safety related systems.

The approach permits differing hazards of the system to be assessed at different levels, making use of the most appropriate assessment techniques. In this way a very high level of confidence can be achieved in a cost-effective way as the most powerful techniques, which are usually the most costly to apply, are only used where necessary.

The approach assesses the system integrity and safety in the context of the complete system and not just the software, although this is an important part of the assessment.

Hazard criteria selection

The first step in an assessment is to establish the Hazard Criteria. These are the criteria to which the system is expected to conform for it to be judged operable. Examples could be:

- no accidental discharge of toxic material from an uncontrolled process.

- no loss of control on the engine management system resulting in sudden acceleration.

- fail-safe under definable operating parameters.

- no drug delivery above a specified rate.

The Hazard Criteria differ for each system, industry sector and application because of the way in which the system is to be used within given legal, economic and environmental constraints.

Assessment levels

For each Hazard Criteria on specific Assessment Level is established. Since there is no universal technique for making this type of selection, the criticality of the system with respect to the application is determined and from this an appropriate level is chosen.

The techniques to be used are selected from those associated with a particular assessment level. The criticality of the application, the user industry and the technological culture of the developer all have an influence on the set of techniques that are chosen.

Evaluation

For each Hazard Criterion, the system will be assessed for conformance with some recognisable standard of safe/reliable operation for the agreed Assessment Level.

The system (the hardware, the software, and the system itself) is evaluated using techniques selected to be appropriate for the chosen Assessment level.

Techniques

The techniques to be used are categorised into a progressively more rigorous hierarchy which will usually be related to the corresponding assessment level.

Level 0: System overview. Formalised design reviews show where safety features may have been overlooked during the development. Design reviews also assist in showing where the system can be improved.

The system configuration, the degree of redundancy or diversity, and the modularity of the system are examined.

Level 1: Structural analysis. Checklists are used to document the safety considerations taken during the design.

On software, techniques such as Code Inspections and Walkthroughs are used.

Level 2: Systematic hazard analysis. Techniques such as FMEA, FTA, ETA, HAZOP, Reliability Modelling and Reliability Block Diagrams, are used.

Some of these traditional techniques may also be applied to software.

Level 3: Rigorous analysis. State Transition Diagrams, Petri Nets and Markov Models are used. These allow the assessment of a system without recourse to a high degree of mathematical ability on the part of the client.

Software may be subjected to Path Coverage Analysis.

Level 4: Formal mathematical methods.

Formal mathematical methods may be applied to establish the correctness of the software. In practice this requires that the project has anticipated such a requirement so that suitable mathematically defined specification languages, such as VDM or Z, are used, together with supporting program construction techniques.

TECHNIQUES USED FOR ASSESSMENT

Fault Tree Analysis.

Fault Tree Analysis(FTA) starts with the identified top level hazards and systematically and logically derives the necessary set of possible conditions leading up to that hazard. The purpose is to determine what pattern of underlying faults are necessary for the hazard to occur. This is used to guide a more detailed examination towards specific aspects of the system. This means that time is not wasted examining sections with no serious consequences.

In many systems it is a requirement that any single fault does not cause a hazard. The assumption is made, but not often stated, that the probability of a single fault must also be low enough that the likelihood of a second fault (which may well cause a hazard) is negligible. Unfortunately this caveat has sometimes been forgotten and entirely probable double faults have been left uncovered.

Fault Tree Analysis has an important shortcoming in that because the analysis starts with known hazards, it will not inherently reveal unknown hazards.

Failure Mode and Effects Analysis.

In contrast to FTA, Failure Mode and Effects Analysis (FMEA) considers the possible low level failures and works upwards through the system to discover what the observable outcome of each failure is. Hazardous outcomes, including those which may not have been anticipated, will therefore be identified.

State transition methods.

State Transition Diagrams and Petri Nets are ways of describing a system in terms of its physical states, based on a model of the system. System phases such as initialisation can be described.

These techniques use a graphical or tabular presentation. Consequently the results are readily understandable by the client, yet have a rigorous foundation,

Safe and unsafe states are identifiable together with the route by which the states can be reached. This makes it possible to establish that unsafe states cannot be reached from normal operation, or that there is a route out of an unsafe state, for example.

Where a state table approach has been taken in the design this aids the assessment. However when used as a design tool without the clarity of a hazard analysis the distinction between desired operating modes and actual system physical states has sometimes not been appreciated.

Mathematical techniques.

Mathematical methods allow a complete and consistent specification to be written, which avoids the loose ends and ambiguities an English language specification often contains.

They also provide the means to apply a mathematical proof which will raise confidence in the correctness of the resulting software.

Despite these apparent benefits mathematical methods have not yet been found useful for any of the systems which we have assessed to date. The main reasons for this are:

(i)   The systems have been coded in Assembly language for which there is no real mathematical support.

(ii) The assessments have been predominately retrospective. Unless the use of mathematical methods has been incorporated in the project from the start applying these methods is not really possible.

(iii) The understanding of the specifiers and users in the relevant techniques is insufficient to communicate the requirements and express them in a form which can be confirmed with all those who are a party to the assessment.

(iv) The cost associated with mathematical methods has been too high for a realistic and acceptable project budget.

Selection of methods.

The new draft IEC standard gives some guidance on the selection of techniques of this kind.

If we consider that the greatest safety confidence

comes from the completeness of the design in the perception of the user then it is possible to argue that a measure of that confidence comes from a combination of the rigour of the technique and the ease of communicating the results of the analysis.

It is likely that for some time the optimum will lie in the use of the readily understood but otherwise systematic or rigorous methods. Simple design review is insufficient. Mathematical specifications are not yet well enough understood.

TYPICAL ASSESSMENT FINDINGS

Software engineering.

Every safety related system design should be undertaken using good Software Engineering practice. At its most basic this means following a methodical development plan, emphasising correctness and completeness at each stage in what is called the software life cycle.

In practice, there is often little or no imposed software QA. Instead reliance is placed on the individual engineer to control himself.

Assembly language is the norm for embedded software. However, coding standards to ensure consistent style and consistent conversion from higher level descriptions are rarely evident.

Code is frequently generated by engineers for whom software has been a self taught skill. There is a willingness to assume that the software will perform correctly. There is therefore a tendency to use unsound structures, for example using a test for equality rather than the safer test for inequality to end a loop.

There is little application of checks on input and output values in modules. Even where these are used the limits for a feasible range check may be too wide to act as a safety trap, particularly for time-out values. Although such checks may yield diagnostic information for repair, over-reliance has lead to hazards not being properly controlled.

Little use is made of other defensive programming techniques to check that data and control flow has not been corrupted by interference, faulty interrupt handling, or other programming errors.

Interrupts are considered undesirable in safety critical systems, because they are inherently complex, and cannot be formally specified or analysed. They can create memory or timing conflicts and some microprocessor interrupt specifications are obscure and perhaps even incomplete.

However, in some applications interrupts are the only method of achieving the required performance, for example in a motor speed control loop.

Where we have been involved in assessment during the design phase of a project we have recommended removal of inessential interrupts. This has then been possible before it affected the functional aspects of the implementation.

Safety Plan.

In a safety related project it is necessary to have a safety plan overlaid on the development plan. This identifies stages and activities which ensure that safety is designed in to the end product or system.

Such a specific recognition of safety is not yet usual in software engineering.

Specifications.

It is important to write down a comprehensive specification. A typical specification frequently only considers the functional requirements in response to expected events and conditions, but beyond these expected situations are unexpected conditions and events. The system response to the unexpected must be specified and controlled.

We have found that specification and requirements deficiencies have been greater than explicit coding errors. This may have been because in a retrospective assessment we are dealing with software which has been debugged, but where original omissions in the specification remain.

There is frequently a lack of any coherent specification document for the software and in some cases for the system as a whole.

One of the difficulties which faces developers of systems is that the specification may not capable of being described objectively. This is not untypical in automotive applications where some functional criteria may be known in terms of driver response but the means for translating these into engineering parameters has not yet been developed.

Hazard Analysis.

Hazard analysis must be as complete as possible.

A serious, but known hazard was totally overlooked in a design because it was a mechanical hazard in a system which was dominated by electronics and software.

A hazard in a piece of equipment was not fully controlled because, although the fault condition was known to the designers, the severity of the resulting hazard was not appreciated.

PRACTICAL EXPERIENCES FROM ASSESSMENTS

There is still an emphasis on hardware reliability and failure modes. This is justified from our experience of actual system deficiencies found. Lack of appreciation of hardware failure modes has resulted in uncovered failures.

There is often a lack of records of past equipment failures in equipment of similar type, so no opportunity can be taken to build on this experience when making upgrades or replacements.

We have found a lack of communication of system features from equipment manufacturer to system integrators. This can lead to integrators relying on a misunderstanding about the operation of the system. There is a reluctance to make specific recommendations to avoid exposure to liability legislation.

There is an over reliance on a single or a few individuals who are not obliged to follow procedures which ensure documentation and maintainability of the system. Poor documentation and traceability also adversely affect the assessment time and cost. Frequently this places the burden on the purchaser commissioning an assessment, rather than the vendor whose documentation and procedures cause the problem.

The findings of an assessment frequently lead to a

need for change. This is costly at the end of a project and may also require significant backtracking in the subsequent reassessment of the changes. As a result it is better to involve the assessors at an early stage.

When involved early in a project the assessment comments have been taken into account during the development programme, resulting in better code, documentation etc. This however raises the question of independence of the assessor from any design decisions which result from such comments.

Where an evolutionary approach to the design has been taken, for example building upon a previous product, there is a tendency for the system design and corresponding code and documentation to retain the unstructured style of the original. This may inhibit some organisations from opening their systems for third party assessment for fear of exposing a lack of good current practice.

Suspicion and skepticism on the part of the software designers/implementers is common at the start of an assessment but seems to ease off and can change to enthusiastic cooperation as the assessment proceeds. This wariness is understandable and is considered to be due to the need for commercial secrecy, a wish to avoid personal criticism, professional concern about weaknesses they are aware of and a belief that deep application knowledge is essential to understanding their software.

Assessment necessarily brings out weak points and can easily bring with it an implied criticism of development staff. This must be recognised and handled with care.

The potential exposure of confidential information also affects the permissible content of reports. We often make separate confidentiality agreements with all other parties in an assessment and keep separate each party's material. The detail which can be included in a report is then limited.

There are usually improvements in QA, Change Control, documentation and coding styles after an assessment. When engineers appreciate the improvements to be gained. Managements need to capitalise on this willingness to embrace a more orderly approach.

There can be difference of opinion in the adequacy of justification for steps already taken in the design which are considered non conforming by the assessors. The frequency of this can be minimised if an adequate hazard analysis has been carried out initially for it is usually an interpretation of the severity of consequences which lies behind the conflict.

DESIGN FOR ASSESSMENT

From our experience during assessments we can offer some recommendations to make an assessment more effective and less expensive. The recommendations also assist the design process and so need not be considered an extra burden.

Design complexity. Eliminate complex designs. The design will be easier to understand and it will be less likely to have hidden problems.

System structure. Segregate safety critical and non-critical components. This limits the scale of the assessment to the critical parts of the system, hopefully few in number. It also makes the critical parts explicit and helps avoid incursions from other parts of the system.

Use Fault Tree Analysis during the design process to guide the structure of the design.

Documentation. Clarify all descriptions. Rigorously identify & justify constants & algorithms. Link these justifications to the supporting design documents and test results. Apart from making an assessment possible, it will also aid team understanding and lead to fewer misunderstandings.

Use Fault Tree Analysis as an aid to documentation of the safety structure.

Standards. Be consistent. Insist on a programming style and documentation standards and stick to them. Again it is more efficient from a project and an assessment point of view if only one scheme has to be learned. It also prevents confusion between team members.

If use of the language 'C' or Assembler, for example, is necessary then the scope for misuse can be reduced by adopting coding standards to exclude obscure constructs and to maintain consistency.

There is a need to make everything explicit, not to depend by default on assumptions by the programmer. There should be guidelines for the use of defensive programming techniques.

Use of interrupts. If interrupts are to be used, their use should be limited, justified explicitly and then recognised as needing special care and attention.

Use of FMEA. FMEA is thorough and it can show up unanticipated hazards. FMEA should be considered for all safety related systems.

THE MEANING OF "PASS"

With current technology it is not possible to know or guarantee that a piece of software is free from errors. The performance of a safety assessment will minimise the probability that errors have been introduced and that safety principles have been incorporated in the design. A "pass" implies that specified criteria have been complied with, to the satisfaction of the Assessors. This may well involve elements of judgment on the part of the assessors.

CONCLUSIONS

Assessment of safety critical systems can be carried out using the framework of the proposed international standard.

It is often not software which causes the problem but the inability to look at the system as a whole which is the root cause of safety problems.

In most industries safety engineering principles have yet to transfer to the computer system and software engineer. This is causing unnecessary expense in correcting specification oversights late in the development cycle. Third party assessment has proved to be an effective learning process for the participants while giving confidence in the system integrity.

Retrospective assessment is limited in depth by the already defined structure of the system and the style and structure of the software and documentation. The ordered methods described for assessment need to be incorporated from the start of the development cycle. By doing this the inherent integrity of the system will be improved and the cost of assessment reduced.

REFERENCES

IEC (1989) *Software for computers in the application of industrial safety related systems*. 65(SECRETARIAT)94.

METHODOLOGICAL ASPECTS OF CRITICS DURING SAFETY VALIDATION

G. List

Department System Reliability and Traffic Electronics, Federal Test and Research Centre Vienna, Arsenal, Vienna, Austria

Abstract. The author's institution is working among others on the field of validating systems used in railway safety applications upon order of users and/or manufacturers. In this connection the critics of a system are very important. In this paper methodological aspects of critics are given by the hand of hardware and software. Together with them selected examples of critics based on carried out validations are presented together with the influences of detected fail-safe errors on the costs of the validation. Because of the close connection of the validation work to the standardizing work within the International Electrotechnical Committee on safety systems relevant problems of the latter will also be discussed, but very briefly.

Keywords. Safety; computer control; computer software; software safety validation; system maintainance.

INTRODUCTION

In this paper methodological aspects of critics found during the safety validation process will be given. These methodological aspects concern the safety validation report, the actions taken by the validation institution in the case of detecting a fail-safe error (in hardware or software) and presenting examples of detected errors with information about the relevant costs and time for the validation of a corrected version of the hardware or the software.

Since a lot of years the Federal Test and Research Centre Vienna-Arsenal works on the field of safety validation for electronic equipments used for railway applications. The validation of an equipment is carried out for hardware and in most cases existing software; at which the number of equipment with software steadily increased during the last years. Especially for safety related software the detection of errors residing in it - errors can hardly be avoided - and the therefrom resulting critics during the validation process is very important. It is not evidently clear in every case of critics of software errors whether the weak point shall be treated as critical or not. This problem will be described later in more detail. In extreme cases, when there arise too many critics during the validation process, the validation of a finalized equipment may turn into a development accompanying validation, with all the advantages and disadvantages.

As concerns the maintainance of an equipment, especially when it has been in use for a long time, similar problems occur as described above for changes of an equipment under test during the validation process. In every case of change there arises the question of how much of the equipment must again be validated. In the sense of fail-safe the whole equipment should be validated but nearly no manufacturer would pay for it without to be forced. Some more thoughts to it will be given later.

In course of the validation works the relevant international standardization is followed very closely by the author's institution and partly active work is carried out for it. Work is done for the IEC[1] committee TC65[2] "Industrial process measurement and control" and especially for its working groups 9 and 10, which are working on standards for safe software respectively safe systems. In the intended standards methods will be given which are related to the design and validation of safety related control systems in order to obtain a dependable system. In this connection it shall be mentioned that the term dependability (see fig. 1.) is a problematic term at the current time, because the IEC committee TC56 "Reliability and Maintainablity" is responsible for this reliability related term and there exists no common opinion about its contents. The topic software is also treated by both IEC TC's and needs to be coordinated as concerns the software life cycle.

THE SAFETY VALIDATION REPORT

The safety validation report is the presentation of results of a work, that the author's institution has done upon an order for a safety validation of an equip-

---

1) IEC = International Electrotechnical Committee, Geneva, Switzerland
2) TC = Technical Committee

ment or system. Such a report may be the result from an analysis of a finalized system or from a development accompanying validation (see also chapter on software validation).

One of the most important points of a safety validation is the presentation of the results of the investigations, which include the critics found during the validation process. Therefore in this chapter a short overview of the contents of such a validation report issued by the author's institution is be given. An example of a shortened table of contents (see also /1/) is presented in table 1.

TABLE 1 Shortened table of contents of the safety validation reports

I. INTRODUCTION, PRESENTATIONS
  A. Presentation of the tested system
    1. Purpose, application and function
    2. General data, system specification
    3. Safety conception of the system and suppositions for the system functions and fail-safe behaviour
    4. Explanation of structure and function of the system in general
    5. Explanation of structure and function of hardware and software in special
  B. Presentation of the used validation methods
II. SYNOPSIS OF THE SAFETY PRINCIPLES TOGETHER WITH CRITICS
III. VALIDATION DESCRIBED IN DETAIL
    1. Hardware aspects
    2. Software aspects
    3. Detailed presentation of the analyses (with the methods given in I.B)
IV. SUMMARY OF THE WORK
    FINAL JUDGEMENT (if possible)
V. ENCLOSURES (containing block diagrams, flow-charts, structure diagrams, measurement results, listed documendation issued by the manufacturer

As can be seen from the above table 1 under point I.A.3 suppositions are named. This is an important chapter, where all those suppositions are named which have an influence on the system respectively on its safety behaviour. Generally it can be stated that the better a system is the smaller is that chapter about suppositions. According to the opinion of author's institution a system that needs a lot of suppositions concerning safety is worse than a system which doesn't need one, although both systems are working in a fail-safe manner provided that all the suppositions are obeyed. Such suppositions which are meant here are e.g. the lighting of a red light of a signalling system has the same function as no light (both mean stop for the train), or suppositions are resulting from specifications for a system e.g. the time of a permissive state of a system output after its switching off is depending from the load at its output, which makes it necessary for the manufacturer to specify a minimum load for the output to obtain a worst case switching off time.

According to the experience of the validation institution the author is of the opinion that such suppositions or overlooked suppositions seem to be more sources of system errors than real system design errors.

HARDWARE SAFETY VALIDATION

The validation of hardware used in safety related systems is a well known practice for the specialists concerned. There exist a lot of methods such as the FMEA (Fault Mode and Effect Analysis), FTA (Fault Tree Analysis) etc. which may be found in the relevant literature and standards /2/, /3/, /4/, /5/ etc. and are therefore not described in detail here. All these methods and useful combinations are applied with success in the author's institution according to their applicability, depending from the system to be validated. In this paper here the relevant points of actions taken by the validation institution upon finding a fail-safe error in the equipment under validation shall be shown; this will be illustrated by hand of some examples as really appeared during safety validations.

The first action after finding a fail-safe error is its exact documentation in the validation working papers. Immediately after repeated internal checks of the correctness of the findings by colleagues of the proving person (proving redundancy) the manufacturer is informed about the problem. In most of the cases the manufacturer takes corrective measures to reach the aim of obtaining a fail-safe system; in other cases (the minority of cases) the manufacturer tries to overcome the problem with specifications for the system and/or its application. This last solution is according to the author's opinion a bad one (see also the chapter before). In the following two examples of errors will be presented.

In the first example a railway equipment for showing the signal state in the drivers cabin is considered. Its task in short words is to receive coded track circuit signals, decode and display them with lamps in the cabin of the driver. The fail-safe error found in this equipment was a wrongly designed digital filter. It had the task to hinder code pulses spaced too tight to be passed to the pulse distance evaluation hardware. The misdesign was in a way that the filter fulfilled its task but freezed the current display information in the case of interferences from the track circuit. So a new signal state would not have reached the display unit as long as the interference would be present. Upon detecting this design error a message was sent to the manufacturer immediately with the description of the misfunction. As reaction upon this the error was corrected, resulting in a more than small design change which in turn required a repetition of a lot of proving work. The amount of additional validation expenditure due to this error has been about 10 % as concerns validation costs and about 30 % as concerns the validation time, the duration mainly due to the required additional development time. This example shows that on the one hand the design was not checked enough and on the other hand the case of interferencies not tested enough.

In the second example, originating from an other equipment, a power-on circuit for a watch-dog built up with discrete components for safety reasons shall be considered. Following fail-safe error could be found: The watch-dog circuit which supervises the distance between pulse packets and has the task to switch off the system in the case of faults is constructed as a self holding circuit (by its own output signal which is fed back to its input and is only present when all the inputs including its own are present) and needs therefore an initial power at the start up sequence for being able to power up itself. The relevant circuit was built by a relay activated manually or externally which transferred a capacitor charge for a certain amount of time via its contacts to the watch-dog output and therefore to its input, starting up the self holding circuit. The fail-safe error in this circuit was that the possibility of breaking relay contacts with following short circuit so that a continuous power up was given and so the function of switching off in the case of other errors was inhibited. After information of this defect the manufacturer changed the circuitry and presented a new solution which he thought would work. The validation work carried out with a FMEA (like for the original circuit) for the new solution again showed a case where the interruption of contacts of the push button for the manual operation resulted in a continuous power up of the watch-dog, again loosing its possibility to switch off its output in the case of faults in the system. This design error was again reported to the manufacturer who didn't see an other solution than demanding the power up voltage to be applied externally and specifying that it is only allowed to apply the power up voltage at the start up of the system for a certain small time. As already mentioned above this was a solution where a technical misfunctioning was overcome by a specification of the manufacturer. In the safety validation report it was stated that under the supposition that the specifications will be observed the system is behaving in a fail-safe manner. The critics in the above described solution is that a technical problem is overcome by a specification. The additional costs of the validation work have been about 20 % of the total costs because the whole FMEA had to be repeated but the date of finalizing the validation had not changed because of a possible flexible personal situation of the validation institution.

SOFTWARE SAFETY VALIDATION

Validation of safe software is a not old theme and therefore hardly standards about the relevant methods exist. In the IEC there is currently tried to produce standards on software in safety systems /6/, /7/. The work is carried out by TC65. About the connected problems see the below following chapter on the standardization work. In this chapter some problems in relation to critics of software validations will be described and an example given.

Because of the quick changing of knowledge of how to validate software not many terms of reference exist and even they are hardly updated to the newest state. The railways that are very much interested in having an "error free" software in their safety equipment have started long time ago with a working group concerned with the setting up of reports on the process of specifying, designing and validating systems that contain safety related software. This work was done in the ORE[3] A155.2 committee and resulted in 5 reports e.g. /8/ and an UIC leaflet /9/. Also some national railway organizations have their own guidelines but due to the development they only may be applied not exactly as written but only in their sense. Along with these facts the problems for the validation are evident. It is not very clear which deviations of a software under test from guidelines etc. are to be considered as not enough to fulfill the safety requirements or are only redactional. The solving of this problem is also one task of the validation institution. As mentioned in the chapter concerning the hardware validation the problem of specification similarly occurs here. A software not specified good enough and/or described in too less details may contain fail-safe errors. This will be illustrated with an example.

The example is taken from a radio equipment receiving data in a fail-safe manner, at which more than one system works on the same frequency. Each individual equipment needs therefore to be addressed for being able to differ between the attached transmitters. The transmission itself is done by repeating two blocks of messages with the same contents, each of which contains 2 bytes (high and low byte) of address. The main part of the specification of the fail-safe receiving software was about to receive both bytes in the two blocks and check the equality with the given address within the equipment. The software was implemented in a way that both received bytes of the 2 blocks have been compared among each other plus both high bytes have been compared to the internal address. Compared to the high bytes, for the low bytes it was implemented that either the one from block 1 or the one from block 2 was identical with the internal address. Although the comparison of each byte with the internal address seems to be redundant, i.e. not necessary, the calculation results of the residual error probability for the data transmission coming from the manufacturer and the validation institution showed differences based on the above described "or" function of the comparison of the low bytes with the internal address. This example shows the importance of the specification and of the description of the software. Less good ones as it was the case here also lead the person(s) carrying out the validation into a wrong direction. Naturally this error was revealed and corrected by the manufacturer. The additional expenses for the detection of the error together with the validation of its correction have been about 5 % of the total validation costs. Time delay for the safety investigations was negligible.

---
3) ORE = Organisation for Research and Experiments (Utrecht, Netherlands) of the International Railways Union (Paris, France)

VALIDATION OF CHANGED SYSTEMS AND MAINTAINANCE

As already mentioned in the introduction, in the case of too many vital and/or redactional critics a validation of a finalized system may turn into a development accompanying validation. Such a validation differs to a not small extent from a validation of a finalized system. Such differences are e.g. proving the specifications with a view to other applications than intended in order to detect eventual restrictions for the system, proving the system on a block by block basis necessarily in more detail than during a conventional validation and some other points. One additional major difference is the price. Compared to a conventional validation, for which the price can be estimated rather exactly and therefore will be given to the our customer as a fixed price, the costs for a development accompanying validation are variable and will be accounted according to the work done. Experiences have shown that a poor designed system, with many changes because of the validation, caused with development accompanying validation roughly about 100 % more of validation costs than a good designed one with the same amount of hard- and software after its finalization with no fail-safe critics.

Concerning the critics related to maintainance of a system it must be pointed out that especially a poor documentation is the reason for introducing fail-safe errors into a safety system. Therefore the critics of system presented in our validation report contains also such points that have influence on maintainance activities. In the validation report these influences are marked there as redactional because they have no fail-safe influence on the system in its present state, but might well have an influence later. Examples for this especially concern software documentation, which is in many cases not as good as it should be. We have found deviations between the actual programmes and the descriptions as well as in the flow charts and/or programme structure charts, a too small number of comments in the programme listings, wrong comments, parts of programmes not used but present (e.g. branches as needed in a design version were not deleted for the final version) etc. For these reasons relevant remarks, which are sometimes not favourable for the manufacturer although the system is fail-safe with consideration of the suppositions, are included in the safety validation reports.

For this case here a positive example shall be described. A railways administration decided to build a second series of an equipment, which has been designed about 15 years ago and validated by the author's institution, because the equipment fulfilles its task very good with no problems. The problem was that some components used at that time don't exist today is the same type or don't exist at all. So this administration consulted the author's institution again to obtain an expertise on the safety of the system for rebuilding it according to the original plans but in some places different components. Because of the documentation contained in the old validation report, the good documents of the old system and the clear lists of changings to be carried for the second series, the validation work for comparing the old and new components and design changes was on the one hand simple and cheap (less than 10 % of the original validation costs) and on the other hand very quickly done. For that case the railways adiminstration has the conviction that also the second series of equipment will be working in a fail-safe manner.

STANDARIZATION AND TERMS

As already mentioned before, the validation work requires to be on the latest state of the relevant techniques and also to develop and apply sometimes new and/or better suited methods and tools than used before. This fact is the major reason of many other reasons for working in relevant international committees that are considering methods and techniques for standardization. As concerns the theme of hardware validation, the IEC technical committee 56 ("Reliability and Maintainability") is working on standards for investigating hardware for reliability purposes, but they are also applicable for safety investigations. This committee, which consists of a lot of working groups has one group that is considering software aspects (WG10). The theme on functional safety of systems is treated by the IEC committee 65 ("Industrial Process Measurement and Control"), which works besides others on standards for software for computers in the application of industrial safety related systems. As can be seen two IEC groups are working on neighboured themes which has resulted in differences. One of the main difference between the groups are the terms and there especially the term "dependability", which should be taken as an umbrella term (see fig. 1.).

Fig. 1. Concept of dependability

This concept, which was pushed by the software, seems to be more adequate to describe, simply spoken, the quality of software. The old term reliability is in principal connected to a decrease of a system behaviour in the run of time which seems not to be appropriate for software because it does not change with time.

This problem is yet an open point together with a difference in consideration of the software life cycle. This last point is only mentioned here because the validation aspects, described in this paper, of a software system is at any rate one step in the software life cycle. The future work within both IEC committees will show the outcome and it remains to hope that both groups will use the same terms and the same model for the life cycle in their standards.

CONCLUSION

This paper has shown some aspects of critics appearing during a validation work regarding as well hardware as software. Some examples of errors have been presented that could be detected due to a very systematic validation work, using a variety of validation methods and the experience gained during many years of practical investigations. Additionally a very short view to the standardization work concerning software was given as the validation of software respectively of a system is a part of its life cycle.

REFERENCES

/1/ Sethy, A. (1990). Testing of Software Safety for Signalling Applications; Methodical Aspects, Proceedings of the 7th International Conference on Reliability and Maintainablity, Brest, France, pp. 320.

/2/ IEC, (1985). Analysis techniques for system reliability - Procedure for failure mode and effects analysis (FMEA). Publication 812, IEC Geneva, Switzerland.

/3/ IEC, (December 1984). Fault tree analysis. Document TC56(sec)192, IEC, Geneva, Switzerland.

/4/ IEC, (January 1988). Analysis techniques for system reliability - General consideration for reliability / availability analysis methodology. Document TC56(co)138, IEC, Geneva, Switzerland.

/5/ Sethy, A. (1988). Research works on Electronics in the BVFA for Rail Applications (in german). e&i, 10, Edition Springer Wien New York, p. 429 ff.

/6/ IEC, (August 1989). Draft - Software for computers in the application of industrial safety related systems. Document TC65(sec)94, IEC, Geneva, Switzerland.

/7/ IEC, (October 1989). Draft - Functional Safety of programmable electronic systems: Generic aspects, Part 1: General requirements. Document TC65(sec)96, IEC, Geneva, Switzerland.

/8/ ORE, (September 1987), Proving the safety of computer-based safety systems. Report A155.2/RP11, ORE, Utrecht, Netherlands.

/9/ UIC, (1989). Processing and Transmission of Safety Information, Leaflet 738, UIC, Paris, France.

SYMBAD: A SYMBOLIC EXECUTOR OF SEQUENTIAL ADA PROGRAMS

A. Coen-Porisini and F. De Paoli

Politecnico di Milano, Dipartimento di Elettronica, Piazza Leonardo da Vinci 32, 20133 Milano, Italy

Abstract

Among existing techniques used to improve software reliability symbolic execution plays an important role. Symbolic testing is potentially more effective than numeric testing, since a symbolic execution represents a class of numeric executions. Symbolic execution can also be used to generate the necessary verification conditions that ensure program (partial) correctness, by adding to the program first order predicates describing its logical properties. Several tools have been built to provide environments in which symbolic execution can be performed. However all of these tools are quite unsatisfactory when dealing with programs that use arrays and/or pointers.

In this paper we present a tool (SYMBAD) that provides an environment in which it is possible to perform both symbolic testing and verification on sequential Ada components. Furthermore SYMBAD can be used to find out which restrictions must be placed on input variables of a program in order to execute a particular path.

SYMBAD has been designed and implemented at Dipartimento di Elettronica of Politecnico di Milano; it runs under Unix operating system and it is written in C and Lisp.

The main SYMBAD features are presented; problems encountered in dealing with arrays and pointers and the adopted solutions are discussed; finally an example of a SYMBAD session is reported.

Keywords: Symbolic execution; Ada; Program testing; Program verification

1. Introduction

Symbolic execution is a powerful technique that can be used for improving software reliability. For instance, symbolic testing is potentially more effective than numeric testing since a symbolic execution represents a class of numeric executions. Moreover symbolic execution can be used to derive constraints on input data that forces the execution flow to follow a given path. Formal verification of programs can be done by means of symbolic execution if appropriate predicates describing their logical properties are provided.

Several tools have been built to provide environments in which symbolic execution can be performed. Among them EFFIGY [6] for PL\1 programs, DISSECT [4] for Fortran programs and UNISEX [5] for Pascal programs.

However symbolic execution still has some open issues; for example symbolic execution of programs containing array and/or pointers is not handled in a satisfactory way by all these tools. Moreover symbolic execution can be performed automatically only if a powerful theorem prover which solves decision points is provided.

In this paper we introduce a tool, SYMBAD, that provides an environment in which symbolic execution of sequential Ada programs is performed. SYMBAD has been built on the top of UNISEX, it is written in C and Lisp and it runs under Unix operating system. SYMBAD consists of a symbolic executor and a theorem prover that resolves most of the decision points encountered during symbolic execution. Furthermore we present an original approach to array handling which has been adopted in SYMBAD.

The paper is organized in the following way: in the next subsection a brief description of basic symbolic execution concepts is provided; Section 2 discusses theorem proving and array handling issues along with the adopted solutions for both topics; Section 3 provides a description of SYMBAD environment along with three examples of work sessions; finally Section 4 draws some conclusions.

1.1. Concepts of symbolic execution

Symbolic execution consists in executing a program with inputs expressed by symbols instead of numeric values. Dealing with symbols requires to the underlying abstract machine the capability of performing algebraic manipulation of symbolic expressions.

Traditional program execution is described by the program counter along with the state of variables, that is, by their actual values. However, during symbolic execution such a description is not enough, since values of variables are symbols that stand for classes of numeric values. In fact when a decision point is reached values of variables may not suffice to select the branch to follow. Hence, some assumptions on values of variables have to be taken. These assumptions on symbolic values are represented by means of a first order predicate which we refer to as *path-condition*. The *path-condition* (PC) is initialized to *true*, i.e. there is no constraint on the initial symbolic values, and then PC is updated whenever a new assumption is made. Every time a decision point is reached PC can drive the execution: the constraints expressed by PC are used to evaluate conditional expressions in order to select the branch to follow. Therefore, a key component of any symbolic executor is a theorem prover which solves the decision points encountered during the symbolic execution. In Section 2.1 a discussion of such topic is provided along with a description of the adopted theorem prover.

2. SYMBAD

SYMBAD deals with every Ada unit but task: it allows any sequential Ada program to be processed provided that its source code is available. This means that standard libraries can be treated only if their source code is provided. A special remark should be done on TEXT_IO package: since symbolic execution deals only with input/output of symbols, SYMBAD has special PUT and GET functions and all the I/O procedures in the program are performed accordingly.

SYMBAD is based on UNISEX, the symbolic executor for Pascal programs introduced above. The major improvements of SYMBAD, besides its capability of dealing with original constructs of Ada, such as packages, generics or the *with* clause, are the original way in which array (and pointers) are handled, and the presence of a theorem prover.

SYMBAD consists of three major modules:
- a translator that translates the Ada source code into an intermediate representation in Common Lisp;
- a set of Lisp functions that forms the run time support of the symbolic executor;
- a theorem prover that solves decision points.

SYMBAD has been developed under Unix operating system. The translator is made of a lexical analyzer and a parser implemented by means of Lex and YACC. The output of the compiler is a single Lisp function containing a prog statement. The prog statement consists of calls to the run time support functions to provide interactive symbolic execution of the Ada program.

In the next two subsections we discuss the most interesting features of SYMBAD: the theorem prover and the array handling mechanism.

2.1. Expression evaluation

Program execution involves the evaluation of conditional expressions in order to choose along which branch the computation has to continue. In symbolic execution such expressions contain symbolic values instead of numeric values. Therefore it is necessary to take into account any constraint on such symbolic values. In other words, each time a decision point is reached it is necessary to check whether the constraints expressed by PC imply the conditional expression or not.

Let us consider the following trivial example:

State:[1] $\{(X, \alpha), (Y, \beta)\}$
PC: $(\alpha > \beta) \wedge (\beta > 0)$

...
if X > 0
then ...
else ...
end if;

The theorem prover has to demonstrate that PC implies the conditional expression X>0 (X≤0), in order to follow the **then** (**else**) branch. Replacing variables by their current symbolic values yields the following theorem:

$(\alpha > \beta) \wedge (\beta > 0) \Rightarrow (\alpha > 0)$

The theorem prover can answer *true*, *false* or *neither*. In this simple example the answer is *true*. *Neither* stands for: the path-condition is not strong enough to imply the conditional expression; for instance, if the test is (X>10) then the answer will be *neither*. In general, theorem proving activities are very hard to perform, both from a theoretical and a practical point of view. So, building a theorem prover that solves any kind of expression is practically not realistic; however, under the remark that most of the expressions involve low degree polynomial inequalities, it is possible to built a useful tool.

Among different approaches found in literature, [2] we have based our theorem prover on Bledsoe's method *sup-inf* [1]. This method deals with linear inequalities; we have extended such method to some general cases of higher degree expressions. The method attempts to demonstrate a conjecture by contradiction. Each variable belonging to the conjecture is viewed as a constant characterized by a subrange (Skolem's constant). The method consists in solving inequalities, i.e. in determining the subrange of each Skolem's constant. If variables belong to an unfeasible subrange, i.e. its inferior limit is greater than the superior, then the conjecture is proven. The method consists of three major steps:

(1) rewrite the conjecture using existential operators only and regarding each variable as a Skolem's constant;

(2) put such expressions in a Disjunctive Normal Form such that only inequalities of type ≤ are involved;

(3) solve the inequalities with respect to Skolem's constants.

The extension of the *sup-inf* method allows to reduce some expressions of higher degree so that the *sup-inf* method is still applicable. Experiences show that most of real programs decisions involve linear or sometimes second degree inequalities, so we believe that the class of problems our theorem prover can solve is large enough to handle most of the cases. Of course this remark does not hold for scientific algorithm where expressions often include operators like *log* or *cos*.

By means of a simple example, let us show how the theorem prover works.

Let us suppose that we want to prove the following theorem:

$(x^2 \geq y) \wedge (y > 9) \Rightarrow (x < 3) \vee (x > 3)$ \hfill (1)

The theorem is true, but let us apply the full method.

The first step transforms in the following way the above theorem in order to prove it by contradiction:

$\exists\, x,y: \neg[(x^2 \geq y) \wedge (y > 9) \Rightarrow (x < -3) \vee (x > 3)]$ \hfill (2)

Introducing Skolem's constants X and Y we obtain:

$\neg[(X^2 \geq Y) \wedge (Y > 9) \Rightarrow (X < -3) \vee (X > 3)]$ \hfill (3)

To prove the initial theorem we need to prove that at least one Skolem's constant in expression (3) (X or Y), has inferior limit greater than upper limit.

Since $\neg(a \Rightarrow b) \equiv (a \wedge \neg b)$ expression (3) is rewritten as follows:

$\{(X^2 \geq Y) \wedge (Y > 9) \wedge \neg[(X < -3) \vee (X > 3)]\}$ \hfill (4)

The next step consists in putting expression (4) in Disjunctive Normal Form, expressing inequalities using only the operator ≤:

$[(Y \leq X^2) \wedge (9 < Y) \wedge (X \geq -3) \wedge (X \leq 3)]$ \hfill (5)

$[(Y \leq X^2) \wedge (9 + \varepsilon \leq Y) \wedge (-3 \leq X) \wedge (X \leq 3)]$ \hfill (6)

Remark that the < operator is transformed introducing the infinitesimal quantity ε ($\varepsilon > 0$).

---

[1] By convention, Greek letters are variable values and Latin letters are variable names

Adding the term $(0 \leq X^2)$ to (6) we obtain:
$[(Y \leq X^2) \wedge (9+\varepsilon \leq Y) \wedge (-3 \leq X) \wedge (X \leq 3) \wedge (0 \leq X^2)]$ (7)
Solving (7) w.r.t. X and Y leads to:
$[(9+\varepsilon \leq Y \leq \infty) \wedge (-3 \leq X \leq 3) \wedge (\max(0,Y) \leq X^2 \leq \infty)]$ (8)
Since Y has inferior limit equal to $9 + \varepsilon$ (8) can be rewritten as follows:
$[(9 + \varepsilon \leq Y \leq \infty) \wedge (-3 \leq X \leq 3) \wedge (Y \leq X^2 \leq \infty)]$ (9)
Therefore the subrange of Y is $[9+\varepsilon, \infty]$; in order to find the subrange of X it is necessary to take into account also the second degree term.

Let us examine the term $(Y \leq X^2 \leq \infty)$: in a general form it can be written as $[(a \leq X^2) \wedge (X^2 \leq b)]$.

The solution for the first inequality is $(X \leq -\sqrt{a}) \vee (X \geq \sqrt{a})$ while the solution for the second one is $(-\sqrt{b} \leq X) \vee (X \leq \backslash R(b))$. Thus in our example we have $[(X \leq -\sqrt{9+\varepsilon}) \vee (X \geq \sqrt{9+\varepsilon})] \wedge [(-\sqrt{\infty} \leq X) \vee (X \leq \sqrt{\infty})]$, that is:
$[(X \leq -\sqrt{9+\varepsilon}) \vee (X \geq \sqrt{9+\varepsilon})]$.

So in order to find the subrange of X we have to solve the following expression:
$[(X \leq -\sqrt{9+\varepsilon}) \vee (X \geq \sqrt{9+\varepsilon})] \wedge (-3 \leq X \leq 3)$ (10)
which can be rewritten as follows:
$[-3 \leq X \leq -\sqrt{9+\varepsilon} \vee \sqrt{9+\varepsilon} \leq X \leq 3]$.

Therefore the type of X is $[-3, -3 +\delta] \cup [3+\delta, 3] \equiv \emptyset$, with $\delta > 0$

At this point it is not necessary to find the subrange of X^2, since we have already found an unfeasible subrange for X; thus we can stop concluding that the theorem is true.

2.2. Array handling

During symbolic execution array indexes are often represented by symbolic expressions. In such a case, the referenced array element may not be uniquely identified. Furthermore, even the array size may be represented by a symbolic value. Solutions that have been adopted in other symbolic executors [6], [4], [5] are not satisfactory: some of them are practically unfeasible even for small-size arrays, while others introduce simplifications that leads to unacceptable losses of information.

In SYMBAD an original approach has been followed. The key idea is to provide an initial description of the array, that is updated each time an assignment involving array elements is encountered. An array is represented by an ordered set of pairs, where each pair contains information deriving from an assignment: the first element of the pair is the value of the array index, while the second one represents the value assigned to the array element. Let us clarify by means of an example. The following procedure:

```
1    procedure FOO (N: INTEGER) is
2      A: array (1..N) of INTEGER;
3      J, K, X: INTEGER;
4    begin
5      for I in 1..N loop
6        A(I) := 0;
7      end loop;
8      GET (J);
9      A(J) := 5;
10     A(J + 1) := 7;
11     X := A(J) + A(2);
12     GET (K);
13     if A(K) >= 0 then
14       X := X + 5;
15     end if;
16   end FOO;
```

declares an array A whom size is determined at run time by the actual value of parameter N. After the execution of the declarative part, (lines 1-4) the value of N is represented by the symbolic value η, and A is initialized by the pair *(any undef)*, where the keyword *any* stands for *any index value* and the keyword *undef* stands for *not yet defined*. Remark that the array representation is independent from array size: only known information is stored. The initialization loop (lines 5-7) involves all elements of array A; such situation is represented by adding the pair *(any 0)*. Thus, after the initialization loop A is represented by *((any undef) (any 0))*. Such representation can be transformed in the pair *(any 0)* by the garbage collection mechanism which deletes the unfeasible pair *(any undef)*. In fact when two pairs have the same first element they refer to the same bunch of array elements; as a consequence the value of the first pair (from left) is not alive since a successive assignment, represented by the second pair, has overwritten the previous one. Therefore the first pair can be safely deleted from the list.

Let us go back to the example: when the assignment at line 9 (A(J) := 5) is executed the pair (γ 5) is added to A; in the same way the execution of the assignment of line 10 (A(J) := 7) adds the pair (γ+1 7). As a consequence, the state of the symbolic execution after line 10 becomes:

A: *((any 0) (γ 5) (γ+1 7))*
J: γ
K: *undef*
X: *undef*

The statement at line 11 (X := A(J) + A(2)) is an assignment to variable X that involves two elements of array A. Let us explain how the evaluation mechanism works. The list representing array A is examined from right to left: the index value of each pair is examined and, if it is sound, the corresponding element value is taken into account. For example the evaluation of A(J) starts with the pair (γ+1 7). Since the value of J is γ, A(J) can have value 7 if the expression $\gamma=\gamma+1$ is sound. In our case $\gamma=\gamma+1 \equiv false$, so, that pair is discarded. The index value of the next pair matches the referenced index; this means that all remaining pairs are overwritten by that pair. Hence the searching algorithm stops returning all found feasible values: in our case only value 5.

In the same way the evaluation of A(2) is performed: however this time, none of the index value of A exactly matches the referenced index, but they are all sound, that is, all index values can represent the value 2. Therefore, all element values should be considered. In other words, A(2) can have the value 0, 5 or 7, depending on the corresponding index value: if γ+1 equals 2 then A(2) has value 7, if γ equals 2 then A(2) has value 5, otherwise it has value 0. Thus the new value of X is represented by an ordered set of pairs where the first element of each pair is a boolean expression and the second element is the corresponding value. The new state of the symbolic execution is:

A: *((any 0) (γ 5) (γ+1 7))*
J: γ
K: *undef*
X: *((true 5) (γ=2 10) (γ=1 12))*

The statement on line 12 (GET(K)) assigns a generic value κ to K. The conditional expression on line 13 (A(K) ≥ 0) involves an array element addressed by the value of K. In this case all feasible values of A(K) must be taken into account. Each feasible value has an associated condition involving κ. However, it is not always necessary to examine those conditions since feasible values may be enough to reach a conclusion. For instance, in our example feasible values are 7, 5, 0, and the expression A(K) ≥ 0 is always true. Thus the **then** branch is selected, and the program terminates with the state:

A: ((*any* 0) (γ 5) (γ+1 7))
J: γ
K: κ
X: ((*true* 10) (γ=2 15) (γ=1 17))

Since no assumptions have been taken during the execution PC is still *true*.
Now let us suppose that the conditional expression of line 12 were A(K) > 5, so that only for value 7 it is possible to follow the **then** branch. Therefore it is necessary to make the assumption (κ = γ + 1), in order to represent the fact that A(K) equals 7. In this case the symbolic execution ends with the following path condition and state:

PC : (κ = γ + 1)
State:
A: ((*any* 0) (γ 5) (γ+1 7))
J: γ
K: κ
X: ((*true* 10) (γ=2 15) (γ=1 17))

Summarizing, any time an assignment to an array is performed a new value is added to the list representing the array; the garbage collection mechanism prevents the list to grow up too much. Each time an array element is referenced, the array list is examined from right to left until either the end of the list is reached or the current index value equals an index value in the list; in both cases, every value belonging to a pair with a sound index value is considered.
In Section 3.3 a more complex example is provided.

3. Description of SYMBAD environment

SYMBAD provides an environment for both testing and verifying Ada programs. After starting the system, the user is asked to choose between test or verify mode. In test mode the user should provide symbolic or numeric input data values. If numeric values are provided, the system acts like a symbolic debugger, vice versa, if variables have symbolic values, SYMBAD can be used either for generating path predicates or for symbolic testing.
In verify mode every variable is initialized by the system, so that SYMBAD executes the program in order to verify it. Moreover SYMBAD keeps track of which paths have been verified and which have not. In order to verify a program, appropriate assertions must be inserted in the program source code. Assertions are written in an assertion language that consists of five major keywords: *entry*, *exit*, *assume*, *prove* and *assert*. Each assertion statement consists in a keyword followed by a list of predicates. Each predicate is a relational expression in which program variables can be referenced. In the assertion predicates the following keywords can be used: *forall* (universal quantification), *exist* (existential quantification) and *implies* (logical implication). Examples of assertions are:

--: entry((z<x),(w<>0))
--: assert(forall x(x>y))

Each assertion statement must be preceded by the special symbol --: ; in such a way it looks like an Ada comment and it will be ignored by any Ada compiler. The symbol ' can be used to represent the old value of a variable, for example the expression (x > x') in a loop invariant means that the value of x after the loop execution must be greater then its value before the loop execution.
SYMBAD provides debug options that may be set by the user during symbolic execution. The most important are the *save* and *restore* options. *Save* let the user save the current state (i.e. the program counter, the values of variables and PC), while *restore* allows him to restore previously saved states. In such a way it is possible, for example, to save the state before an **if-then-else** statement, then execute the **then** branch, restore the old state, and finally execute the **else** branch. This allows to execute both paths without restarting the program execution from the beginning. Other options allow the user to set break points, step by step execution, verbose execution, etc.
The next two subsections discuss two executions, the former performed in test mode and the latter in verify mode. Finally the last subsection shows a program execution involving arrays.

3.1. An example of program test case generation

By means of the next example we want to clarify how SYMBAD behaves and how it can be used for defining test cases for branch testing. Let us consider the following procedure which calculates the power of a real number.

1. **procedure** POWER
2. (BASE: **in** FLOAT; EXPONENT: **in** INTEGER; RESULT: **out** FLOAT) **is**
3. EXCHANGE : BOOLEAN;
4. L_RES : FLOAT;
5. L_EXP : INTEGER;
6. **begin**
7. L_RES := 1.0;
8. **if** EXPONENT >= 0 **then**
9. EXCHANGE := FALSE;
10. L_EXP := EXPONENT;
11. **else**
12. L_EXP := -EXPONENT;
13. EXCHANGE := TRUE;
14. **end if**;
15. **while** L_EXP /= 0 **loop**
16. L_RES := L_RES * BASE;
17. L_EXP := L_EXP - 1;
18. **end loop**;
19. **if** EXCHANGE **then**
20. RESULT := 1.0 / L_RES;
21. **else**
22. RESULT := L_RES;
23. **end if**;
24. **end** POWER;

We want to symbolically execute this program to find out which constraints should be imposed on input parameter in order to obtain a set of data for branch testing. Until line 8 there are only assignments, therefore the state and the path-condition are:

State: ((BASE, α), (EXPONENT, β), (L_RES, undef), (L_EXP, undef), (RESULT, undef), (EXCHANGE, undef))
PC: true

On line 8, a decision point is reached. The theorem prover has to prove whether PC implies EXPONENT >= 0, i.e. *true* ⇒ (β >= 0). The theorem prover answer is *neither*, thus the system asks the user to choose between the answer *true* and the answer *false*. Before making a choice we save the current state using the save option. Let us assume *false*, so that the **else** branch is executed and the corresponding assumption ($\beta < 0$) is added to PC. Before the execution of the **while** loop, line 15, the state and the path-condition are:

State: {(BASE,α),(EXPONENT,β), (L_RES,1.0), (L_EXP, β), (RESULT, undef), (EXCHANGE, true)}
PC: $\beta < 0$

The theorem to evaluate on line 15 is ($\beta<0$) $\Rightarrow$ (-β/=0). The answer is *true*. Therefore, the execution goes on with statements on lines 16-17. Then the test at line 15 is evaluated again. This time the theorem to consider is ($\beta < 0$) $\Rightarrow$ (-β-1 /= 0). The answer is *neither*, and we assume *false* since we have already executed the loop body. Therefore PC and the state are:

State: {(BASE,α),(EXPONENT, β), (L_RES, 1.0), (L_EXP,β), (RESULT,undef),(EXCHANGE, true)}
PC: $\beta < 0 \wedge -\beta - 1 = 0$

The next statement to execute is the test on line 19. Since PC implies the conditional expression, the **then** branch is followed, and the program ends.
So far, we have discovered that giving to input parameter EXPONENT values less than 0, statements 6,7,8,11,12,13,14,15,16,17,18,19,20,23,24 are executed.
We restore the state before the loop execution. We can do that using the *restore* option to restore the previously saved state. In such a way we go back to the state and path-condition after the execution of lines 6,7 (see above) and we choose to execute the **then** branch, i.e. we assume *true* ($\beta \geq 0$). After execution of statements 8,9,10,14 PC and the state are:

State: {(BASE,α),(EXPONENT,β), (L_RES, 1.0), (L_EXP, β), (RESULT, undef), (EXCHANGE, false)}
PC: $\beta \geq 0$

The theorem to evaluate on line 15 is ($\beta \geq 0$) $\Rightarrow$ (β/=0); this time the answer is *neither*, and we assume to not execute the loop; as a consequence the next statement to execute is the test on line 19. Since PC implies the opposite of the conditional expression, the **else** branch is followed, and the program ends. with the following PC and state:

State: {(BASE,α), EXPONENT, β), (L_RES, 1.0), (L_EXP β), (RESULT, undef), (EXCHANGE, false)}
PC: $\beta \geq 0 \wedge \beta = 0$

The described execution let us to define two test cases: EXPONENT < 0 and EXPONENT $\geq$ 0 and, we can use this information for testing, the program.

3.2. An example of partial correctness verification

SYMBAD can also be used for formal verification of programs partial correctness, provided that suitable assertions are inserted in the source code. The next example shows how to verify a program: function ROOT takes as input an integer A and computes the greatest natural number smaller than or equal to $\sqrt{A}$, i.e. the following property is intended to hold:

$\forall A ((A \geq 0) \Rightarrow (ROOT(A)^2 \leq A < (ROOT(A) + 1)^2))$

The code of function ROOT is the following:

```
1   function ROOT (A: INTEGER) return INTEGER is
2     X,Y,Z : INTEGER;
3   --: entry ((A >= 0))
4   --: exit ((X**2 <= A'),(A' < (X+1)**2))
5   begin
6     X := 0;
7     Y := 1;
8     Z := 1;
9   --:assert ((X**2 =A),(Y=(X+1)**2),(Z = 2*X+1),(A=A'))
10    while Y <= A loop
11      X := X + 1;
12      Z := Z + 2;
13      Y := Y + Z;
14    end loop;
15    return X;
16  end ROOT;
```

Before the function body, an entry and an exit assertion have been inserted; the entry assertion is used to initialize the path-condition while the exit assertion must be proved at the end of the execution, in order to prove the partial correctness of the function. These two assertions must always be inserted in the source code in order to verify a program. Other assertions, like the loop-invariant at line 9, are evaluated when they are encountered.
In order to verify a program containing a loop statement the proof is divided in three steps:
1) starting from the beginning, with PC set to the entry assertion, the value of PC before the loop statement must imply the loop assertion;
2) starting from loop statement, with PC set to the loop assertion, the value of PC after loop body execution must verify loop assertion;
3) starting from the end of loop statement, with PC set to loop assertion, the value of PC at the end of the program must imply the exit assertion.
When SYMBAD works in verify mode, every variable is automatically initialized with a symbolic value formed by the variable name preceded by the special character $. The state and the path-condition after execution of lines 5-8 are:

State: {(A, $A),(X, 0),(Y, 1),(Z, 1)}
PC: ($A >= 0)

Before continuing the execution, the assertion at line 9, representing the loop invariant, must be proved, that is the following theorem has to be proved:

($A \geq 0$) $\Rightarrow$ (0**2<=$A) $\wedge$ (1=(0+1)**2) $\wedge$ (1=2*0+1) $\wedge$ ($A=$A)

which can be reduced to ($A >= 0$) $\Rightarrow$ (0 <= $A), that is obviously true. Therefore the path 1,2,3,4,5,6,7,8 is verified with respect to the assert statement.
Now we execute the loop starting from the following state and path-condition:
State: {(A, $A),(X, $X),(Y, $Y),(Z, $Z)}
PC: ($X**2<=$A) $\wedge$ ($Y=($X+1)**2) $\wedge$ ($Z=2*$X+1) $\wedge$ ($A=$A)
where each variable has been initialized and the path-condition has beet set to the loop assertion. Now the conditional expression of while (line 10) is evaluated. The theorem to be proved is:
($X**2<=$A) $\wedge$ ($Y=($X+1)**2) $\wedge$ ($Z=2*$X + 1) $\wedge$ ($A=$A) $\Rightarrow$ ($Y<=$A)
The answer is *neither*, thus an assumption should be taken: the system automatically assumes *true*, since we want to execute the loop body. The loop body is executed and then the assertion at line 9 is evaluated again. The new state and path-condition are:
State: {(A, $A),(X, $X+1),(Y, $Y+$Z+2),(Z, $Z+2)}
PC: ($X**2<=$A) $\wedge$ ($Y= $X+1)**2) $\wedge$ ($Z=2*$X+1) $\wedge$ ($A=$A) $\wedge$ ($Y<=$A)
The predicates on line 9 evaluates to:
(($X+1)**2<=$A) $\wedge$ ($Y+$Z+2=($X+1+1)**2) $\wedge$ ($Z+2=2*($X+1)+1) $\wedge$ ($A = $A)

which is implied by the current value of PC. In such a way we have verified also the path 10,11,12,13,14. Now, the system re-initializes every variable and set PC to the loop assertion; since the loop should not be executed the expression ¬($Y <= $A) is added to PC.
  State: {(A, $A),(X, $X),(Y, $Y),(Z, $Z)}
  PC: ($X**2 <= $A) ∧ ($Y = ($X+1)**2) ∧ ($Z = 2*$X+1) ∧ ($A=$A)∧ ¬($Y <= $A)

The execution goes on with statement at line 15 which does not modify neither the state nor PC. Since PC implies the exit assertion the program is verified

3.3. An example of array execution

Let us consider the following procedure SORT that sorts an array A by selection. The array index bounds 1 to N and the array elements are integer.

```
0.   type INTEGER_ARRAY is array<> of INTEGER;
1.   procedure SORT(A:inout INTEGER_ARRAY(1..N))is
2.   TEMP, M: INTEGER;
3.   --: entry(true)
4.   --: exit (FORALL K (1<=K and K <= N-1 IMPLIES
                         A(K) <= A(K+1)))
5.   begin
6.     --: assert (FORALL K (1<=K and K <= I-1 IMPLIES
                             A(K) <= A(K+1))
7.     for I in 1..N-1 loop
8.       M := I;
9.       --: assert (FORALL T (I+1 <= T and T <= J-1 IMPLIES
                               A(I) <= A(T))
10.      for J in I+1..N loop
11.        if A(J) < A(I) then
12.          M := J;
13.          TEMP := A(I);
14.          A(I) := A(M);
15.          A(M) := TEMP;
16.        end if;
17.      end loop;
18.    end loop;
19.  end SORT;
```

Furthermore let us suppose that we want to verify the program; as shown in Section 3.2 the proof of correctness will be split into several parts.
Remark that this example contains two nested loops; therefore it is necessary to verify first the inner loop and then the external one.
At the beginning (line 5), the state and PC are:
  State: {(A, ((any, unknown))), (N,$N), (M,$M), (I,$I), (J,$J), (TEMP,$TEMP)}
  PC: true

The representation of A is given by the pair (*any, unknown*), that stands for: all values of A are unknown valid values. The assert statement of line 6 is implied by PC[2]; thus path 1,2,3,4,5 is verified.

At this point variables are re-initialized and PC is set to the loop assertion. Since we have to execute the loop body the expression ($I ≥ 1 ∧ $I ≤ $N) is added to PC. At line 9 PC and state are:
  State: {(A, ((any,unknown))), (N,$N), (M,$I), (I, $I), (J,$I+1), (TEMP, $TEMP)}
  PC: (FORALL K(1<=K ∧ K<=$I-1 IMPLIES A(K)<=A(K+1))) ∧ ($I ≥ 1 ∧ $I ≤ $N)

The assertion of line 9 is implied by the current value of PC[3]; the path 7,8 is verified, that is, the initial condition of the inner loop. At this point variables are re-initialized, PC is set to the assertion of the inner loop and the loop body is executed. After completion of inner loop verification SYMBAD will restore the state and PC of line 9 to continue the verification of the external loop. Thus at line 11 PC and state are:

---

[2]Since before loop execution the value of I is 1, the theorem is:
true ⇒ FORALL K (1 ≤ K ≤ 0 IMPLIES *unknown* ≤ *unknown*)
that is
true ⇒ FORALL K (false IMPLIES *unknown* ≤ *unknown*)

  State: {(A, ((any, unknown)($I,unknown1))), (N,$N), (M,$I), (I, $I), (J, $J), (TEMP, $TEMP)}
  PC: (FORALL T($I+1 <=T ∧ T <= $J-1 IMPLIES unknown1 <= A(T))) ∧ ($J ≥ $I+1 ∧ $J ≤ $N)

The evaluation of the test at line 11 leads to the following theorem:
  PC ⇒ *unknown2* < *unknown1*

The answer is *neither*, SYMBAD automatically saves the state and assumes that the answer is *true*. Thus, after the execution of statements 12-17 the state and PC become:
  State:{(A,((any, unknown)($I,unknown2)($J,unknown1))), (N, $N), (M, $J), (I, $I), (J, $J+1), (TEMP, unknown2)}
  PC: (FORALL T($I+1 <=T ∧ T <= $J-1 IMPLIES unknown1 <= A(T))) ∧ ($J ≥ $I+1 ∧ $J ≤ $N) ∧ (unknown2 < unknown1)

Now we should prove that PC implies the assert statement of line 9:
(FORALL T($I+1<=T∧T<=$J-1 IMPLIES unknown1<=A(T)) ∧ ($J≥ $I+1 ∧ $J ≤ $N) ∧ (unknown2<unknown1) ⇒
(FORALL T ($I+1<=T ∧ T<=$J IMPLIES unknown2 <= A(T))

The theorem is true, therefore the path 10,11,12,13,14,15,16,17 is verified. To goes on with verification, the system restore both state and PC previously saved and it assumes that the answer to test at line 11 is *false*.
The state and PC at line 17 are:
  State: {(A, ((any, unknown)($I,unknown1)($J,unknown2))), (N, $N), (M, $I), (I, $I), (J, $J), (TEMP, $TEMP)}
  PC: (FORALL T($I+1 <=T ∧ T <= $J-1 IMPLIES unknown1<= A(T))) ∧ ($J ≥ $I+1 ∧ $J ≤ $N) ∧ ¬(unknown2<unknown1)

Also in this case the loop assertion is implied; therefore the path 10,11,16,17 is verified.
The inner loop has been completely verified, thus we can go on with the verification of the external loop restoring the state and PC of line 9; moreover this time assertion of line 9 and the expression $J > N are added to PC.
  State: {(A,((any,unknown)($I,unknown1))), (N,$N), (M,$I), (I, $I), (J, $J), (TEMP, $TEMP)}
  PC: (FORALL K (1<=K ∧ K<=$I-1 IMPLIES A(K)<=A(K+1))) ∧ ($I ≥1∧$I≤$N)∧($J>$N) ∧ (FORALL T ($I+1 <= T and T <= $J-1 IMPLIES unknown1<= A(T)))

The current value of PC implies the assertion of line 6 and therefore also the external loop is verified. Finally variables are re-initialized and PC is set to the external loop assertion:
  State: {(A, ((any, unknown))), (N, $N), (M, $M), (I, $I), (J, $J), (TEMP, $TEMP)}
  PC: (FORALL K (1<=K ∧ K <= $I-1 IMPLIES A(K)<=A(K+1))) ∧ ($I > $N)

Since PC implies the exit assertion the program is verified.

4. Conclusion

The tool described in this paper can be usefully exploited both for program validation and program testing. The current version is still a prototype but we are working to obtain a more effective tool. Some parts, such as the user interface or the theorem prover should be improved. Moreover, the tool may be included in a more general environment for software development. In such a way the programmer could specify, design implement, test or validate his/her programs in an integrated environment.

---

[3]The theorem is:
(FORALL K(1≤K≤$I-1 IMPLIES A(K)≤A(K+1)))∧(1≤ $I ≤ $N)⇒
(FORALL T ($I+1 ≤ T ≤ $I IMPLIES A($I) ≤ A(T)))
that is
(FORALL K(1≤K≤$I-1 IMPLIES A(K)≤A(K+1)))∧(1≤ $I≤$N)⇒
(FORALL T (false IMPLIES A($I) ≤ A(T)))

At the Politecnico di Milano we are working on a project concerning with the definition of an environment for software development and reuse; a program transformation technique that allows to specialize existing Ada components has been studied [3]. The goal is improve software reusability by means of a tool that customizes existing components in order to match more restrictive requirements. This kind of transformation is accomplished by means of symbolic execution. A tool performing such transformation in a (semi)automatic way has been implemented starting from SYMBAD.

5. References

[1] Bledsoe, W.W. "The Sup-Inf method in Presburger Arithmetic." Tech. Rept. Department of Mathematics - University of Texas, 1974.

[2] Chang, C.L. and Lee, R.C. *"Symbolic logic and mechanical theorem proving",* Academic Press (1973).

[3] Coen-Porisini, A., DePaoli, F., Ghezzi, C., and Mandrioli, D. "Reusing Software Components by Specialization." Tech. Rept. 89-062, Internal Report, Politecnico di Milano - Dipartimento di Elettronica, December, 1989.

[4] H., W.E.H. "Symbolic testing and the DISSECT Symbolic Evaluation System." *Transactions on Software Engineering SE-3*, 4 (July 1977).

[5] Kemmerer, R. and Eckmann, S. "UNISEX a UNIx - based Symbolic EXecutor for Pascal." *Software-Practice and Experience 15*, 5 (May 1985).

[6] King, J.C. "Symbolic execution and program testing." *Communications of the ACM 19*, 7 (1976).

TOOLS AND METHODOLOGIES FOR QUALITY ASSURANCE

U. Anders, E.-U. Mainka and G. Rabe

Technischer Überwachungsverein Norddeutschland e.V., Hamburg, FRG

Abstract.

Nowadays the qualification and assessment of microprocessor based safety related systems cannot be done manually. Therefore a set of tools has been developed which consists of disassemblers for an important group of microprocessors, a control flow analyser and a data flow analyser. To do the analysis unaffected by errors in the documentation, compilers or linkers it starts from the object code.
The disassemblers supply direct metric survey information about the programs. Further result files of the diassemblers are processed by the control flow analyser and data flow analyser. All tools give hints on formal errors on the one hand, on the other hand they deliver accounts that allow an efficient semantic analysis of the control and data flow. A large amount of safety relevant applications can be ascertained by this method.

Keywords.

Computer control; computer evaluation; programmable controllers; quality control; software tools.

INTRODUCTION

Faulty safety relevant appliances have again and again led to accidents with serious consequences for employees and environment. To avoid such accidents producers and managers have to pursue Quality Assurance (QA), which has gained in importance because of the new development in product liability. Additionally, the legislation has passed a large number of regulations and technical rules according to which such installations have to be inspected and approved.

Due to the development in control and instrumentation microprocessor based systems are more and more responsible for the safe and reliable function. This means that the same quality measures (precautions) have to be applied and the same inspections and acceptance tests have to be taken as for the other components of the system.
These precautions are necessary in any safety related application on the one hand for the qualification and safety measures of the applied production lines, on the other hand they are necessary for the qualification of the products.

The TÜV Norddeutschland has been engaged in quality assessment of safety related systems and particularly their software since the early seventies, when we were first confronted with the problems that arose from this. In the course of this work a large spectrum of methods has been applied. Approaches accompanying the development and after completed development tractable methods such as
- reliability growth models
- black box tests
- code inspection and reviews
- white box tests
- static programme analyses
have been tried.

Experience with the differing models has shown that for safety related appliance with some risk potential the following combination of methods produces the most economic results:
- qualification for standard parts: via reliability growth models
- single components: static program analyses supplemented by whitebox tests
- complete systems: black box tests

Analysing the error occurence in different parts of computer systems it becomes obvious that the errors occur more frequently in the software than in the hardware and that most software errors occur in user programs rather than in standard ones. For this reason we have concentrated our work on Quality Assurance and assessment of user programs.

Beginning with a method described by Ehrenberger in 1973 we developed a tool set for testing safety and reliability of software in safety related applications. We will now proceed to describe the requirements of this tool set, its characteristics and the results.

REQUIREMENTS AND CHARACTERISTICS

The following four observations are basis for the requirements on the tool set for QA and assessment of software:
- the well defined field of coding (well defined compared to requirement specifications and system design) seems to be the most efficient starting point to put QA and assessment of software on a defined methodical and tractable basis
- even today software for microprocessor based systems is often written in assembler or machine close meta languages, e.g. PL/M
- documentation and realized program often do not correspond fully.

Beginning with the machine code and not, as is usual, with the program source has advantages that prove important for systems with a larger risk potential:
- independance of the quality of documentation
- independance of linker-/compiler errors
- possibility of evaluation of time consumption for object programs.

Due to the characteristics of compilers the generated code may seem to be not well structured. But these disadvantages are due to the present methods of programming (assembler, machine close languages) presently of no great importance, at least for the important group of small 8/16 bit processor based systems.

Following these observations we start from the object code for QA and assessment of software.

This approach requires as an absolute **must** as first step the disassembling of the object code. For this purpose a family of disassemblers has been developed with the following requirements:
- display of results independant of the processor via meta language by tools for static analysis of control- and data flow
- traceability from restart and interrupt addresses to differ between code and data
- ability to solve indirect addressing problems in an interactive way.

We have developed disassemblers for ZILOG Z80, INTEL 8031/8051, INTEL 8048, INTEL 8086/8088, Motorola 6802 and Motorola 68000.

In the second step the results of the disassembling are processed by the static control flow analyser (CONY). CONY is not much different in its main characteristics from other commercial static control flow analysers. Main feature is however its independancy of language and its ability to analyse almost any program.

CONY has following layout requirements:
- processing independant of language or processor type
- display of global information on the tested software
- display of informations that allow evaluation of efficiency for QA and assessment via
  * white box test or
  * more profound static analyses
- finding hints on weak points of realized code
- display of basic informations for the understanding of the code
- display of control flow informations necessary for data flow analysis.

The development of the control flow analyser is finished, it processes the results of above mentioned disassemblers.

Finally the results of the disassembling and of the static control flow analysis are processed together with a static data flow analyzer (DANY). Again it is not much different than other static data flow analyzers. As CONY, it can analyse almost any program and works independantly of the used language. This is possible by integrating the disassemblers and CONY.

DANY has following requirements:
- processing independant of language and processor type
- display of global informations on the used data
- display of further informations that allow evaluation of efficiency for QA and assessment via
  * blackbox test or
  * more profound static analyses
- finding hints on weak points of realized code
- display of further informations for the understanding of the code.

The data flow analyser is being developed at the time.

RESULTS OF TOOLS

The interaction between disassemblers, CONY and DANY is displayed in Fig. 1.

The three tools communicate in such a way as to send each other data via interface files. At the same time each tool outputs partial results that can be interpreted separately.
All three tools have been developed in the project 'Software Safety Tools' SOSAT that was subsidized by the German Ministry of Research and Technology (1500 679/A "SOSAT 2").

Results of Disassembling

The disassemblers generate four result files. The first one contains (amongst other items):
- number of programbytes in object file
- number of disassembled instructions
- list of start addresses
- number of sub-routines
- number of sub-routine calls
- nesting depth of sub routines
- call and nesting of sub routines
- list of recursively called sub routines
- list of not disassembled address areas

The second result file contains the disassembled and formatted program as a text file.
The third result file is the interface to CONY. It is independant of processor type anhd language. It contains for each basic block a start and end address, addresses of direct and jump sucessors.

The fourth result file is the interface to DANY and is also independant of processor type and language. With reference to basic blocks it contains the assignment function for each major output item. The usage of the corresponding minor output item, e.g. flags, is simply mentioned here.
Fig. 2 shows the hierarchy of calls as the disassembler displays it - based solely on the object code.

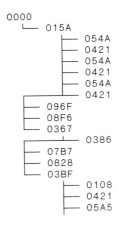

Fig. 2: Calling hierarchie

Results of Control Flow Analysis

Input data for CONY are the hardware definition and, if applicable, the linker list apart from the already mentioned interface file of the disassembler.
The hardware definition contains information about the hardware on which the tested program is run, e.g. the physical memory storage, interrupt addresses and the usage of ports. Some plausibility tests are hereby possible, for example whether all interrupt addresses have been used as starting addresses while disassembling.
The linker list is used to allocate symbolic names to the physical routine addresses that were found in the object code. It supports the communication with the developer as well as the reference to the documentation of program development.

The first result file of CONY is a static overview with information such as
- number of basic blocks per routine
- nesting depth of forward and and backward jumps
- number of jumps that violate the

rules of structured programming
- a rough estimate of processing time and number of paths.

Additionally, hints are given as to formal peculiarities of the control flow:
- jump from one routine into another (code sharing)
- unconditional jump to succeeding address
- conditional jump to directly succeeding basic-block
- RETURN in main program
- RETURN before the end of a routine

Some sheets for commenting peculiarities can be printed on option for each routine with peculiarities.

The control flow structure of the analysis program is displayed as 'Programm-Ablauf-Diagramm' (PAD). They contain the basic blocks in the middle, on the left the backward jumps and on the right the forward jumps. An example is given in Fig. 3.

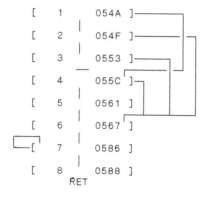

Fig. 3: Example of a PAD

Last, not least, CONY generates a pseudo code representation of the control flow structure. The syntax is according to PASCAL rules to enable the use of commercially available Nassi-Shneiderman-Diagram generators.

Results of Data Flow Analysis

The data flow analyser is still being developed. It starts from the interface file of the disassembler and the control flow analyser. In its present form it analyses only few formal data flow characteristics of the test piece, e.g.
- number of input and output variables of the single routines and of the the complete program
- number of indirectly addressed variables
- list of indirectly addressed variables
- list of possible input and output variables.

To simplify the understanding of the function of the test program DANY fills up the pseudo code generated by CONY with data flow informations. This output (Fig. 4), oriented on the PASACAL notation, is the basis for manual handling and additionally the input for the structogram generator.

EXPERIENCES IN THE APPLICATION OF THE TOOLS

We think it important to concentrate on the following questions:

- Are the tools suitable for actual usage?
- Can we derive statements on the quality of the tested software with the results of the tools?
- Do the results suffice for assessment?
- Can we make general statements on software by applying the tools?

We have analysed nineteen programs with the tools. These programs consist of an object code length of altogether 285 kilobytes in 2089 routines.
Two programs were written for the Z80, two for the 8048, five for the 8051, three for the 8086, four for the 6802 and three for the 68000. Ten programs were written in assembler, five in compiler languages and four in a mixture of assembler/compiler. Three different compilers were used.
The tools were run on all programs and delivered all of the above mentioned results. This also applies to examinees, in which indirect addressing was used. With only one exeption the tool behaviour did not make any differences between assembler and compiler programs nor between the programs for the different processors. The actual velocity of analysis was 20 bytes/second. Only the object code generated by a certain compiler led to velocities of more than two orders of magnitude above this value. Obviously the code generated by this compiler is structured worse than the usual manually generated codes.
All in all, the tools were apllicable in quality assessment procedures without any problems.

The usage of the tool results showed that based on the report files of the disassemblers and the statistical overview of CONY it is possible to gain an overview on the tested program. It was possible to estimate roughly the amount of work needed for QA and assessment, due to the calling hierarchy generated by the disassembler as well as to the statistical overview and the hints on formal peculiarities of CONY. Decisions whether to certify the correct function rather by analytical methods or rather by tests could now be made guided by the number of hints on formal irregularities of the control flow and the number of paths. Hints on formal weaknesses of the tested programs were factually given in the output of all three tools. They are far more complete than we could so far find with manual handling. The examiners that have actually used the tools found it very helpful to use the result outputs as PADs, Pseudo code and structograms as a basis for the understanding of the tested software.

Wheras up till now the tool-created results sufficed for a formal evaluation, for the evaluation of the contents it is necessary to tighten the data flow related results. For efficient processors, e.g. Motorola 68000, the problem of the indirect adressed data must be solved. The often mentioned opinion that assembler programs are usually worse structured than meta language programs can, to our present knowledge, not be obtained for the object code. About 80% of the routines that have been examined consist of less than 100 paths. But these 80% of routines include only 30% of the code. Nevertheless it is a large number compared with the general expectation. For this amount a complete path test (C0-Coverage) is manageable with relatively little work.

CONCLUSION

In some fifteen years a workable methodology for approving software quality has been developed. The methodology is designed mainly for safety aspects but can also be successfully applied to other software quality attributes, e.g. software reliablity.

For the application of this methodology we have developed a set of tools. Reference point for the tools is the object code. This ensures that they are not influenced by drawbacks in documentation, compilers or linkers.

The tools set has been designed to give an overview, hints on quality relevant weaknesses and to provide quality assurance staff with some basic information for semantic analysis. The tool set consists of a set of disassemblers, a static control flow analyser and a static data flow analyser. The application of the tool set is restricted to 'small' 8/16 bit microprocessor based systems with program sizes of some 10 Kbytes.

The practical use of the tool has shown that they enable an evaluation on expected results and amount of work necessary at an early stage of development due to their generated overviews. The tools generate hints on quality relevant formal weaknesses of the testes software much more complete and in a much shorter time than by manual analysis. On top of this they deliver the basis information for the functional analysis, and this again much more complete and of a stable high quality that is hardly achievable by manual work.

Irrespective of the practical use of the tools they assist or even enable following tasks:
- post documenation of assembler programs
- support of the calculation of processing time for the operating system based real-time applications
- proof of formal realization characteristics
- optimisation of machine programs

REFERENCES

Ehrenberger, W. (1973).
Zur Theorie der Analyse von Prozessrechnerprogrammen
Laboratorium für Reaktorregelung und Anlagensicherung
München

TÜV et al. (1988).
Werkzeuge für den standardisierten Software-Sicherheitsnachweis (SOSAT2);
1. Technischer Fachbericht, Rev.1, July 1988.

TÜV et al. (1989).
Werkzeuge für den standardisierten Software-Sicherheitsnachweis (SOSAT2);
2. Technischer Fachbericht, Oktober 1989.

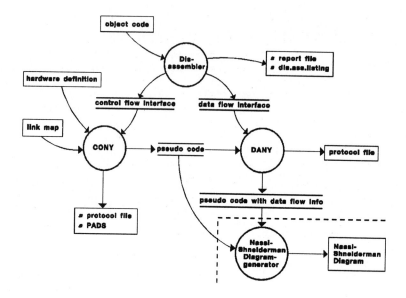

Fig. 1: Tool set

```
PROCEDURE   H_054A;
BEGIN
        Block 054A.a
        acc:= r1C;
        IF (bE7=#0) THEN BEGIN
        Block 054F.a
        acc:= ANL(acc,#E0);
        IF NOT (ANL(acc,#E0)<>0) THEN BRANCH TO 0567       !!!;
                Block 0553
                r1C:= #1F;
                r1D:= #FF;
        END
        ELSE BEGIN
                Block 055C.a
                acc:= CPL(ORL(acc,#1F));
                IF (CPL(ORL(acc,#1F))<>0) THEN BEGIN
                        Block 0561
                        r1C:= #E0;
                        r1D:= #00;
                END;
        END;
  x?: Block 0567
        mem[#A003]:= r1C;
        mem[#A002]:= r1D;
        8251dat:= #00;
        acc:= pmem[ANL(r1E,#03)+#546];
        dptr:= #A001;
        mem[#A001]:= pmem[ANL(r1E,#03)+#546];
        r02:= #50;
        REPEAT
                Block 0586
                r02:= r02-#1;
        UNTIL (r02-#01=#0);
        Block 0588
        acc:= #0F;
        dptr:= #A001;
        mem[#A001]:= #0F;
        R E T U R N ;
END;
```

Fig. 4: Example for a pseudo code

A COMPARISON OF STATIC AND DYNAMIC CONFORMANCE ANALYSES

M. A. Hennell* and E. Fergus**

Statistics and Computational Mathematics, University of Liverpool, UK
**Liverpool Data Research Associates Ltd., 131 Mount Pleasant, Liverpool, UK*

Abstract

The paper examines the strengths and weaknesses of Static and Dynamic Analysis when used to show conformance with a set of assertions.

Keywords

Static Analysis, Dynamic Analysis & Conformance.

Introduction

The objective in this paper is to compare Static and Dynamic Analysis as tools for the purpose of software validation. In particular their suitability for the evaluation of critical code is examined.

Static analysis is an analysis performed on a representation of the software to be implemented (e.g. a specification or the source code listings), either manually or preferably automatically. The type of analysis which is to be performed depends on what characteristics of the software are under scrutiny.

Dynamic analysis is an analysis performed during or after the program has been executed but it can also be performed earlier by, for example, simulation or interpretation. Its objective is to find how the software performed its task.

Firstly let it be noted that Dynamic Analysis requires a considerable measure of Static Analysis to be performed in order to determine the control and data flow structure of a program. The difference which this paper explores is in the use of Static Analysis to probe the program structure in order to detect errors. The paper will show that both theoretical and experimental work has demonstrated that in practice Dynamic Analysis is more powerful than Static Analysis as an error detector. The technical reasons for this will be explored in order to highlight areas where future research is required.

To distinguish between Static Analysis performed for the purpose of establishing program structure and metrics, and Static Analysis performed for the purpose of validation, the latter will be referred to as Static Conformance Analysis. The reason for this nomenclature is that the analysis purports to show that the software conforms to some criteria. Similarly, the essence of Dynamic Analysis is that the correct outputs shall be generated, and hence there is an implication that something or someone (the oracle) checks that the outputs conform to the expected outputs. It will be shown that the same conformance criteria used in the Static Analysis can also be used in the Dynamic Analysis and hence we will use the term Dynamic Conformance Analysis for this case. Clearly the two techniques can be directly compared.

Benefits of Conformance Analysis

Conformance analysis is the process in which a program or procedure can be shown to be consistent with an assertion or set of assertions. Hereafter the term program will mean either program or procedure. An assertion is a constraint expressed in terms of variables which are present in the program. Such assertions usually occur as one of four kinds:

a) an assertion in terms of the input variables to the program,

b) an assertion in terms of the output variables of the program,

c) an assertion in terms of any variables which are in scope at a particular point, and

d) a generalised assertion involving a set of rules e.g. as in Data Flow Analysis. Where variables should be defined before they are referenced etc.

These assertions may be local, global or even intermittent.

The ultimate object of any conformance analysis is to ensure that all paths through the code conform to one or more assertions. That is, the actions are inconsistent with the assertions are not violated. Specific examples of generalised assertions are a set of rules [5], or a specification or an allegation [3] (the terms flavour [1] and sequencing constraint are of this type [2]).

An assertion may be a condition (or set of conditions) which is derived directly from the program specification, i.e. the statement of what the program does rather than how the program is to do the task. An assertion could also be a statement that a violatory function must yield a specific value. A validatory function [11] is a function which is not referenced in the specification but is introduced solely to establish that same component of the system is correct e.g. a checksum for validations data movements, or a standard known integral of an integration routine.

An allegation is a weaker condition which summarises a belief about the program variables. It is possible for an allegation to be incorrect in at least some context. They are used primarily to increase understanding about a program or sharpen up the analysis. In some circles these additional conditions are called axioms.

Conformance analysis may be performed either statically (i.e. by analysing the representation) or dynamically (i.e. by animation or either tree executing the code). This paper compares the two approaches.

Both Static and Dynamic Conformance Analysis are path based techniques. They depend ultimately on directed graph representations. In both there is an assumption that all the paths are not independent. Indeed there are many representation of basis sets of paths. The assumptions with the basis sets is that, if the basis can be shown to be correct, then all combinations are also correct. This assumption can of course be weakened and certain combinations of the basis set can be explored. No technique is valid for all paths for any arbitrary program. Static Data Flow Analysis is valid for certain types of program (those with connected directed graphs).

Deriving constraints from a specification is itself a weak activity: it is difficult to construct an adequate specification [12], and it is difficult to establish that a desired goal has been achieved. This second difficulty - the construction of an "oracle" - can be seen with the following examples:

1) Consider a calculation to compute the energy content of the sun. The specification will state the equations that have to be solved. It does not state what the actual answer is (there would be no point in the calculation if it did).

2) Consider a graphics package. The package is specified as needing to draw curves of particular types (perhaps circles). How could one check that what appears on a screen is indeed the required curve? The quality of the display terminal would need to be considered, or even the eyesight of the viewer.

Thus a specification alone is frequently inadequate for the purposes of testing.

The problem may be very widespread, particularly in certain areas. For those areas the use of allegations is a possible way of improving the situation.

Static Conformance Analysis

In Static Conformance Analysis a program or procedure is analysed statically to yield a directed graph representation of the program. The directed graph is traversed in some way by a path traversal procedure.

Once a path has been chosen then this path is extracted and the source code corresponding to this path is examined in detail. The predicates or conditions which occur on this path are collected together to form the set of path constraints. Any intermediate variables are removed by replacing each occurrence with the last value assigned to that variable. This is the process of symbolic execution (symbolic replacement). In any representation there is the equivalent activity.

The symbolic execution process usually produces predicates which are complex algebraic expressions in terms of the input variables to the path.

To the set of path constraints can be added any other constraints such as assertions which one expressed in terms of the input variables. Assertions which involve intermediate or output variables must be included in the symbolic execution activity.

If this set of path constraints are consistent, i.e. there are no contradictions for any values of the input variables, then the path is feasible and satisfies the assertions. In some circles this is referred to as proving or establishing correctness.

It is the claims which are made following an analysis of this type which provides the motivation for this paper.

The principle technical problems which occur in static conformance analysis are:

1) For general problems non-trivial loops cannot be summarised accurately by means of loop assertions. The principle difference between Static Analysis and program proving is that program proving systems attempt to replace loops by loop invariants. These invariants are assertions which summarise in some ways the actions of the loop. They extract the essence and discard the house keeping aspects.

2) The choice of paths. Static analysis cannot resolve computed loop indices and hence the unfolding of a loop will be arbitrary. For example: in
for i: = m to n do .... od ;
neither the lower nor upper limits will be known in general since their values must first be computed and hence they are not available for Static Analysis. This problem is usually solved by an arbitrary decision such as to traverse the loop twice only. This same problem occurs in a higher level representation whenever the selection from a set of values is part of the problem.

3) The presence of infeasible paths. An infeasible path occurs when the path condition X is unsatisfiable. To demonstrate infeasibility one must show that not-X is a theorem. This is very difficult to achieve by static methods when the evaluation of X involves terms which are computed dynamically (e.g. Choice clauses, loops indices, pointer arithmetic). For example: in the previous example m < n may not occur and hence the loop will never be executed, or in
if (exp (x) < sin (x)) then......
where exp(x) may always be greater than sin(x)

In high level representation this may not be a problem because the mathematical properties of functions etc., are well defined. The problem can arise when the results of the functions etc. cannot be represented in a simple closed form. At the code level, with the use of approximate arithmetic this problem is very common.

The properties of infeasible paths in code have been studied extensively in [6,7]. In general it has been shown that the probability of any chosen path being infeasible is highly likely. Reports of strategies to generate predominantly feasible paths can be found in [9]. Nevertheless the problem is still highly significant.

4) Indexed variables such as arrays must be treated as simple variables because in general the indices are computed and cannot be resolved statically. For example:
i: = gam (x)
a [i]: = 10;
Statically the value of I is unknown and hence no statements about individual elements can be made. This problem is always present as an ambiguity in any representations.

5) Pointer variables which are dynamically determined cannot be analysed. In some circumstances the possibilities can be reduced to a small set. For example pointers to procedures can only point to the members of set of procedures. Usually this is a small set.

6) Floating point arithmetic errors cannot be detected. Thus round-off and truncation errors are uncontrolled. Interval arithmetic can produce bounds but these are usually useless. Essentially the conformance of any computation involving floating point numbers cannot be established statically.

7) Metrics to record the state of progress (i.e. static coverage) are usually unreliable. For example whilst all the paths in a given class (e.g. once round each loop) may have been analysed (and the metric has reported all such paths analysed), not all the executable statements in the program will have been analysed. This is because some of the chosen paths will be infeasible and if these paths were to be rejected and removed from the analysis then the above problem might be exposed. Basically infeasible paths make the metrics optimistic. This problem is particularly dangerous because statements of the type "all paths executing the loops twice have been analysed" is believed by some to have relevance to reliability. In fact all the paths may have been infeasible and the analysis is then worthless.

It is the previously mentioned ambiguities and infeasibilities which are the basis of this problem.

8) Static techniques only detect anomalies. All such anomalies currently require a human oracle to resolve whether an error exists. Many anomalies give rise to equivalent programs which are notoriously hard to resolve. An equivalent program is one which is computationally (or functionally) identical to a given program. The anomaly is therefore benign if it leads to an equivalent program. The number of anomaly messages can be extremely high which in turn reduces the possibility of a human checker detecting the error or an equivalent program. A rate of 1000 anomaly messages per 1000 lines of code is typical.

9) Since there is no measurable certainty in the choice of control flow structures as described in 7) no reliability metrics can be constructed in general. Only in specific cases such as Data Flow Analysis for strongly connected graphs can the contribution to reliability be accurately stated.

10) The oracle problem mentioned in [8] is compounded by the need to compare two expressions i.e. the program computations and the constraint, and show that they are equivalent. Even with sophisticated theorem provers this is extremely difficult and in the face of finite arithmetic beyond modern technology.

11) Called procedures (utilities, operating system calls) cause ambiguities. The reason is that in general it is not possible to accurately summarise the actions of the called procedure.The call cannot therefore be replaced by the summary. This weakens the analysis of higher order procedures to the extent that almost nothing can be said at the highest level. Only if the variables are independent can any sensible analysis be performed.

12) Ambiguities arising from exception handlers and interrupts can in general only be resolved with worst and best case analysis. In general such constructs give rise to disconnected graphs.

13) The indexed variable problem causes expression ambiguity. For example:
a [i]:= 10;
a [j]:= 12;
IF (a [i] = a [j]) THEN a [i]:= a [i] + inc.;

The final values of a [i] and a [j] can not be predicted. Their final value depends on whether i was equal to j, which may not be known statically.

This again is a manifestation of a higher level ambiguity.

Finally note that exceptions and gotos are not themselves the cause of any problems in Static Analysis. Any construct can be abused by bad design.

The issue of the contribution to reliability referred to in [10] is a complex one. In general, reliability is influenced by three factors:

> The usability of the code.
> The testedness of the code.
> The use patterns.

Ignoring the first factor, the aim of the static analysis must be to generate and analyse a set of paths which in some sense is a basis set for the use patterns. To date, very little work on this aspect has been reported [see Veevers].

More usually the practice is to cut loops arbitrarily at 2 traversals, analyse the resultant paths and compare the path constraints with a set of assertions. Clearly relating this to actual use patterns is somewhat difficult. This then is a major deficiency in a method which is targeted towards the achieval of reliability.

Dynamic Conformance Analysis

For Dynamic Conformance Analysis the assertions and other constraints can be translated into executable code with a mechanism for reporting constraint violations. Thus an assertion violation will result in a run time message.

The resultant program is then run with test data sets which achieve a known level of coverage. The measurement of the coverage can be performed with the constraints inserted, or pre-measured test data can be used. Since the assertions do not affect control flow or data flow patterns the two approaches are equivalent.

The Dynamic Conformance Analysis may be contrasted as follows:

1) The choice of paths is determined by the test data. Therefore the unfolding of loops is irrelevant.

2) Infeasible paths only appear indirectly as noise in coverage metrics. Certain statements or branches may not be executed because all the paths which lead to these statements or branches are infeasible. However, the nuisance value of this is considerable.

3) Indexed variables are explicitly resolved and hence there is no problem.

4) Pointer variables are resolved explicitly. Whether all possibilities are explored is determined by the class of test data. In general there is no surety on this point. Because the Static Analysis cannot always list the possibilities.

5) Loops are treated explicitly.

6) Floating point errors are detectable. Sensitivity depends on the nature of the oracle.

7) In this case the metrics are reliable, for example, "every statement executed" cannot be weakened by any infeasible path. This is because the definition requires that paths have succeeded in reaching each statement and by definition these paths are feasible.

8) Dynamic Analysis primarily detects errors not anomalies. This position depends to some extent on the nature of the oracles (see below). The most potent oracle is that of an observer examining the software outputs

9) Reliability metrics can and have been widely used.

10) The dynamic analysis oracle can be composed of:
a) an assertion checker
b) inspection (by humans) of the outputs
c) comparison (by humans or automated tools) with expected results.
d) the use of validatory functions.

11) In dynamic analysis the calling of procedures causes no additional problems.

12) The presence of exception and interrupt handlers only adds to the path explosion problem.

13) All indexing ambiguities are explicitly resolved.

Both Static and Dynamic Analysis are seriously affected by the quality of the Oracle. In both techniques humans can play a large part and the drive towards automation is seeking to reduce this dependency. The difference in the methods, primarily centres, around practicality and the nature of the oracles.

Static analysis uses assertions as adjuncts to the code to produce a more meaningful analysis. In Dynamic Analysis the same assertions can be inserted into the code as executable statements with a test to see if they are satisfied (see figure 1). Then, the actual traversal of a path ensures that the path constraints are satisfied (for one or more sets of input values) and also that the assertions are or are not satisfied.

Curiously one technique widely used in Dynamic Analysis has yet to be tried in static analysis. It is the concept of the validatory function [11]. This is an introduced function whose correct functioning implies that some housekeeping code, which is common to both the validatory function and the actual function is correct. Clearly statically, the highly complex actual function could be removed and the analysis could be performed with the housekeeping code and validatory function. The correctness of this substitute system would imply that at least the housekeeping code for the actual system was correct.

Thus the efficiency of dynamic analysis can be substantially increased by the inclusion of these executable assertions. This is directly comparable to the same process in static analysis. In static analysis the demonstration of consistency is for all input whilst in dynamic analysis it is for the set of data points. Depending on the structure of the path computation the correctness of the executions for these data points can imply that the paths are correct for all data points in the valid range. Techniques for establishing the truth or otherwise of this possibility are still in the laboratories.

Theoretical Comparison

In Static Analysis the aim of automation is to sharpen up the quality of the analysis in order to reduce the number of anomaly messages. Nevertheless, most of the anomaly messages are benign, they never cause any problems and are merely the consequence of algorithmic style.

The above criticisms apply to the analysis of any representation of the requirements but to different degrees whether it is called a specification or code. This is recognised in the specification arena by the drift towards animation which is another term implying a limited form of interpretation or dynamic analysis.

Dynamic Analysis establishes correctness point wise through the input space. Under certain conditions this can establish correctness for all the input spaces. Any assertions which can be formulated can be checked dynamically. The principle weakness lies in the sensitivity of the outputs and the oracle to errors.

Both Static and Dynamic Analysis are primarily path based testing methods. Both suffer from the lack of an inference argument that, having tested a given set of paths and found them to be 'correct' in some sense, it is not possible to deduce that all paths are therefore 'correct'.

If one seeks the reverse argument i.e. "is there a class of programs for which the above inferences are valid?" then the answer is that indeed there is. However, the constraints which determine this class are highly restrictive and tend to exclude all useful programs (they would, for example, exclude real numbers).

Experimental Comparison

One of the most comprehensive experiments to be performed is reported in [8]. Six testers evaluated six testing strategies. It was found that overall branch testing (a particular case of dynamic analysis) was twice as effective as static techniques. This result is in line with experiments performed by Howden and others.

Static Techniques, because they are currently highly intensive have tended to be applied either to specific types of software or to small examples of more general types of software. In contrast dynamic techniques are used widely and occasionally on very large software systems (100 000 lines and bigger)

Conclusions

Dynamic Analysis is firmly based on the results of Static Analysis. The sharper the Static Analysis the sharper will be the corresponding Dynamic Analysis. Moreover, the two are nicely complimentary, that which is difficult to resolve statically is usually straightforward dynamically.

Both methods require an oracle to examine output and detect errors. The building of automated oracles is expensive and difficult. Much of the future research work will involve the construction of automated tools for this task.

Dynamic analysis scores much higher when it comes to regression testing. It is usually very significantly cheaper to revaluate a Dynamic Analysis than a Static Analysis.

Figures

```
#include <stdio.h>

typedef int state;
#define initial 0
#define final 1

t(s)
int s;
{
  return(final);
}

main()
{
state current = initial;
state next_state;

next_state = t(current);
/*assert*/
/*
    current = initial && next_state
!= final
 */
/*end assert*/
current = next_state;
}
```

Figure1

An example of an assertion expressed as a comment in a program implemented in C

```
#include <stdio.h>
typedef int state;
#define initial 0
#define final 1
t(s)
int s;
{
  return(final);
}
main()
{
state current = initial;
state next_state;
next_state = t(current);
/* TESTBED assertion 1 */
   if ( ! ( current = initial &&
            next_state != final )
   { assert_fail(1);
   };
/* end TESTBED assertion */
current = next_state;
}
```

Note:

The C routine that implements the T function does not comply with the specification and will cause the instrumented assertion to fail.

Figure 2

The assertion automatically transformed to executable code. In this case a procedure assert_fail is called to report the violation

References

[1] W.E. Howden, "A General Model for Static Analysis", Proc. 16th Hawaii Int. Conf. on System Sciences, 1983.

[2] K.M. Olender and L.J. Osterweil, "Specification and Static Evaluation of Sequencing Constraints in Software", Proc. Workshop on Software Testing, Banff, July 1986.

[3] L.J. Osterweil, "Allegations as Aids to Static Program Testing", Dept. of Computer Science Report, University of Colorado, 1975.

[4] C.V. Ramamoorthy, S.F.Ho and W.T. Chen, "On the Automated Generation of Program Test Data", Proceedings of the IEEE International Conference on Software Engineering, San Francisco, 1976.

[5] L.G. Stucki and G.L. Foshee, "New Assertion Concepts for Self-metric Software Validation", Sigplan Notices, Vol. 10, No.6, pp. 59-71, June 1975.

[6] M.R. Woodward, D. Hedley and M.A. Hennell, "Experience with Path Analysis and Testing of Programs", IEEE Transactions on Software Engineering, Vol.6, No. 3, pp. 278-286, May 1980.

[7] D.Hedley and M.A.Hennell, "The Causes and Effects of Infeasible Paths in Computer Programs", Proceedings of the 8th International Conference on Software Engineering, London, August 1985.

[8] L.Lauterbach and W. Randall, "Six Test Techniques Compared: The Test Process and Product", Proc. NSIA 5th Ann. Nat'l Joint Conf. and Assn., Washington, D.C. 1989.

[9] D. Yates and N. Malevris, "Reducing the Effects of Infeasible Paths in Branch Testing", Proc. ACM SIGSOFT '89, Third Symposium on Software Testing, Analysis, and Verification (TAV 3), Key West, Florida, 1989, Ed. R. A. Kemmerer.

[10] A. Veevers, E. Petrova, A. Marshall, "Statistical methods for software reliability assessment", in "Achieving safety and reliability with computer systems", ed. B.K. Daniels, ISBN 1-85166-167-0.

[11] M.A. Hennell, P. Fairfield and M.U. Shaikh, "Functional Testing" Proceedings of workshop on "Requirements Specification and Testing", 1984, pub. Blackwell, ed. T. Anderson.

[12] W. Swartout, R.Balzer, "On the inevitable intertwining of specification and implementation", CACM, V25, N7, July 1982.

COMPUTER BASED TRAINING FOR CONTINGENCY DECISIONS

K. H. Drager, H. Soma and R. Gulliksen

A/S Quasar Consultants, Harbitzalléen 12, 0275 Oslo 2, Norway

Abstract. The paper describes an EDP-simulation program package SIMLAB (Simulation Laboratory). SIMLAB is a simulation package developed as a Computer Assisted Learning tool for personnel involved in safety and contingency planning and those who have operational responsibility for such matters.

Successful handling of an emergency depends on the efforts of the personnel both as individuals and as groups. It is therefore important that any training given, allows the opportunity of developing the skills of the individual as well as the group. The SIMLAB development is trying to meet both these requirements.

Contingency training is normally based upon simulation of an accident scenario. The objective is to establish a realistic scenario, without passing the danger threshold which separates a real accident from a simulated one.

In order to perform a simulation, an offshore platform is modelled with respect to accident development, escapeway system, personnel distribution, etc. Its surounding fields and platforms with their rescue resources are also included. SIMLAB contains the following program modules:

ADS	- Accident Development Simulation
EVADE	- Simulation of the mustering phase of an evacuation
PTRACK	- Personnel tracking of every individual involved in the evacuation process
LBL	- Simulation of the launching and escape phase of the lifeboats
HEVAC	- Simulation of helicopter evacuation
ORS	- Simulation of a SAR operation offshore

The purpose of using these simulation programs in training situations is to show the effects of decisions that might influence the development of an emergency situation. By such a simulation it is possible to include extreme scenarios, which are difficult or impossible to test out in real life.

In addition to being a training tool, the programs also represent a valuable planning tool for safety and contingency matters.

The paper describes the EVADE modelling and computer program in detail, and how all the computer programs form together the SIMLAB system, and finally how the intended use of the system will be.

Esso Norway's platform ODIN, located in block 30.10 on the Norwegian Continental Shelf is modelled in the prototype version.

Keywords. Expert systems; computer-aided instruction; safety, computer simulation; educational aids.

INTRODUCTION

The SIMLAB training concept is intended to be used for training groups or individuals. In the group exercise, the contingency staff (trainees) are supposed to meet in the emergency room/control room as soon as possible after an emergency occurs, and from there manage the emergency operation. Each individual is supposed to play the role he would have if a real emergency was to occur. By use of different types of communication tools, like PA-system, telephones etc., the contingency staff have to act upon information in order to bring the emergency under control, or minimize the losses/damage.

The information fed to the contingency staff will be given by a simulation staff (instructors), taking on the roles of personnel outside the control room. A situation as sketched in Fig. 1 is established.

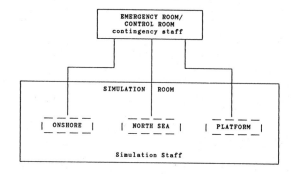

Fig. 1. The SIMLAB training concept

The SIMLAB is primarily intended to be a tool for the simulation staff, to keep track of the simulated emergency situation onboard. This will help them to act realistically when their "roles" are relayed into the emergency room.

In addition SIMLAB also logs events and the time history of the emergency simulation in order to facilitate evaluation of the performance of the trainees during debriefing.

After the training has been performed a debriefing and exercise session follows. In this session the instructors may rerun the simulation in order to demonstrate the trainees decisions and their outcome. The purpose of using SIMLAB in the debriefing exercise is to improve the trainees ability to make correct decisions.

In the debriefing session the trainees are not faced with the same complex situation as during the accident simulation, nor do they face the same time pressure. The trainees will therefore get feedback on their performance and have time for reflection without added stress.

Individuals may also use SIMLAB, and can "play" their own scenarios and test out the results of their choices.

Before starting up the simulation, a thorough briefing session is necessary to clarify the issues for the trainees, and the goal of the training/exercise.

THE SIMLAB COMPUTER PROGRAMS

The intention of the SIMLAB computer programs is to simulate the development of an emergency scenario offshore. The simulation comprises the following

- Accident development
- Evacuation and personnel tracking
- Rescue

performed by the following computer programs:

ADS	- Accident Development Simulation
EVADE	- Simulation of the mustering phase of an evacuation
PTRACK	- Personnel tracking of every individual involved in the evacuation process
LBL	- Simulation of the launching and escape phase of the lifeboats
HEVAC	- Simulation of helicopter evacuation.
ORS	- Simulation of a SAR operation offshore

In the following, the basis for the EVADE program and its modelling are described, and how this program is linked to the others making up the total SIMLAB system.

OFFICIAL REQUIREMENTS, RULES AND REGULATIONS

The main requirements concerning evacuation time as stated in "Regulations for Lifesaving Appliances etc. on Drilling Platforms and other Mobile Installations in the Sea" (February 3, 1982) will apply to the approvement of offshore platforms. This regulation reads:

> 'The number and location (of lifeboats) shall further be such that it is possible in a safe and realistic way in all thinkable accident situations to evacuate all persons on board the facility 15 minutes after the alarm has been given. Tests of this may be required.' (para 4.1.1, last item).

Two main types of impact will then have to be accounted for:

- the accident renders certain escapeways inaccessible
- the evacuation time must meet the requirement under a list condition

This regulation was the basis for the EVADE development.

ESTIMATION OF EVACUATION PERIOD

Preferably, the time to evacuate should be measured in real situations to give a correct estimate. Fortunately, accidents are rare, and detailed measurements of evacuee performance during accidents are non existing.

Another possibility to get real data is to observe mustering drills. Although it is an exercise situation, and the evacuees are aware of this fact, such data can be valuable for the identification of the time needed to evacuate the installation. Therefore, the regulation mentioned above states that such tests may be required.

Actual observations of evacuation performance of new concepts on the planning stage are not possible. However, the new concepts will only be realized if there is a high probability that they will comply with the requirements. Thus, there is a need to use analytical tools to estimate the evacuation performance, based upon the number and distribution of people to be evacuated, the escapeway lay-out, procedures used, as well as other factors that are known to influence evacuee performance.

There are two possible ways to assess the evacuation period analytically.

First, a simple approach would be to calculate the time to muster from the most distant part of the installation to the most distant lifeboat. Given a certain walking speed in corridors and stairs/ladders, and measuring the distances as described in the General Arrangement drawings, the mustering time can be calculated. Adding the time to enter and launch the lifeboats will then give the total evacuation time. On some occations, the time needed for shut down or quick release operations must be considered if applicable.

However, the simple approach described above does not take care of the fact that queuing might occur. Such an effect will typically be present when a number of persons are gathered in the same place, e.g. the dining room, cinema, etc. at the time of mustering alarm. During the evacuation process, queuing might also occur at escapeway confinements.

As a response to this problem, the analytical simulation approach has been followed to arrive at the computer simulation program EVADE.

BASIC MODEL PARAMETERS

There are two main parameters concerning escapeway lay-out:

1) Escapeway geometry

2) Characteristics of escapeway elements

The escapeway geometry describes which elements are interrelated, e.g. how corridors and stairs are interconnected, where the exits and lifeboats are, etc. The geometry can then be described with a reference to input and output elements to each node.

There are two types of escapeway elements:

- personnel centres
- route elements

The centres are characterized by:

- the number of persons present at start of simulation
- the time period for the first evacuee to reach the exit
- the maximum time period to evacuate the centre if no queuing

The escaperoute elements are characterized by:

- length
- minimum width
- average width
- type: corridor, down stair or up stair

There are three basic human parameters in the model:

1) Evacuee walking speed, measured in meters per second.

2) Evacuee ability to pass constrictions (e.g. doors) along the route. This ability is denoted the 'throttle effect', and is measured as the number of persons that can pass a door opening of one meter per second.

3) The maximum acceptable density of persons, i.e. the maximum number of persons per square meter that is allowed before queuing occurs.

A higher value in any of the human parameters above will normally result in a reduced mustering period, i.e. faster escape.

SELECTION OF BASIC PARAMETER VALUES

Some uncertainty exists as to what basic human parameter values should be used in a simulation of evacuation of offshore installations. Studies of evacuation of office buildings are reported in the literature, indicating parameter values in the range of:

Evacuee walking speed: 0.8 - 1.5 meter/sec
Throttle coefficient : 1 - 1.5 persons/meter/sec
Maximum density : 3 - 5 persons/m2

The vertical component, v, of evacuee walking speed in downgoing stairs are:

$$v = 1.0 - a/125 \text{ meter/second} \quad (1)$$

and for up-going stairs:

$$v = 1.0 - a/100 \text{ meter/second} \quad (2)$$

where a is the stair angle with the horizontal in degrees.

It is to be noted that for medium stair angles (30-45 degrees), the vertical component may be considered constant, having the values:

Down stairs: 0.38 m/s
Up stairs : 0.28 m/s

Data concerning walking speed in stairs were collected on previous exercises on an offshore drilling rig.

However, some of the other data are reported from studies in Japan where the evacuees were not dressed for outdoor winter activitiesl and their validity or applicability in North Sea conditions could be questioned. In order to clarify this point, validation studies has been performed on an offshore rig.

PERSONNEL TRACKING

When developing SIMLAB, the need for tracking each individual was evident, and a further development of EVADE into a personnel tracking of individuals was performed. The program named PTRACK gives the possibility of a 3D look into the platform escapeway system seeing the movement of personnel.

The following operations are possible to simulate with PTRACK.

- Position the personnel at different locations
- Simulate mustering of the personnel
- Give order to personnel to move to a given position
- Simulate injured or dead personnel as result of accident impact
- Simulate a rescue-team for transportation of injured personnel
- Simulate a fire-fighting team
- Simulate a search team
- Simulate panic and apathy occurence
- Show accident impact result for the escapeway system

LIFEBOAT EVACUATION

The motion of the lifeboat with respect to the collision hazard from start of the launching operation until the lifeboat is at a safe distance away from the platform is simulated.

Figure 2 shows a simulated launching operation from the ODIN platform.

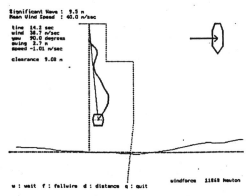

Fig. 2. Lifeboat/raft launching simulation

HELICOPTER EVACUATION

Based on the accident development and mustering process, a number of persons may need to be evacuated from the platform. The HEVAC program will cover the evacuation performed by helicopters.

The scenario needs to be described in terms of the status of the personnel involved:

- The total number of people to be evacuated.
- The number of people to be evacuated by the means of stretchers.
- The number of people who need medical assistance during the transport.

These factors influence the passenger capacity of the helicopters and time to bring the people in distress abroad (time from when the helicopter lands to take off).

OFFSHORE RESCUE

The program ORS simulates the rescue operation and the results will be displayed on the data screen as shown on Fig 3. One graph shows the number of people picked up from the sea as a function of time, another shows the number of people rescued alive. A comparison of the two graphs will then indicate whether the operation should be considered "successful".

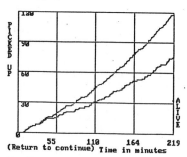

Fig. 3. Rescue history

SIMLAB APPLICATION

The programs described in the preceeding, form together the SIMLAB system, and the SIMLAB application sequence is shown in Fig. 4.

In the briefing phase the trainees will be introduced to how the simulation will be performed and the simulation environment. It is assumed that the trainees are familiar with the platform in question. If not, a platform layout demonstration also has to be performed.

The trainees are after the briefing session supposed to be familiar with the simulation environment, and a demonstration of the lifeboat launching program will be performed. The intention is to demonstrate the effect of launching boats or rafts into the given environment. The program will calculate the probable success rate for all lifeboats and rafts, and this information will be available for the trainees when the simulation starts.

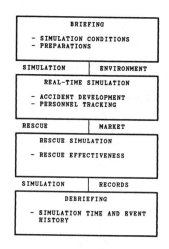

Fig. 4. SIMLAB application sequence

The trainees are now ready for the simulation and leave for the emergency/control room, and the instructors start the real-time simulation.

The real-time simulation performed by SIMLAB is illustrated in Fig. 5.

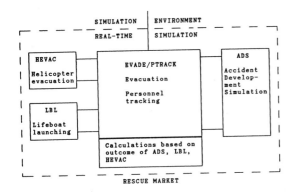

Fig. 5. Real time simulation

The main program in this sequence is EVADE/PTRACK which gives the instructor an updated picture of the situation on the platform. The instructor may choose the updating frequency. PTRACK gives the status of the platforms internal situation concerning personnel, where they are and their condition, and the status of escapeways and platform based rescue means. Due to the stochastic feature of the program and decisions taken by the trainees, the outcome of the simulation will vary.

The instructors have to feed the computer with the trainees decisions, and may introduce into the simulation scenario different equipment failure, etc. in order to put a strain on the trainees in the emergency situation. The trainees themselves only communicate through ordinary communication means with the instructors.

ADS (Accident Development Simulation) gives information to EVADE/PTRACK on accident effects, and EVADE/PTRACK calculates the consequences with respect to personnel i.e. if an explosion should occur, which personnel are within the danger zone and probabilistic selection of the consequences to them.

The response from the trainees will also have impact on the accident development, i.e. use of fire fighting equipment, activating manual deluge, etc.

If the trainees decide to start lifeboat/raft launching or helicopter evacuation, the programs LBL or HEVAC will be initiated. Based on the outcome of the launching or helicopter evacuation, EVADE/PTRACK will summarize the consequences with respect to personnel.

When the outcome of the simulation is clear, either the accident is brought under control or the platform has to be evacuated, the simulation ends, with a potential rescue market. The rescue market is the number of persons that have to be saved either from the sea or lifeboat/rafts. The real-time simulation stops here, and the instructors and trainees meet to run a rescue simulation of the potential rescue operation.

THE RESCUE SIMULATION

This is a joint session between instructors and trainees, and SIMLAB is used to calculate the probable outcome of a rescue operation after the accident, given the non-platform based rescue means available.

Based on the non-platform based rescue means available, and the rescue market, the ORS program will calculate the rescue effectiveness, or probable survival rate of personnel.

It is now possible to vary the assumed parameters and rescue means used during a potential rescue operation, in order to find the right match of rescue means to achieve the best probable outcome.

All records from the simulation runs are saved by SIMLAB, to be used during the debriefing session.

During simulation, time and event history is recorded so that it is possible to rerun the simulation in order to evaluate the decisions made by the trainees, and judge if alternative decisions could lead to a better handling of the emergency.

The SIMLAB system is very well suited in the debriefing phase in teaching the trainees to take the right decisions in order to arrive at the best handling of an emergency situation.

CONCLUSIONS

SIMLAB is a training tool under development and no experience has yet been gained of SIMLAB in use. However, SIMLAB has several advantages as a training tool, which cannot really be met by todays manual systems.

- Repetetive possibility
  * The simulation runs can be repeated exactly at any time

- Easy to change
  * Any changes can be easily implemented

- Accumulation of knowledge
  * The simulation system can be updated continously with gained experience, and thus be updated to represent the state of the art at any time.

- Training of groups and individuals
  * Both individuals and groups can use the system without alternations, and achieve the same level of training.

- Recording advantages
  * Every communication with the system can be recorded and thus be repeated exactly

- Debriefing advantages
  * With an exact recording, the debriefing analysis will be even more valuable.

- Easy implementation both offshore and onshore
  * the system requires only a micro computer system which is easy to transport, and can be quickly installed anywhere

Moreover, EDP has not been used in contingency training, and represents innovation and a challenge in this field. Esso Norge has been willing to take up this challenge, and thus make the development of SIMLAB possible.

Finally, all the programs which make up the SIMLAB-package have been developed for safety and contingency planning and evaluation purposes. The SIMLAB development makes them more valuable as such tools.

SIMLAB also allows companies the chance to model all their platforms with potential accident situations, personnel and surroundings, and thus have easy access to a live picture of the platform when safety and contingency matters are to be discussed.

SIMLAB is developed for the offshore industry's particular needs. However, the concept is usable for other applications as well, which also will be exploited in the future.

REFERENCES

/1/ Drager K. Harald, Helge S. Soma and John-Arne Skolbekken
"Simulation of an evacuation process of ships and offshore structures"
European Congress of Simulation
Prague, Czechoslovika, Sep. 1987.

/2/ Drager, K. Harald and Helge S. Soma
"Optimalisering av beredskaps-ressurser" NIF-seminar, Trondheim, Norway, Feb. 1987.

/3/ Skolbekken, John-Arne og K. Harald Drager: "Trening i beredskapssituasjoner, SIMLAB" NITO-seminar, Oslo, Norway, April 1987.

/4/ Drager, K. Harald and Helge S. Soma
"Modelling and simulation of an offshore evacuation and rescue operation."
IMACS symposium on AI, Expert System and Languages in Modelling and Simulation
Barcelona, Spain, June 1987.

/5/ Drager, K. Harald, Helge S. Soma, Roy Gulliksen and Cort Holtermann:
"SIMLAB - Simulated Safety and Contingency Training"
OCC-88, Aberdeen Scotland, March 1988

/6/ Norwegian Maritime Directorate ('The red book')
"Regulations for Lifesaving Appliances etc. on Drilling Platforms and other Mobile Installations in the Sea"
(February 3, 1982) (para 4.1.1, last item).

/7/ Wright, J.F.
Critical behavior patterns in an evacuation operation. Det norske Veritas Report No. 82-1144.

/8/ Rimeid, B.E.
Offshore emergency preparedness. Base accident scenarios in case of blowout and process fire. Det norske Veritas Report No. 82-1143.

/9/ Soma, H., Karlsen J.E., Solem, R.R., Wright, J.F.
Mustering analyses for the cruise vessel M/S Phoenix. QUASAR Report No. T-03-85-1. 1985.

/10/ Kuwabura, H. et.al.
A fire-escape simulation method by describing actions of evacuees as probabilistic phenomena.
Musashino Electrical Communication Laboratory. Nippon Telegraph and Telephone Public Corporation (NTT).
Musashino - City, Japan.

/11/ Muta, K. et.al.
Study on a total fire safety system.
Kajima Institute of Construction Technology, Tokyo, Japan.

QUALITATIVE KNOWLEDGE IN A DIAGNOSTIC EXPERT SYSTEM FOR NUCLEAR POWER PLANT SAFETY

I. Obreja

Department of Databases and Expert Systems, Technical University of Vienna, Paniglgasse 16, A-1040 Vienna, Austria

Abstract. This paper presents partial achievements in an on-going research for detection and diagnosis of Nuclear Power Plant faults. A new approach to the field of Nuclear Power Plants, qualitative reasoning, was tested. Qualitative reasoning has become a widely discussed field of advanced artificial intelligence research.

The paper describes an artificial intelligence implementation, a diagnostic expert system. A small prototype has been implemented in C-Prolog and is currently running on a Sun386i workstation. We have tested the expert system on a qualitative model of a part of the Emergency Feedwater System at the Seabrook Nuclear Station and have achieved satisfactory results.

Keywords. Artificial intelligence; diagnosis; expert systems; failure detection; nuclear plants; qualitative modeling.

INTRODUCTION

Research for possible expert systems software applications to nuclear power plants (NPP) has substantially increased in the last two decades. The dynamically complex system of a NPP is a most challenging topic for artificial intelligence (AI) specialists. Malfunction diagnosis of NPP's systems generally uses shallow knowledge incorporated in fault models. This derives from the fact that most NPP's systems receive signals from sensors and that their possible malfunction causes and effects on system variables are well known.

In recent years many important results have been obtained about representation and reasoning on structure and behaviour of complex physical systems using qualitative causal models. The current AI trend trying to cover this aspect is qualitative reasoning using deep knowledge representation of physical behaviour (Herbert, 1987).

This paper describes the implementation and results obtained in a research effort devoted to study the suitability and the limits of a qualitative model based on deep knowledge for fault detection and diagnosis of the emergency feedwater system (EFWS) in a NPP. The EFWS has been chosen because of its importance for the safe functioning of the NPP.

The first part of the paper sets the basis for describing such a complex world as a NPP in a model apt to be exploited for system building. It provides a short general information on the EFWS stressing its importance and concluding with some details on the functional description of the EFWS. After giving a description of the modules of the expert system: Fault Detection Module and Fault Diagnosis Module, the paper continues by giving an overview of the Diagnostic Process in which the main feature of the resulting expert system (making diagnostic knowledge explicit) is underlined.

The paper then goes on with a detailed presentation of the Knowledge Base Information. The deep knowledge, i.e. formal knowledge about the system structure, design principles and physical laws underlying system operation is desribed in detail.

In the conclusion we analyze the efficiency of this approach as well as of the chosen implementation language, i.e. PROLOG. Finally, the future development of the system is outlined.

THE REAL-WORLD PROBLEM: OVERVIEW OF THE EFWS

Generalities

Figure 1 shows the Seabrook NPP EFWS.

The EFWS is a standby system which would not be operated during normal plant operation. The role of the EFWS is to provide full cooling of the Reactor Coolant System in emergency conditions.

The EFWS is automatically activated in three cases of NPP malfunction:

1. loss of offsite power (LOOP)

2. low-low level in any steam generator

3. loss of alternative current (LOAC)

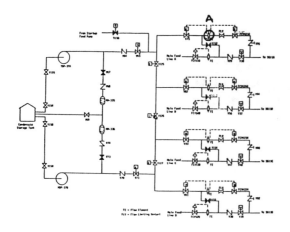

Figure 1: Schema of the Seabrook station EFWS

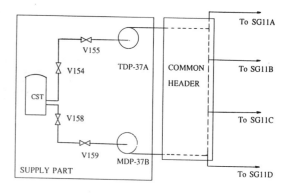

Figure 2: Simplified sketch of the Seabrook NPP EFWS

Functional Description

For our purpose we divided the EFWS in three subsystems (see Fig. 2):

- supply part
- common header
- supply lines to steam generator

In this paper only the supply part subsystem is diagnosed. Therefore, a detailed description of the other subsystems of the EFWS is given in (1984).

The supply part subsystem consists of two pumps (TDP-37A, turbine driven, and MDP-37B, motor driven), each supplied by individual suction lines from the Condensate Storage Tank (CST). Each pump suction line to the CST contains two manual isolation valves (V154, V155, V158 and V159).

During operation of the EFWS, both pumps discharge into a common header, which in turn supplies four individual supply lines to each of the four steam generator main feed lines (SG11A-SG11D). The value of the fluid pressure obtained at the output has to be constant. Any deviation from a constant value corresponds to a system fault.

The possible malfunctions which can occur in the EFWS, their causes and effects on system variables are well known. They are associated with cracks in the pump or CST casing, pipe or valve ruptures, and pump or valve operation failure.

As the EFWS is working only in emergency conditions, the occurrence of a malfunction in the EFWS would lead to catastrophic results. Safety insurance is an acute problem in NPP. We consider that expert systems can contribute to the improvement of flexibility and man-machine communication in NPP.

THE EXPERT SYSTEM

The expert system in its current stage is designed to operate off-line. The diagnostic process is performed by a forward-chaining inference engine that operates on the knowledge base.

The inference mechanisms adopted in deep modeling techniques (deKleer, 1986,1987) generally consist in:

a) identifying that a fault exists by comparing the simulated system behaviour with the actual one;

b) picking up a list of candidate faulty components by reasoning on system structure;

c) identifying a set of new test points in order to further discriminate among the candidate set;

d) completing the discrimination process through examination of further test cases.

The Diagnostic Process

When describing the Diagnostic Process we will refer to Fig. 3.

The diagnostic process module consists of two modules: Fault Detection and Fault Diagnosis Modules.
The process starts with the Fault Detection module which detects a symptom of malfunction by observing the qualitative variation on the output parameters. Several information sets (see Fig. 3) are instantiated in the initialization phase.

The process then continues with the identification of the causes of malfunction by exploiting the information contained in the "Model". The "Model" actually represents the Knowledge Base. It contains descriptions on the Physical System (generic components, initial measurements, connections, possible measurements, actual components). Note that only the correct system behaviour is described in the model.

The Fault Diagnosis module then propagates the observed qualitative variation through the system model using a constraint propagation method (Reiter, 1987;deKleer, 1987). Thus, all possible fault models are generated.

This step ends when some input parameters (i.e. parameters in the left-hand-side (LHS) of the rules) are unknown thus making further propagation impossible. The qualitative reasoning process can continue only if new measurements are taken.

The choice of the optimum measurement to do is taken according to heuristic criteria, i.e. probabilities of component failures. From these probabilities we compute candidate probabilities and Shannon's entropy function (deKleer, 1987).

After the most appropriate point to measure next has been identified, the measurement is taken, and the qualitative propagation is continued for this measurement.

The goal is reached when the best diagnosis is found.

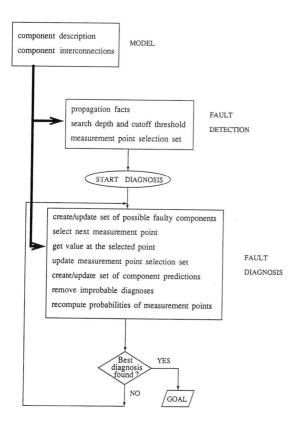

Figure 3: The diagnostic process

KNOWLEDGE BASE INFORMATION

The Knowledge Base contains qualitative information derived from the EFWS model. This information is used by the Diagnostic Process a presented above.
The EFWS model is described by:

- components
- connections
- equations involving:
    - process variables
    - design parameters

Components (considered up to now) are manual isolation valves, pumps, tanks and "t_connections" (Friedrich, 1989), i.e. pipes.
Components are connected together by process variables. The components' behaviour involves process variables and design parameters. Design parameters have nominal values stated by design. Qualitative analysis considers design parameters as constants.

Figure 4 shows an example for the component type *valve*. The global system has a hydraulic behaviour. Therefore, pressure and flow were analysed among representative process variables.

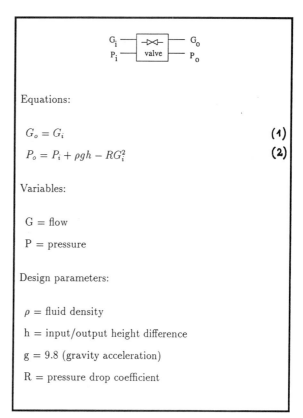

Figure 4: Valve description

A symptom of malfunction is detected when a variation on the variables' qualitative values is observed.

We show below the variations of affected variables caused by EFWS malfunctions:

	P_o	G
tank malfunction	dec	std
pipe break	inc	inc
valve malfunction	dec	dec

where

- dec = decreases
- inc = increases
- std = steady

Each possible malfunction affects in a known way (inc or dec), but in an unknown measure, some variables. Thus, the model and its analysis are intrinsically qualitative.

Other information. The knowledge base also contains information on possible measurements. I.e. we consider possible component ports (measurement points) and then define possible values.

Measurement points can be viewed from two different aspects:

1) points in the schema where instrumentation devices could be placed (distinguished by the design, e.g. point A in Fig. 1)

2) "qualitative" points defined by us.

Not all variables can be directly measured. Moreover, the measurement instrumentation is subject to frequent decalibration. Thus, the supplied qualitative information is not always reliable.

Therefore, we defined "qualitative" points and values. I.e. we considered each component port as a possible measurement point and assigned it the values [1,0,-1] (for *inc, std, dec*).

CONCLUSIONS

In this paper we have outlined on-going research for detection and diagnosis of NPP faults. We used a new method for the NPP field, i.e. qualitative reasoning using deep knowledge representation of physical behaviour. For this purpose we have developed an expert system.

The expert system has been designed such as to maximize flexibility and completeness needed in deep level diagnosis systems. Therefore:

- an inference engine which is independent of the plant model undergoing diagnosis has been developed. Due to the facilities of the Prolog programming language, backtrack and cutoff procedures have been implemented to find possible diagnosis candidates. The cutoff procedures are most useful for finding only minimal or most probable candidates.

- the knowledge base can be updated simply by adding (changing) specific knowledge according to the plant model. We have shown that the use of qualitative knowledge leads to savings on the cost of expensive and unreliable measurement instrumentation.

However, our expert system in its current stage proves to have several disadvantages, e.g. it does not include time as a variable. The plant system is modelled only statically, and thus, unrealistically. We consider this to be the most critical problem for our future research.

Moreover, the expert system has been tested only on a part of the EFWS. It remains to be seen if it will be equally efficient when extending it to the whole EFWS model.

REFERENCES

deKleer, J., and J.S. Brown (1984). A qualitative physics based on confluences. Artificial Intelligence., 24, 7-83.

deKleer, J. (1986). An assumption-based TMS. Artificial Intelligence., 28, 27-128.

deKleer, J., and B.C. Williams (1987). Diagnosing multiple faults. Artificial Intelligence., 32, 97-130.

Friedrich, G., and I. Obreja (1989). Model-based decision tree generation for diagnosis and measurement selection. IFAC/IFIP SAFECOMP Workshop., 109-115.

Herbert, M.R., and G.H. Williams (1987). An initial evaluation of the detection and diagnosis of power plant faults using deep knowledge representation of physical behaviour. Expert Systems., vol. 4, no. 2.

Reiter, R. (1987). A theory of diagnosis from first principles. Artificial Intelligence., 32, 57-95.

(1984). Review of the Seabrook units 1 and 2 auxiliary feedwater system reliability analysis. NUREG-3531.

Copyright © Controller HMSO London 1990. IFAC Symposium on Safety of Computer Control Systems. Published by Pergamon Press, Oxford, 1990

CASE STUDIES IN INDUSTRY PRACTICE

MANAGEMENT OF COMPUTER-AIDED CONTROL SYSTEM DESIGN FROM CONCEPT TO FLIGHT TEST

B. N. Tomlinson, G. D. Padfield and P. R. Smith

Flight Management Department, Royal Aerospace Establishment, Bedford, UK

Abstract. Active control technology (ACT) has brought a dramatic change to the way aircraft can be designed and flown. The challenge for flight control research is, given the potential of ACT, to define what is required. The problem is that to answer this question a 'flyable' implementation is needed, whether for a piloted flight simulator or for full-scale flight. The need for an implementation introduces issues of software design and management and a possible 'conflict' with the needs of research. This paper describes an environment for flight control law research being developed in Flight Management Department at RAE Bedford, to provide a rigorous yet flexible framework. A comprehensive life cycle is defined for the evolution of flight control laws from concept via piloted simulation to flight test. In its current form, the life cycle has four major phases: conceptual design, engineering design, flight clearance and flight test. Conceptual design covers off-line simulation - modelling, design and analysis - and some pilot-in-the-loop simulation using RAE's Advanced Flight Simulator. The outputs from this stage include information on the response types and system characteristics required. Engineering design is the process of full control law design and includes further refinement of control system architectures via modelling and more detailed piloted simulation. Flight clearance consolidates results from earlier stages and achieves a verified implementation for the target flight control computer. Flight test evaluates the control system in full scale flight and appropriate operational conditions, with further comprehensive data collection. A structured description of all these phases is presented, with emphasis on the conceptual phase; some examples of the control law life cycle are given with particular application to helicopter flight control. The verification and validation issues associated with the attendant software development are also addressed.

Keywords. Control system design; Flight control; Active Control Technology; Helicopter control.

INTRODUCTION

RAE Bedford has been conducting flight control research for many years and has created a number of special facilities for this purpose, both on the ground and in the air. The objectives of RAE control law research are to define requirements - what modes of flight control are feasible and desirable - and how they should be expressed, and to create design tools and a design environment to enable concepts to be explored. A current activity is aimed at establishing a comprehensive life cycle definition for the evolution of flight control laws from initial concept via piloted simulation to flight test, using computer-based aids and tools as much as possible.

Two flight test vehicles serve as the focus for flight control research at RAE: the specially adapted two-seat Harrier known as VAAC (Vectored Thrust Aircraft for Advanced Flight Control) (Nicholas, 1987) and the RAE ACT Lynx project (Padfield and Winter, 1985). The latter project, still at the proposal stage, is the main driver for this paper and poses particular challenges as its proposed low level, high speed flight envelope requires that its experimental flight control system be designed to be flight critical.

Significant improvements in helicopter handling qualities are possible through the use of active control technology (Agard, 1984; Buckingham and Padfield, 1984) tailoring the response characteristics to the flight condition. This increased versatility presents a new challenge to the control law designer. Response types that confer Level 1 handling (acceptable with low workload) vary with flight conditions, mission task element (MTE) and useable cue environment (UCE). The UCE (AVSCOM, 1989) is classified as 1, 2 or 3 depending on how difficult it is, with the prevailing visual cues, to control attitude and velocity precisely (1 being the easiest).

Figure 1 illustrates how the required (pitch/roll) response types change according to the currently proposed requirements. In the low speed regime below 45 kn, to achieve Level 1 handling (Cooper and Harper, 1969) as UCE degrades from 1 to 2 and 3, the preferred response type would need to change from rate command to attitude command/attitude hold and finally to translational rate command with position hold. In forward flight, rate command is preferred for all UCEs. Also indicated in Fig. 1 are generic soft and hard flight envelope limits corresponding, for example, to fatigue and static strength loadings. An active control system (ACS) needs to be quite complex to take account of these varied requirements, with automatic mode blending or pilot-operated mode switching together with appropriate limiting at the envelope boundaries.

For a production application, such an ACS would evolve from a requirement specification through design and implementation to flight test and certification. Through this life cycle, design freedom and knowledge typically vary as shown in Fig 2 (Schrage, 1989). The greatest freedom coincides with the least knowledge at the beginning of the process. When the system is ready for flight test, the knowledge has increased but the design freedom is very limited. This interchange of freedom and knowledge is inevitable but it does highlight the need to accumulate as much correct knowledge as possible while the design freedom is high. This requirement is further emphasised by the curve showing typical committed risk and cost during the evolution. Typically, 80% of the commitment is made by the end of engineering design. All the functional details, including errors, frozen at this stage will have considerable impact on risk and cost in future phases. With a sound and valid requirement, good quality design tools and coherent verification and validation (V&V) procedures, the process should deliver a successful control system with a high probability of error-free software.

In a research context, several aspects of the above process need modification. In most cases, the requirement is poorly understood or even becomes the objective of the research itself. The engineering design tools may be immature, because of the novel application and again, development of these (eg vehicle simulation model, control law optimisation method) may be an objective of the research. Implementation and the ensuing verification will, in principle, be similar to the production application although the demand for rapid change at this stage may place more emphasis on tools and automated aids than in the production environment.

Flight tests are the ultimate validation, provided relevant operating conditions can be found, but the need to explore a range of configurations, categorising good and bad features, requires considerable flexibility. To emphasise the need to raise both design freedom and knowledge in order to confer this flexibility, the life cycle has to expand to include an iterative mechanism feeding knowledge back to the requirements capture and design phases. In research, it has been found to be appropriate and productive to introduce a conceptual design phase as a requirement capture activity. The introduction of this phase highlights the need for design during the phase of greatest leverage on the final research results. Introducing conceptual design also acknowledges the iterative nature of research, as illustrated in Fig. 3. Features of this iteration are:

    a. at all stages, the discovery of a fault, design error or uncertainty will require the return to a previous stage.

    b. for safety, changes to control laws made in the flight phase should cover incrementally only those regimes already mapped in ground simulation throughout the previous stages.

    c. the iterative cycle accumulates knowledge and this has to be documented in a consistent and coherent manner.

    d. passage from one phase to another should only be allowed following a satisfactory outcome of procedural tests.

Considering on the one hand the highly interactive nature of these development phases and on the other the safety issues associated with flight critical software, the need for a management support environment is paramount. Important attributes of such an environment are:

    i each development phase should consist of a defined set of activities with distinct documented outputs.

    ii capture and retrieval of design knowledge should be emphasised, together with audit trail data.

    iii clearly defined tests, consistent throughout the phases, are required.

    iv procedures should enhance, not inhibit, creativity.

    v management support and design tools should form an integrated, computer-based environment.

This paper describes the development of such a management support environment at RAE Bedford, in the form of a control law life cycle model. Examples are drawn from the helicopter flight control domain.

OUTLINE OF THE LIFE CYCLE

General

In its current form, the RAE Control Law Life Cycle has four major phases (Fig. 4): conceptual, engineering design, flight clearance and flight test.

The <u>conceptual</u> phase evaluates the basic concept in a form that can encapsulate the operational requirements. It includes off-line simulation - modelling, design and analysis using control system design and analysis packages such as TSIM (Winter, 1983) - and some pilot-in-the-loop simulation using RAE's Advanced Flight Simulator. The outputs from this stage include information on the response types and system characteristics required.

<u>Engineering design</u> is the process of full control law design in conjunction with a representative vehicle model, and includes further refinement of control system architectures via detailed modelling and extensive piloted simulation.

<u>Flight clearance</u> consolidates results from earlier stages and achieves a verified implementation for the target flight control computer. Validation of the design to include a loads and stability analysis will also form part of this phase.

<u>Flight test</u> evaluates the control system in full scale flight and appropriate operational conditions, with further comprehensive data collection.

Each phase is described with the aid of structure diagrams, as used in the Jackson JSP and JSD methods for software and system design (Cameron, 1983). The diagrams are complemented by a process dictionary containing text describing the sub-processes and activities.

Formalised structure diagrams of this nature possess many benefits

    - they clarify the overall process;

    - they identify the sequence of possible and necessary activities;

- they require the basis of decisions (whether of selection (o) or iteration (*) - Fig. 4) to be defined;

- they provide a helpful means of communication and a focus for discussion among team members.

A computer-based tool is available to generate them (MJSL, 1987).

The whole life cycle model and method is currently in a prototype stage at RAE. Activities in the first two design phases are now discussed briefly, together with aspects of the design knowledge that needs to be captured. The structure of activities within the clearance and flight test phases are still evolving and will be reported on at a later date.

Conceptual Phase

As noted above, the emphasis in the conceptual phase is to establish the design requirements and criteria for the engineering design phase. This phase is creative, the principal stages being problem expression, design (consisting of modelling and evaluation) and review; the full process structure is illustrated in Fig. 5, emphasising the specific activities of each sub-phase, including documentation.

Within problem expression, a simple but non-trivial sub-phase, is a choice of activities that have considerable influence on the value of the research. Problem expression can take the form of text and diagrams; the entry in the process dictionary may look like

> 1.1.2 Express: create new expression of the problem - using text and diagrams express the high level statement of the problem in sufficient detail to initiate the conceptual phase.

An example might be:

> design a full authority active control system for a Lynx helicopter to achieve Level 1 handling qualities in air combat.

Necessary subsidiary problem expressions would be (AVSCOM, 1989):

> a. Determine location of (pitch/roll/yaw) Level 1/Level 2 handling boundary on the bandwidth/time delay diagram for rate/response type rotorcraft in tracking phase of air-to-air combat MTE.

> b. Determine location of Level 1/Level 2 handling boundaries on the attack-parameter diagram for acquisition phase of air-to-air combat MTE.

> c. Determine minimum level of various pitch/roll/yaw cross-couplings necessary to guarantee Level 1 in air-to-air combat MTE.

Design criteria can be expressed in various ways. In the frequency domain, bandwidth and phase delay metrics are defined by AVSCOM (1989) indicating the nature of the system response. These two metrics can be plotted against one another as shown in Fig. 6, which identifies boundaries between the three handling qualities levels. Hence, at the design stage, a control law can be tuned for good closed loop characteristics. Desirable locations for the system roots are also specified in the 's' plane for different vehicle axes and tasks. To cater for large amplitude motions, time domain criteria are specified as shown in Fig. 7. Ratios of peak angular rate over peak attitude change (the attack-parameter) are used to characterise handling qualities levels in the time domain.

Eventually this part of the research is complete when the criteria are validated and their range of application established but, to begin with, many gaps may exist. For example, whereas the basic format for (a) and (b), as illustrated in Figs 6 and 7, may be established, that for cross-coupling may not be and another level of problem statement needs to appear. In the conceptual phase, problem expression is the key to 'starting on the right track', but it is the modelling and evaluation sub-phases that give substance to the overall phase and generate new knowledge. Activity 1.2.1 may be described in the process dictionary as

> 1.2.1 modify existing model: an appropriate conceptual model for developing the required criteria already exists, so select and modify as required.

The modelling requirement must clearly be traceable to the problem expressions. For example, expression (a) above will require models that allow bandwidth and time delay effects to be evaluated independently and in isolation from other interferences. Simulation activities occur in the second of the design sub-phases, as part of evaluation, and are composed of non-real-time TSIM and real-time piloted simulation activities. The piloted experiments must be designed to the same level of detail as a more comprehensive engineering simulation. In principle, the tests here should be identical to those conducted later, in Phase 2, in terms of tasks, UCE, pilots, etc. Full documentation is crucial at this stage; the dictionary entry might be as follows:

> 1.2.9 document results of piloted evaluation - should contain a complete description of tasks, simulation environment (eg cues) plus supporting validation documentation, pilot details, together with the simulation results, eg pilot comments plus ratings, results of data analysis presented in format established in Phase 1.1.

A key feature of the conceptual design phase is the iteration, allowing several passes through the modelling-evaluation sequence if required, to derive the required knowledge. This may, for example, be necessary to establish a suitable format for pitch/roll or roll/pitch cross-coupling criteria.

Most recently a good example of the success of conceptual simulation has been the exploration (Massey and Wells, 1988) of the potential of helicopter carefree handling systems in the Bedford Advanced Flight Simulator. Several configurations were evaluated, including visual warnings, tactile warnings via collective and pedal shakers, and direct intervention systems. Dramatic improvements were observed (Fig. 8) particularly with the highest level of augmentation. These experiments demonstrated the power of direct intervention control and were only possible because it was relatively straightforward to add carefree handling features to the conceptual model. Devising such techniques for a full model of a helicopter and its advanced control system is currently in progress but is expected to take several man-years of effort. Robust and realisable requirements have, however, been established.

A review sub-phase is included in the conceptual phase, as in all four major phases, acknowledging the need to make a decision at each point as to whether the results are satisfactory, and sufficiently promising to proceed further, and if so, to deliver a specification for the next phase.

Engineering Design

As in the conceptual phase, problem expression, design and review, cover activities in the engineering design phase - Fig. 9; however, the level of detail will generally be considerably greater and elapsed activity times considerably longer. The problem expression sub-phase takes as a starting point the specification output from the Conceptual phase, representing, in part, the design criteria for the control system.

Greater detail will be required, however, to reflect the depth to which the problem is tackled in this phase. Environmental constraints and robustness criteria will form part of the expression, as will requirements on uncontrolled modes eg structural. Internal control system loop performance requirements may also be defined in terms of gain, admissible interactions, structure etc. The design sub-phase contains more substantial activities within modelling, synthesis and evaluation. Fig. 9 is expanded as far as the leaves only for the synthesis activities: these include method selection, control law structure and parameter optimisation and documentation. Control law design method selection is emphasized as an activity; the approach taken here, whether time or frequency domain, classical single-input/single-output (SISO) or multiple-input/multiple-output (MIMO), will depend on a number of factors. Experience of the engineer is important but a method that matches the way the problem is expressed will always have clear advantages. The optimisation activity involves craft-like skills, trading off performance against robustness, to achieve the best controller. On-line documentation during the activity is crucial to avoid the perpetration of the 'black-box syndrome', ie a unit whose internal functions are not known in detail. A working practice that emphasises rational choice and decision making and the associated recording is favoured.

The design method highlighted in this paper is that of H-Infinity, which enables the designer, as part of the controller synthesis phase (Fig. 9), to specify frequency-dependent weighting terms that characterise performance and robustness requirements. The detail of the method will not be elaborated here. Yue (1988, 1989) describes the use of this design method in a helicopter application, which resulted in a successful piloted simulation trial on the RAE Advanced Flight Simulator.

As an example, Fig. 10 shows the locus of phase delay versus system bandwidth in the design scale that can be obtained from use of the method, by incrementally changing the required bandwidth design parameters. The handling qualities at each point can then be evaluated in piloted simulation to check compliance with the criteria.

Control design methods are usually based on linear systems theory, hence the maximum performance that can ultimately be obtained depends upon the extent to which non-linear effects cause the closed-loop repsonse to be degraded. This can arise from non-linearity either in the vehicle model employed in the simulator or in the actual aircraft aerodynamics experienced in real flight, as well as in the actuators and sensors. In particular the actuators are prone to authority, rate and acceleration limits which must be accommodated.

Introduction of the controller into the more realistic non-linear model may therefore introduce difficulties in meeting the required performance, which will need to be resolved. It may be that the initial specification was too optimistic, or poses unacceptable demands on the aircraft actuation, engine, or structure. If so, this would require an iteration back to the conceptual phase to quantify the impact of a reduction in system performance.

The engineering design phase is completed when 'acceptable' closed-loop performance is obtained from the combination of control law and vehicle model. The output of this phase is a definition of a non-linear control law, in the form of a set of state-space matrices and block diagrams that specify the controllers, to enable the required performance to be achieved over the full flight envelope of the aircraft. Discrete switching between the controllers evolved for each design point may be necessary, using so-called 'bumpless transfer methods'.

Clearance and Flight Test Phases

Activities within these later phases are the subject of current research at RAE. The clearance activities will include software verification and a degree of validation using more comprehensive vehicle mathematical models than in earlier phases. Flight test represents the ultimate research evaluation although, ironically, this phase offers no scope for design innovation and creativity; flight test is essentially a knowledge gathering exercise but there is considerable scope for innovation in experimental design and interpretation of results. Such activities will be emphasized in this phase as the experiment design will, above all else, determine the success of the research and utility of results. The iterative nature of the whole life cycle (Fig. 3) is again emphasized. Most concepts are expected to have several iterations before yielding mature knowledge, fit for use in a procurement requirement specification or in definitive handling criteria.

TOOLS AND FACILITIES FOR THE LIFE CYCLE

Tools and Facilities

In progressing through the phases of the control law life cycle, a control law designer will employ a number of tools and facilities. Tools are generally computer-based software packages. Facilities are major resources such as aircraft and flight simulators. Those that exist within the RAE environment at present are described briefly here.

During the conceptual phase, a modelling and prototyping tool, such as the TSIM (Winter, 1983) or MATLAB (Moler, 1986) software packages, will be employed to express and synthesize the initial concept and provide a first level of analysis. TSIM, for example, enables a model to be defined and analysed by classical frequency response and root locus methods. Time response behaviour can also be generated and examined, with and without disturbances such as turbulence.

A fully engineered control law model can also be analysed in TSIM, in conjunction with an aircraft mathematical model, using the same software implementation that is used in the simulator. Portable control law and aircraft models are created to work both under TSIM (on VAX/VMS computers) and on the AFS (on Encore Concept-32 computers).

Software design is aided by tools which support the Jackson design method (JSD), eg Speedbuilder and Program Development Facility (PDF). The latter has the ability to generate compilable Fortran, Coral66 or Ada source from the design description expressed as process structure diagrams. Static analysis tools are also available.

Control of developed software is also an important task, to ensure that what is used is approved, and to provide a mechanism to manage change. The specific tool in use for this purpose is Lifespan (Yard, 1984).

The major facilities are the Advanced Flight Simulator (AFS), and, for flight test, the research aircraft - VAAC Harrier and Lynx as noted in the Introduction.

The progress of all simulations using the AFS is monitored using software (Tomlinson, 1988) built on a database management package (Ingres). This enables definition of what is flown to be closely controlled, and the results of experiments to be captured, so that, after a sortie, the user can identify, through an automated sortie journal, all the main actions and the precise circumstances in which data were gathered. Data acquired from flight tests is also collated and managed by an Ingres-based system, but using a simpler approach (Foster, 1987).

Computer Implementation

The control law life cycle outlined in this paper generates in each phase information and knowledge about the control law design and its implementation that need to be retained. A potential problem is that the total procedure could involve a new and significant overhead; in practice, the procedure must be a help to the designer. The computer can help here by automating the procedure and by engaging the designer in a dialogue at the end of each phase to ensure that all design knowledge is captured properly. Design information accumulates progressively: the state of the design, with descriptions at various levels of detail, both functional and structural; goals and goal structure; design decisions and the rationale for the design, eg related to sensor availability, choice of feedbacks, assumed motivator authority; and results of analysis. Initially, this would be assembled in a database as an information source. A second phase could involve creation of a 'design associate' to provide another level of interpretation or transformation of the knowledge. Such an associate, on which research is in progress elsewhere (Myers and McRuer, 1988), could, for example observe that the designer has varied a parameter during the design activity and selected a final value, and could ask why. A design associate is relevant not only to the design process but also to future 're-design' within the system, in real-time (Boulton, 1988).

CONCLUSIONS

Flight testing controllers for enhancing flight performance provides the ultimate test of the viability of an ACT system. In flight critical conditions, control laws have to be correct with a very low risk of failure or of the occurrence of a design fault. This emphasises the need to establish a coherent and consistent requirements capture and design cycle prior to test. The evolution of the control law through these early stages is likely to be considerable. In a research environment, when many ideas are developing in parallel, a disciplined working practice needs to be established. This paper has covered the topic from the perspective of research at RAE and has outlined the prototype of a suitable life cycle model. A number of observations can be made in conclusion:

1 Given the 'freedom' of a software-based control law, the challenge today is to define the requirement. Research is needed to identify -

what is required - ie the character of the response;

how the requirements might be achieved ie effective methods to design and implement the requirements;

how well the requirements are actually achieved in an implementation.

The last of these constitutes validation of the initial concept and is the aim of taking a control law through to flight test.

2 Research into flight control laws poses particular problems because the emphasis is not on the technology of the hardware implementation or even on the software but more on the concepts being evaluated. For the concepts to reach the stage of being suitable for evaluation through flight test, an implementation still has to be achieved

to high standards. Control law design in a research environment thus requires a more rigorous approach than used in the past, and demands a coherent and consistent requirements capture and design cycle. How to impose such discipline without inhibiting creativity is the management challenge RAE is attempting to meet.

3 A control law life cycle has been defined for research at RAE consisting of four main phases:

Conceptual
Engineering design
Flight clearance
Flight test

4 The conceptual phase includes conceptual simulation. This kind of simulation has an important part to play in the identification of appropriate response types and design criteria for control laws. In effect, it delivers the requirement and should be performed before any detailed design begins. This is a special feature of the RAE approach.

5 Requirement capture and expression should be separated from any consideration of representation in software but requirement definition and conceptual design are interactive.

6 Throughout the control law life cycle, capture of design knowledge is vital, not just what but why. Some control law concepts will be abandoned during the overall process. In a research environment, it is important to record the details, otherwise, the same path may be followed again at a

later date. Furthermore, unsuccessful systems are also a contribution to knowledge and may identify the need for further research.

7 Concepts that complete the life cycle through to flight evaluation will only do so on the basis of 'audit trail' information that keeps track of what decisions were taken, why and by whom.

8 Computer-based methods to capture this design knowledge and 'audit' information are essential, to reduce overheads and encourage designers to record this information.

9 Tools used on computers, (eg design packages) need to be implemented with 'quality' in order to be trusted; their implicit methods and algorithms must be visible and their range of validity must be declared. This does not do away with the condition that tools still need intelligent users.

The life cycle model under development at RAE utilises features of the Jackson software design method to formalise the multitude of activities and sub-phases within the process of control law research from concept to flight test. This paper has highlighted the conceptual and engineering design phases within the model and has provided examples of how previous research activities fit into the proposed structure. Work is continuing to detail the clearance and flight test phases in readiness for ACT flight research in flight-critical helicopter applications.

REFERENCES

AGARD (1984). FMP Symposium on 'Active Control Systems - Review, Evaluation and Projections'. Toronto AGARD CP 384.

AVSCOM (1989). Handling Qualities Requirements for Military Rotorcraft. United States Army Aviation Systems Command, Aeronautical Design Standard, ADS-33C.

Boulton, C.B., M.E. Bennett, and J.N. Clare (1988). A study of the feasibility of using artifical intelligence in aircraft flight control systems. Cambridge Consultants Ltd.

Buckingham, S.L., and G.D. Padfield (1986). Piloted simulations to explore helicopter advanced control systems. RAE TR 86022.

Butz, B.P. (1987). An autonomous associate to aid in control system design. IEEE Montech 87 Conference.

Cameron, J.R. (1983). JSP & JSD: The Jackson approach to software development. IEEE Computer Society Press.

Cooper, G.E., and R.P. Harper (1969). The Use of pilot ratings in the evaluation of aircraft handling qualities. NASA TN-D-5153, AGARD Report 567.

Foster G.W. (1987). A system for analysing flight trials data. RAE TM FS(B) 672.

Massey, C.P., and P. Wells (1988). Helicopter carefree handling systems. Proceedings of Royal Aeronautical Society Conference on Helicopter Handling Qualities and Control. London.

MJSL. (1987). Program Development Facility. Michael Jackson Systems Ltd.

Moler, C., et al (1986). PRO-MATLAB Users' Guide. The Maths Works, Inc.

Myers, T.T., and D.T. McRuer (1988). Advanced piloted aircraft flight control system design methodology - the FCX flight control expert system. Systems Technology Inc.

Nicholas, O.P. (1987). The VAAC VSTOL flight control research project. SAE Paper 872331 Proc International Powered Lift Conference P-203, Santa Clara, CA.

Padfield, G.D., and J.S. Winter (1985). Proposed programme of ACT research on the RAE Bedford Lynx. RAE TM FS(B) 599.

Schrage, D.P. (1989). The impact of total quality management (TQM) and concurrent engineering on the aircraft design process. AHS Conference on Vertical Lift Technology, San Francisco.

Tomlinson B.N., R. Bradley, and C. Flower (1988). The use of a relational database in the management and operation of a research flight simulator. AIAA Paper 87-2573 CP (1987) Also RAE TM FM 1.

Winter, J.S., M.J. Corbin, and L.M. Murphy (1983). Description of TSIM2: a software package for computer aided design of flight control systems. RAE TR 83007.

YARD (1984). Lifespan - YARD Software Systems Ltd.

Yue, A. (1988). H-Infinity Design and the Improvement of Helicopter Handling Qualities. PhD Thesis, Oxford University.

Yue, A., I. Postlewaite, and G.D. Padfield (1989). H-Infinity Design and the Improvement of Helicopter Handling Qualities. 13th European Rotorcraft Forum, Sept 8-11 1987; Vertica, Vol 13, No 2, pp 119-132.

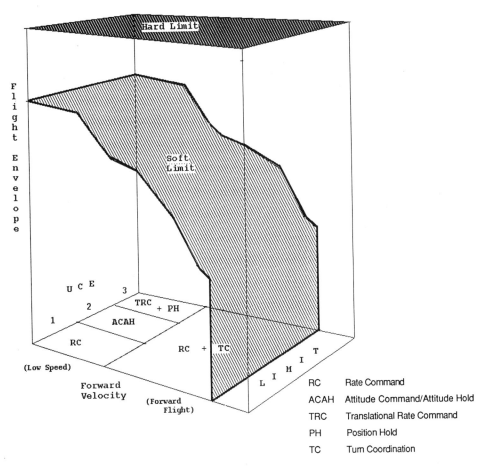

Fig 1 Rotorcraft (Roll/Pitch) Response Types for Level 1 Handling in Specialised Mission-Task-Elements eg Sidestep, Slalom, Air Combat

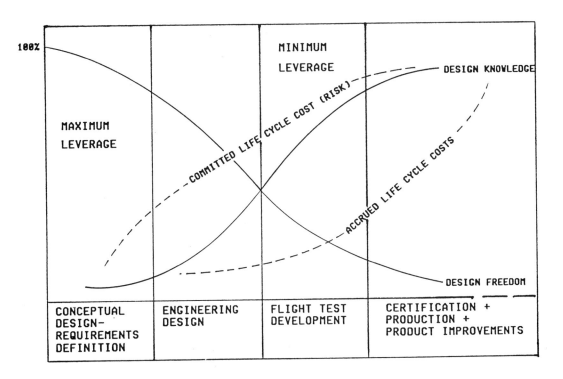

Fig 2 Trends During Procurement Cycle

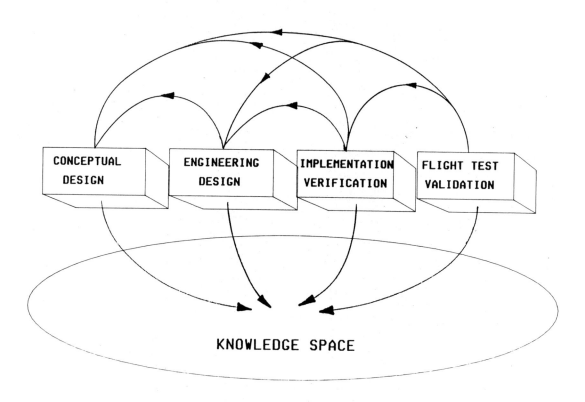

Fig 3 Sources of Knowledge

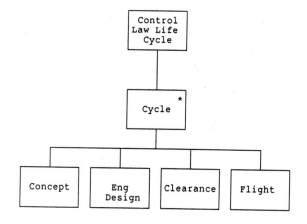

Fig 4 Control Law Life Cycle

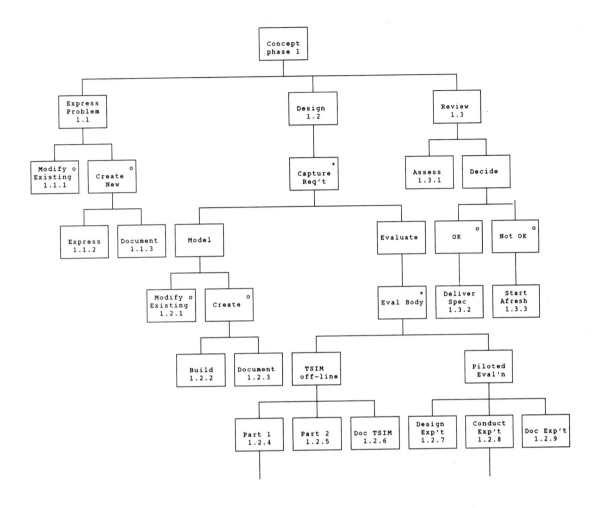

Fig 5 Conceptual Phase Structure Diagram

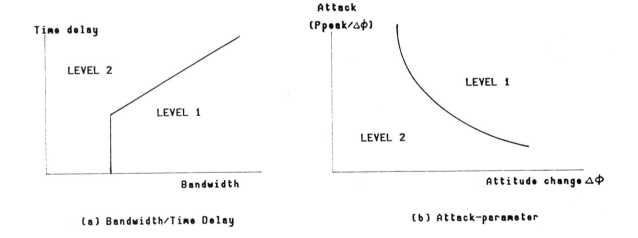

Fig 6 Bandwidth/Time Delay Handling Qualities Diagram

Fig 7 Attack-Parameter Diagram

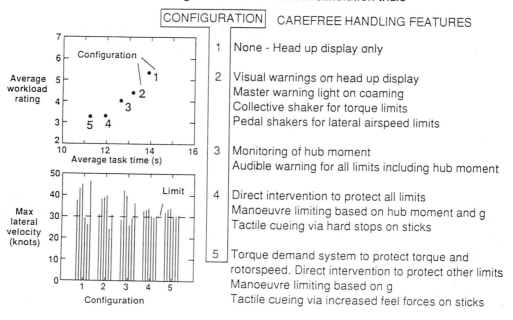

Fig 8 Sidestep Task Results from Carefree Handling Simulation

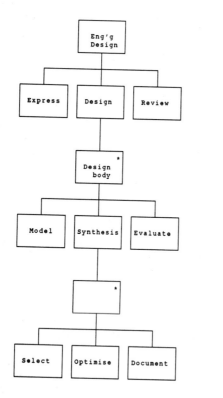

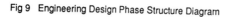

Fig 9 Engineering Design Phase Structure Diagram

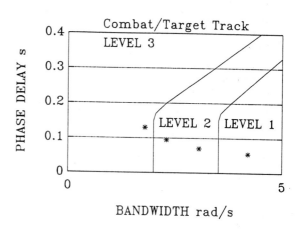

Fig 10 H-Infinity Design Results for a Lynx Helicopter

THE HISTORY AND DEVELOPMENT OF COMPUTER BASED SAFETY SYSTEMS FOR OFFSHORE OIL AND GAS PRODUCTION PLATFORMS FROM THE SIXTIES TO THE PRESENT DAY

C. J. Goring

August Systems Limited, 1-5 Kelvin Way, Crawley, West Sussex, UK

1. INTRODUCTION

To understand how computer based safety systems have developed over the past few decades and how they have been applied, often in a leading role, in the Offshore environment of Oil and Gas Production Platforms in the North Sea, we must first provide an understanding of types of requirements of Safety and Protection Systems needed for these applications.

In this paper I have attempted to give the reader a general understanding of these requirements before proceeding to describe how these requirements have been met during this interesting and exciting period.

Safety Systems existed well before the development of microprocessor and, of course, these older technologies were applied with good engineering practice and safety concepts to the early offshore facilities and when the application is relevant, continue to be applied. To show how, why and where microprocessor based systems have superseded these technologies, I have described their applications and have shown where their application is relevant today.

The growing awareness of safety concepts throughout the industry has encouraged the production of a number of informative guidelines, leading standards embodying good engineering principles and safety concepts. In the UK the Health and Safety Executive's (HSE) Programmable Electronics Systems (PES) guidelines assisted the industry to form a correct understanding of how programmable systems should be applied in safety application. In todays applications some of the examples that this paper provides, show how these guidelines have been interpreted in the offshore environment.

2. **SAFETY RELATED SYSTEMS FOR OFFSHORE PLATFORMS**

A large offshore Oil and Gas Production Platform has to provide all of the facilities of a typical land based site production facility, all compacted together in what could be termed a relative pin-head. The production facilities range from drilling and wellhead control through to oil, gas and water separation and where necessary, gas compression and re-injection or export facilities.

As a self contained site in an hostile environment all of the following utilities must also be provided, these include power, environmental protection, heating and ventilation, personnel accommodation and safety protection systems.

When we analysis the systems needed to provide these functions, the requirement for a number of primary safety systems becomes evident.

2.1 Process Plant Protection

The production and processing areas of the platform use, as their "feed stock" the highly combustible hydrocarbons in both liquid and gaseous form. The platform handles large volumes of these combustible products, often under high pressures. The production and process plant is monitored for process parameters being outside their normal operational range and if these parameters are exceeded, the relevant part or the whole of the process and production equipment is shutdown.

This plant shutdown philosophy is termed the process shutdown, the system providing this type of protection may also cover, by the use of additional inputs, the Emergency Shutdown of the production and processing plant. In the shutdown and therefore, safe state, a number of interlocks are provided preventing the automatic re-starting of the plant.

There will be a number of defined levels for Emergency Shutdown, the highest priority of which will be configured in a belt and braces manner, to ensure shutdown. Typically, as well as commanding the system to shutdown the plant, the power to the shutdown solenoids will be removed, guaranteeing shutdown.

2.2 Personnel Protection

Working in such a confined, but potentiality hazardous environment, great store is set on providing the best personnel protection systems that are commercially available. In the terms of computer based safety systems, the Fire and Gas Detection and Protection System fulfils a primary role.

The Fire and Gas Systems utilises field mounted fire detectors and combustible gas detectors and if required, additionally toxic gas detection is also provided. The information from these detectors is used by the system to determine whether a hazard exists in each operational module +location.

If it is determined that a hazard exists, a number of possible actions are taken by the Fire and Gas Systems, depending on where and what the determined hazard is. For instance a command may be sent to shutdown all or part of the production process, alarms will be raised, commands will be sent to close or open appropriate dampers and to switch on or off appropriate fans. In the case of confirmed fire, deluge or halon release may be operated with the appropriate warning time delays.

2.3 System Interaction

As it can be seen from the above sections it is often appropriate for the Fire and Gas Systems to send commands to the ESD system to shutdown appropriate plant areas. These commands are normally hardwired between the systems. Additionally, alarm management information systems, often based on colour graphic display system, need information updates to occur. The CRT interface may be dedicated to the safety system or alternatively may be part of a separate system, such as the process control system. To ensure the correct operation and information availability, hardwired mimic matrix panels would also be provided ensuring that the safety officers would at all times be fully aware of the status of the platform.

3. ENVIRONMENTAL CONSIDERATION

The offshore environment associated with the North Sea is potentially relatively a severe environment. This is certainly the case for the parts of the safety systems located in areas such as the drilling module.

The main safety system panels would however, normally be mounted in 'safe areas' with a controlled environment, although in the worst case the possibility of environmental control system failures will have to be taken into account when specifying the safety system. The mechanical construction of the equipment would be considered rugged with an ability to survive typically up to 20G shocks in any plane and to be continuously subjected to vibration levels of up to 1G.
The equipment temperature range under worst case would fall into the -10 C to 50C range with the possibility of up to 100% humidity being present for short periods.

Salt spray protection would only be considered for areas such as the drilling modules and deck mounted equipment, these areas often require panels sealed to IP65/66, manufactured out of stainless steel.

In certain overseas environments where humidity and temperature play a significant role, most PCB assemblies would need to be covered with a protective anti-fungal coating.

4. OPERATIONAL AND MAINTENANCE PROCEDURES

The certification by the regulatory authority of the safety systems enables production to start, all maintenance actions being under strict control.

In principle, a permit to work systems is operated with senior management personnel providing the logged documentation that enables the maintenance staff to work on particular areas of the equipment.

If for instance, a major maintenance operation requires the loss of fire detection equipment in an area then fire watches would need to be on duty for 24 hour coverage.

Maintenance actions cover three main categories:-

a) Preventive Maintenance
b) Corrective Maintenance
c) Perfective and/or adaptive maintenance

The safety system design should minimise the need for a), make b) easy to achieve without loosing plant safety coverage or production up-time and ensure that if and who c) is required, then changes are simple in their implementation, testable before re-certification and readily documented.

Percentive maintenance would normally be categorised into degrees of influence on the safety of the system. For instance, changing or cleaning cabinet ventilation filters with no required access to the system cabinets would be considered a low risk category, where as calibration of analogue inputs or gas modules inputs a high risk.

The operational procedures ensure the appropriate level of access, supervision and quality inspection of maintenance actions are covered.

Corrective maintenance is normally a higher level of risk then preventive maintenance. It usually requires the on-line changing of active safety elements and therefore, the risk of false trips or false alarm is obviously higher. Again the level of access is strictly controlled by the procedures, additionally the safety system will be engineered in a hierachical manner to minimise the chance of errors.

Adaptive maintenance, due to increased safety requirements or plant changes, requires that strictest level of control. The procedures that cover adaptive maintenance therefore, ensure that changes are analysed for safety implication that changes are first implemented in an off-line test environment, that the changes are independently verified for safety and correct operation prior, to implementation on-line.

Documentation procedures are thorough and all but the most minor of changes will need re-certification through the safety authority.

5. EARLY MICRO SYSTEM

With the advent of microprocessors in the early 1970's, the safety industry began considering how these units could be utilised for safety and critical control. The good engineering practices applied ensured that hardware single points of failure were eliminated or made fail-safe. However, the majority of the early real time programming was completed by hardware engineers, self trained in software and invariably used the microprocessors assembly language.

To compound the limitations of this approach, the tools available were very limited, the successful micro manufacturers of the 1980's being those who provided reasonable, although expensive, tools in the form of microprocessor workstations (eg. Intel/Motorola) in the late 1970's.

A typical example of an interlock system using microprocessor is shown in Figure 1. The advantages over and hardwired interlock systems of the past, included an increased flexibility, limited diagnostics, lower cost and smaller size. The flexibility was limited by the fact that programming was completed in assembly and therefore, only a limited number of personnel were available to make changes.

In retrospect, a number of areas of concern, with respect to safety, have to be reviewed. These mainly occur around the question of maintainability of these systems. The majority of safety risk areas occur during the maintenance cycle and therefore the correctness and ease of maintenance is of paramount importance. The lack of formal methods in the approach to design on the early systems increased significantly the risk, when changes are to be made.

During these early days in the North Sea, safety, although of paramount importance was not as formally controlled as in the present day, therefore, the overall procedures for modifying safety systems produced further potential risk areas.

The man-machine interface provided with these early systems was also limited to LEDs/lamps and switches and could therefore, not provide the sophisticated alarm and maintenance data that is required from present day systems.

6. EARLY TRIPLE REDUNDANT MICRO SYSTEM

Majority voting (2 out of 3) has always been recognised as probably the most cost effective solution to both system correctness of operation and availability. During the early 1980's TMR micro systems were developed to provide these facilities on Fire and Gas Detection and Protection Systems.

These systems were designed using three identical processing and I/O scanning systems, running totally asynchronously with the resultant digital control action being determined from a six element hardware voter. A typical system architecture is shown in Figure 2.

Although this architecture did provide increased system availability, a number of minor problems had to be overcome, both for design and maintenance functions.

6.1 Communication Problems

When information had to be transmitted over communication links to other systems or to the CRT man-machine interfaces, the only practical solution was to allow one of the selected healthy processors to provide all of the communication interfaces. This highlights two problem areas, firstly the selected processor would become significantly slower due to the extra work load, slowing down communication and CRT response time. Secondly when the selected processor failed, providing this could be detected, the changeover mechanisms become a potential single point of failure.

6.2 Maintenance Problems

When an item in one of the system signal paths failed, generally the only mechanism to determine which path had failed was by the result of the majority vote and the resultant discrepancy raised. It then often become a significant maintenance task to find the fault, hence increasing maintenance involvement and total Mean Time To Repair (MTTR).

7. DUAL REDUNDANT PLC

With the advert of the Programmable Logic Controller from a number of large manufacturing organisations, the approach of configuring these units in a redundant manner for safety applications has been utilised in a number of applications over the past few years. An example of a simple ESD system is shown in Figure 3 and as it can be seen, the configuration is similar to Figure 1.

The primary advantages of this configuration results from the use of well proven products with many millions of hours of operation experience. However, the safety configuration must be a choice between either complete fail to safe or availability.

The I/O shown in Figure 3 is for a typical ESD which would be configured fail safe.

The other problem area is once again with respect to maintenance. The PLC's main on-line diagnostics is a self test configuration, which relies upon the PLC proving itself correct with the small but inherent dangers of a missdiagnosis.

Additionally I/O diagnostics will be limited to comparisons in the on-line mode with extensive off-line test equipment needed for full coverage.

8. SAFETY PLUS AVAILABILITY

During the early 1980's, as a result of fly by wire research in the 1970's, Triple Modular Redundant (TMR) PLC's utilising Software Implemented Fault Tolerance (SIFT) were developed. The design of the SIFT Kernel was formally validated as part of the original research project and the units now have many millions of operational hours in safety critical applications.

The typical architecture for these PLC's is shown in Figure 4. The applications range from Subsea Control, Nuclear Safety and Complete Oil and Gas Platform Safety Systems. One of the primary advantages of the TMR/SIFT architecture is provided by the continuous voting and validation of data which ensures that not only are faults detected without delay, they are also pin pointed to the exact maintenance location.

With the correct design allowing on-line replacement of all major modules, system reliability and hence, availability can be provided to the highest level.

8.1 Typical Application Configurement of TMR SIFT

The example shown by Figure 5 is for the complete safety system for a major North Sea Offshore Platform. The system provides cover for both the Emergency Shutdown and Fire and Gas Detection and Protection Systems, providing system availability in excess of 99.99%.

9. CONCLUSION

There is little doubt that safety systems are getting both safer and providing the user with high availability. National and International guidelines and specifications are now readily available, allowing the professional engineer with experience, flexible solutions to meet safety requirements.

Typical Hazard Studies and Fault Schedules developed from Fault Tree analysis are now capable of determining the critical application areas of a safety systems. National standards provided by organisations such as the NII and the TUV's of Germany, provide risk categorisation to allow fault schedules and Hazops to be interpreted and the level of redundancy and diversity to be determined. I believe therefore, we can look forward to a continuing improvement in the safety record of a major industrial operation.

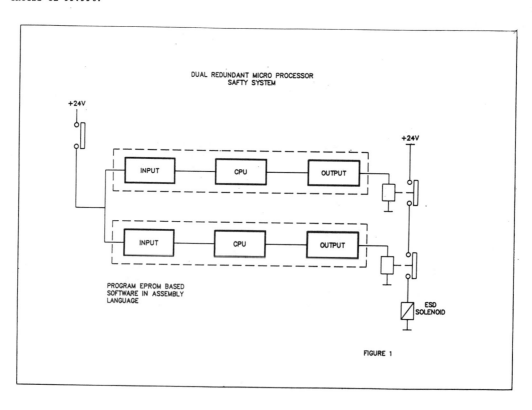

FIGURE 1

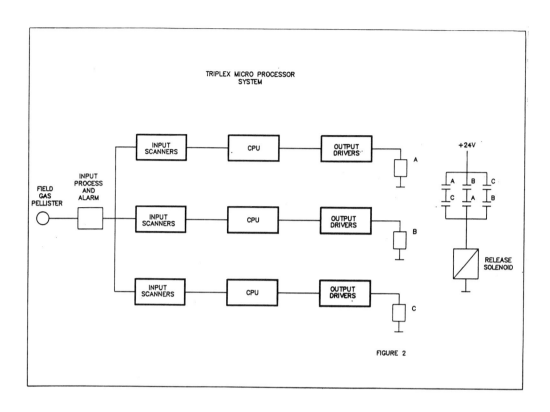

FIGURE 2

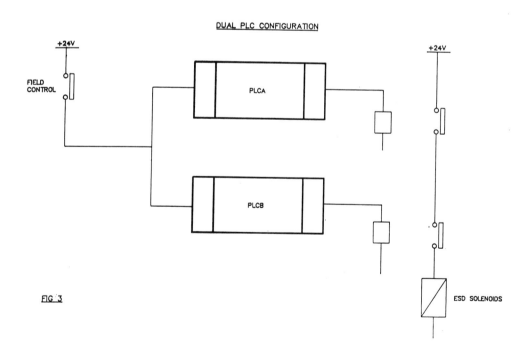

FIG 3

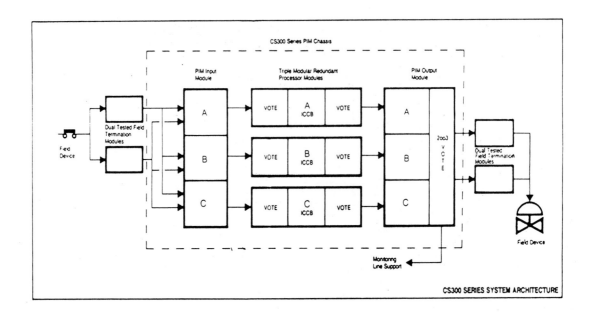

Figure 4

Figure 5

CONTROLLING SOFTWARE PRODUCTION, FROM A CUSTOMER POINT OF VIEW

F. Ficheux*, Y. Mayadoux** and C. Pain***

*Electricité de France, Direction des Etudes et Recherches
**Service Informatique et mathématiques Appliquées
1, avenue du Général de Gaulle, 92141, Clamart Cedex, France
***Service Ensembles de Production, 6, quai Watier, 78401 Chatou Cedex, France

Abstract. EDF/DER, as a customer, must control several development teams on many software projects. This control is done on three directions : management, tools, effective production of software.
We divide development in two agreements, one addressing the specifications, the second the production of software.
The software workshop must be realistic and based on already known products, to avoid risks during development.
The results of code analyzers give good indications of the quality level achieved, but can not be used as formal metrics.

Keywords. Software development. Maintenance Engineering. Quality control.

A CONTEXT TO PRODUCE AND TO CONTROL SOFTWARE.

The Research and Development Division of EDF (EDF/DER) is notably in charge to deliver or to control software products to the other Directorates. This may be either scientific codes, ie numerical calculus, or software to control systems, ie software connected to hardware equipments. In any case, EDF/DER is faced with an increasing amount of software that are subcontracted for their production. The duration of development as well as the long time of use of these products (over ten years) implies that EDF/DER must take into account not only the development aspects but also the maintenance ones from the beginning of the software life cycle, in order to restrain costs after the delivery of products.

Inside this organization, the Software Quality Group is in charge to promote methodologies, to provide advice and to control when requested.

THE STATE OF THE INDUSTRY.

An ideal answer to the quality problem is to be able to apply the Software Development Plan and Software Quality Plan that may be established and standardized, like the ISO 9000 series, or the French AFNOR and AFCIQ Plans. It provides frameworks to set up quality systems using well structured methods
We are faced with two kinds of situations. EDF/DER may be its own producers, and then must control itself, often when new areas are explored; this is illustrated by the research side. On the opposite, EDF/DER may contract agreements with suppliers, and is well illustrated in computer based systems side. We are now

going to explain some of the major difficulties encountered in the two areas then we will propose a strategy.

The Research Side.

The goal of research engineers is to experiment new techniques, and so the first version of a product is often without the quality attributes that may be expected; then, the set up of a new qualified version implies a major modification in the mind of developers. The request of a well planned system is sometimes difficult in a research project, where the objective can not be established in terms of Q.A. attributes, but is determined by the experiment itself. The system looks like a set of prototypes, the results gained from each one having an influence on the next step of developments.
Some of these softwares remain as prototypes and do not need any quality procedure; in other cases, the softwares are requested by industry and then must have all the required quality attributes.

The Computer Based Systems Side.

In computer controlled systems, software is part of a larger problem, and can not be completely defined, while the hardware is not entirely chosen. The managers of this type of projects are scarcely software engineers, and the software aspects must be delegated to specialized teams, increasing the cost of communication and so bringing up new constraints between people having different interests.
To be able to control, the quality team need a well established reference and the access to the project informations.

The EDF/DER Strategy.

The following procedures to improve quality are now adopted for computer based systems.

- A strategy to separate the Requirements from the following phases of a project in two separate agreements; using such a technique, the customer gets a much better view of the specific software aspects of his problem. It is recommended to use two separate agreements, a first one to provide the Specifications from the Requirement, and a second one to develop the software according to the Specifications. A possibility from this two time production is to change of supplier between the two agreements.
This strategy gives different benefits :
    - a better scale in terms of cost and delay;
    - a more detailed second agreement, according to a accurate specification; it must also include a detailed Software Quality Plan and a detailed Software Development Plan.

- The introduction of generic Software Quality and Software Development Plans which provides frameworks to guide the redaction or the evaluation of the supplier plans. A guidance is provided for each section of the plan, and highlights the possible pitfalls; a description of different types of tools existing at each phase of a project allows the customer to verify the veracity of the supplier proposal.

As many project managers are not software engineers, a set of course sessions informs of the specific aspects of software production, and emphasizes the necessary knowledge to understand the process.

BEFORE CONCLUDING AN AGREEMENT.

A good "supplier-customer" relationship is preponderant to achieve the quality of the product, either the specifications or the final delivery. The customer must carefully select his supplier within different directions.

Evaluation of the Supplier Team.

The experience shows that it is extremely difficult to rectify a development process with a "unreliable" supplier team; in such an occasion, the A.Q. rules become ineffective, or inapplicable.
We must note that a particularity of the software area is the important turn over of the staff; it is difficult to be insured of the stability of a supplier team. On small projects, we must be care of some human behaviors, where people tend to be too much confident on one person, sometimes a "hacker" unable to follow formal A.Q. rules.

Evaluation of Methods and Tools.

The development system to be used must satisfy some confidence to the stability and duration of the chosen tools: the cost of maintenance is already affected by these choices. The possibility of immature tools, that would affect the schedule by the delay introduced in waiting for a valuable release of them, implies the greatest care in their selection. In order to allow maintenance, it is also crucial to determine the property of the tools and of their environment, the training of the maintenance team.

The development Plan must be correct in terms of scheduling, manpower allocation.

The connection between the different phases and the tools associated must be verified. It does not mean that an integrated workshop is better than the use of toolkit, it only points out that a definite production of documents is planned.

The configuration management must reflect the completeness of the process. The size of this environment must be connected to the size of the project and to the knowledge of developers.

All these matters addressed, we must remind that all that intentions may become ineffective if the actual developers, who usually do not know this development plan in advance, do not agree with them and are not ready to apply theses rules.

Subcontracting.

It is recommended to avoid any subcontracting activity. Otherwise, the same quality procedures must be applied to the subcontractors. A control of subcontractors must be laid down in the agreement.

Forecast of Controls.

All the quality control procedures must be included in the agreement. We have to mention what documents will be controlled, when it will be done. It is very important not to forget the consequences in terms of delay and cost. Otherwise, it will be dificult to perform any action, either to control or to correct, as no budget and time have been dedicated to the task.

CONTROLLING THE REQUIREMENTS: THE SPECIFICATION PHASE.

In EDF/DER life cycle model, this phase is essential, as its results, the specifications become the references to judge the final delivery.

The following issues must be addressed :

- The specifications must not be merged with some design components.

– A pretty way to insure a good communication between supplier and customer is to share a common tool, to define requirements and specifications. This practice implies that the customer get enough knowledge about the tools. No tool is currently widely agreed within the software community, so it appears difficult to select one of them and to force suppliers to use it.

GOING ON THE PROJECT.

Once the realization phase of the project started, the job of the customer is mainly to verify the actual mapping with the scheduled one. We are going to focus on two points : the tool usage and the analyzers.

Tool Usage.

Our main advice is to survey the effective usage that is made of the tools. Roughly, there is no major problem during the Design and the Coding phase : tools like high level editors are now well established, and easy to use. Situation is very different during the test and integration phases; tools are practically not used. We may sometimes think that the obtained results of the tools are only produced to demonstrate the real usage of the tool (but on small portions of software), justifying the invoice. To answer this problem, we must insure, by agreement, that the results of tools are available to the customer.
This point becomes crucial if the maintenance is going to be done by a customer team, and not the development one. In particular, a description of the regression test system is mandatory to enable the maintenance team to reuse the same materials as during development. Unfortunately, few commercial products are available; so the goal is to set up a system based on an integration of standard tools provided by the test system.

We also must insure that recovery training from the configuration management are done from time to time, to avoid the waste of time in case of a disk failure.

The Analyzers.

Many static and dynamic analyzers are now available, for most of languages. Several aspects must be clearly set to establish the rules.

– The responsibility to achieve the analyze: is the work done by the supplier or the customer?
In the first case, the supplier is associated to the Q.A. process.This method provides good results if the tools are available to the coding team. Unfortunately, the high cost of analyzers forbids such a technique for small projects.
In case of an analyze by the customer, the role of Q.A. may become ineffective, if only late versions are analyzed, at a time where no significant modification can be done. The interest is to be closer to a real situation, where the maintenance team must take into account the (hidden) structure of a software.
On the opposite, the analyzed version should not be a too earlier one, where it is claimed that the next versions would erase all the detected problems, and setting the control out of the scope of a Q.A. validation process.

– The expected results must be clearly defined. The goals may be considered only as highlights and warnings, or may be hold points to achieve. In the first case, the analyze tends to be only a registration time (the production process is then supposed to be concluded), but in the second case, no metrics are today universally agreed. The current usable metrics (Mac Cabe and

Halstead complexity metrics) gives results for each module of a system; it must be care that a not too modular software has been produced, in a way to reach "a perfect unit metrics", to the prejudice of a very complex system, when integrated.

Practically, we are not using derived metrics, but only the basic ones, like the cyclomatic number, the number of operators and operands. This should always be verified against the structure of control graphs, and the mandatory actions. It must be noted that most of these tools are only concerned with the process control, and not with the data usage, like the transmission of parameters between procedures. Anyway, the results from analyzers always gives indications of the quality level of the software, and the acceptance of their usage is also an indicator or an alarm of the confidence that the supplier team may have of itself.

CONCLUSION.

The division of a software production in two contracts, one for the specification and one for the development, allows the customer to have a better understanding and a more accurate anticipation of its software.

The software workshop must be realistic and based on already known products, to avoid risks during development.

The static and dynamic analyzes give indications but the software metrology is still not enough mature to produce laws that can never been trespassed.

THE IMPACT OF SOCIAL FACTORS ON ACCEPTABLE LEVELS OF SAFETY INTEGRITY

I. H. A. Johnston

The Centre for Software Engineering Ltd, Bellwin Drive, Flixborough, Scunthorpe, DN15 8SN, UK

Abstract. The paper emphasises the inherently subjective nature of the concept of safety, whilst accepting the need for some objectively based means of measuring it. The concept of levels of safety is introduced. The combination of risk and consequence as a means of determining required safety integrity as described in IEC 65A(Secretariate)-94I is referred to. Possible means of combining values for risk and consequence to give a meaningful measure of safety integrity requirement are discussed. The potential benefit of the application is introduced as a contributor to the required safety integrity and possible means of incorporating this factor are discussed. A range of social factors which contribute to risk/benefit/consequence and a means of combining these to produce an 'objective' measure of the level of safety integrity which should be required of a system are proposed. This combination draws on the ideas presented in DIN V 19 251.

Keywords. Safety; computer software; computer selection and evaluation; social and behavioural sciences; human factors; risk; consequence; benefit.

INTRODUCTION

Public concern and interest in the extent to which people are safe in the course of their daily lives has been increasing for many years. Early efforts in the United Kingdom provided legislation governing the operation of shipping, mining and automobiles. Some of these efforts have helped provide the basis for levels of safety of which the United Kingdom has been proud. Some, such as the man with the red flag in front of the automobile, have long since been abandoned. In more recent years the growth of consumer organisations and the setting up of the United Kingdom's Health and Safety Executive (HSE) have seen increasing improvements in the degree of safety which the general population expects in all kinds of activity. With the emergence of the green movement major environmental issues are increasingly seen as involving the safety of our planet.

In the world of software engineering and computer control these recent developments have been paralleled by an increasing concern with the safety aspects of automated control systems. This has been paralleled by an increase in the use of such systems in all areas of life. In particular much effort has been devoted to the development of suitable means of ensuring that computer software is safe.

SUBJECTIVITY

There is no clear cut point at which we can say something becomes safe when before it was not. Safety is essentially a subjective concept which depends on current social values and which concerns the level of risk which the population is prepared to tolerate in different situations (HSE, 1988; IGasE, 1989). Many considerations affect the level of risk which is accepted as safe. These include the number of people affected when a hazard occurs, the degree of harm caused to them and the degree to which individuals may, or may not, choose to subject themselves to the risk. The benefits to be derived from the application should also be taken into account.

The conclusion as to whether or not something is safe, whilst being based on the same or similar considerations, may well differ considerably depending on time and geographical location. Indeed different individuals have different viewpoints on, for example, the acceptability or non-acceptability of the risk to safety posed by nuclear power generation. Often such differences may be explained by differences in the individuals knowledge or understanding. Nevertheless, differences of opinion on such matters can and do arise between people whose knowledge and understanding of the matter in hand are at least broadly comparable, if not identical.

Safety cannot therefore be absolute, it is essentially a qualitative rather than a quantitative characteristic, and by its nature cannot be considered to be 100% (HSE, 1988; Bennett, 1984). In practice objectivity of some sort is required and this has been provided by the adoption of standards, frequently backed up by legislation. Such standards usually apply to particular industries or to particular categories, of product or system. They provide a means by which a manufacturer, user or assessor can say "Yes, this product is safe. It meets the requirements laid down by the standard."

In order to be useful standards of this sort must allow such a clear cut decision to be made. Furthermore there must be some rational basis for the implied correlation between "meeting the requirements" and providing a level of safety which will prove acceptable to the public in the context in which the product or system will be used.

In order to achieve such correlations different standards are often applied to systems which might otherwise be regarded as the same, depending on the environment in which they are to be used. Examples of this approach are the differing requirements for electrical appliances intended for use indoors and outdoors or the military and civil versions of electronic components. In a slightly different way the maintenance requirements placed on public transport vehicles reflect the greater public concern with the safety of such vehicles, in comparison with the private car.

SOFTWARE SAFETY

The safety of systems involving software (Programmable Electronic Systems or PESs) has been of growing concern within the computer and control communities. In the last few years this concern has become more widespread and is now shared by many in the more general population.

Software has certain characteristics which make the creation of standards similar to those used for other products rather difficult. Software is used for many different types of application and whilst apparently simple in small doses is exceedingly complex in its operation for most practical applications. There is therefore no easy application criterion which can be applied to software. Furthermore the complex logical structures embedded in software together with its digital nature severely restrict the usefulness of testing.

The nature of software is such that its failures are systematic rather than random, though in practice they may appear to be random. Conventional approaches to defining safety requirements often involve the specification of required figures for availability or failure on demand. There has been much work on the development of models which allow such figures to be calculated for software. Whilst this work is clearly of use the systematic nature of software failures means that from the point of view of safety such figures must be treated with extreme caution. Existing work tends to stress the importance of quality rather than measurement.

Safety Integrity levels have been used (IEC, 1989) in order to fit such qualitative criteria into the regime of different degrees of safety for different types of application. These safety integrity levels are currently identified simply as;

   Very high
   High
   Medium
   Low
   Normal

COMBINATION OF RISK, CONSEQUENCE AND BENEFIT

Safety Integrity is a continuous quantity which is derived from a combination of many different components and which is essentially subjective in nature. These Integrity levels allow a degree of objectivity to be brought to bear provided suitable mappings can be generated from the various factors which contribute to required safety integrity.

The IEC (1989) has proposed a structure for such mappings which I have extended (Johnston, Wood, 1989) and which is shown in Fig. 1.

Risk is defined by the IEC (1989) as;

"The combination of the frequency, or probability, and the consequence of a specified hazardous event".

The means of combination of the frequency, or probability, and the consequence is undefined. The following means of combination could be considered given suitable values for the consequences of hazards;

   Frequency of   )   (Consequence
   occurrence of)   x   (of the
   a hazard   )   (hazard

There are a number of problems with this approach;

Values

Frequency of occurrence and Severity of hazard must be expressed as appropriate numeric values (or by some other means for which a suitable "multiplication operator" is defined). Ideally these values should be scaled in such a way as to provide for an "easy" transformation from Risk Level to Integrity Level.

Effect Of Integrity Level On Frequency

Since the system configuration, amongst other things, affects the reliability of the system we have a potentially circular situation.

The Frequency is used in determining the required Integrity Level which is used to drive the design of the system (software) which affects the Frequency. Clearly this problem is not insuperable, but it must be taken into account.

Benefit

No account is taken of Benefit. Given the approach suggested benefit would be most naturally included as a divisor;

$$\frac{\text{Frequency} \times \text{Severity}}{\text{Benefit}}$$

Benefit could be incorporated, as shown above, directly in Risk; or separately in Integrity as shown in the Fig. 1. Alternatively Benefit may contribute in both ways.

Subjectivity

Frequency of occurrence is a "fairly" straightforward value in some ways, though it may be difficult to justify a specific figure. Severity and Benefit are much less clear cut. They depend on a number of factors some of which are rooted in the opinions and beliefs of society.

POSSIBLE APPROACH

As a starting point I have taken the following approach, which whilst raising further problems does provide a means of reasoning about the combination.

Assume frequency is expressed as a figure within the range [0,1] where;

   1 is "occurs all the time"
   0 is "never occurs".

In other words expressed as something like a probability.

There must be some question as to whether a hazard which occurs all the time is something which should be considered, remember that we are looking at the overall system. Perhaps a complete redesign of the system is appropriate. Equally is a hazard which never occurs really a hazard?

Assume severity is expressed as a value within the range [0,1] where;

    0 is "completely benign"
    1 is "some catastrophic event".

Consideration must be given here to quite what is meant by "catastrophic". If we wish the range [0,1] to encompass all possibilities then 1 must represent a "worst imaginable catastrophe", for example collision between Earth and Sun. From the human point of view this would be the "end of the world" and nothing could be worse. This gives a very wide range. I propose a restriction under which 1 represents "the worst catastrophe which is possible as a result of the failure of a system which we are prepared as a society to countenance the existence of". This approach is discussed in Appendix A.

A straightforward multiplication;

Frequency x Severity

now gives some number to which I believe we can assign a realistic meaning;

S = Severity of hazard $\in [0,1]$
F = Frequency of Occurrence $\in [0,1]$

$S \times F = R$ = Risk $\in [0,1]$

$R = 0 \Rightarrow$ There is no risk.
    $R = 1 \Rightarrow$ There is the highest possible risk.
        (Unless some protection is provided
        the worst imaginable catastrophe
        will occur now)

In real situations we would expect to get results for R such as $R = 1/2$ or $R = 0.132$. Figure 1 depicts a mapping from the risk level (R) to a system integrity level (SI). We could define this mapping (f) as

$f(x)$ =   1 for $x \in [0, 0.2]$,
          2 for $x \in (0.2, 0.4]$,
          3 for $x \in (0.4, 0.6]$,
          4 for $x \in (0.6, 0.8]$,
          5 for $x \in (0.8, 1]$

Such a mapping provides a means of combining Frequency and Severity to derive integrity level. The intervals which the mapping uses to select an integrity level may well need modification, but the method is defined.

THE EFFECT OF CONSIDERING BENEFIT

The means proposed above of combining frequency and severity to derive a risk and thence to map to a system integrity level is certainly attractive. Clearly there are some systems where the risk is too great for any level of safety integrity to be appropriate (unless we postulate a 6th level of;"system shall not be implemented in this form"). A Risk level of 1 on this basis implies a certainty of the hazard occurring and a catastrophic severity. This can only be acceptable if the system is such as to reduce the frequency of occurrence to some very low level, which results in the Risk not in fact being at the level identified.

There is then a boundary level at which Risk maps to the 6th level. Beyond this boundary the system is not, or should not be, implemented.

Any system is implemented in order to achieve some objective. This objective and possibly side effects of the system are the benefits to be had from its implementation. The degree of benefit clearly has an effect on the mapping from Risk to required system integrity level.

A level of safety integrity which is acceptable for a power plant may vary depending on the application for which the power is being generated. Power required for essential services might be considered to constitute a greater benefit than power required for leisure activities.

SOCIAL FACTORS

This modification of safety integrity requirement relates to the mapping between Risk and integrity level as shown in Fig. 1. The mapping is also conditioned by the ways in which society views risk taking. The HSE (1987) points out that it is natural to prefer one kind of hazard to another and cites ethical objections as being amongst the reasons for such differences.

Both the HSE (1987) and the IGasE (1989) discuss the public's perception of risk. In their terms risk is not quite as discussed here, instead it is level of fatalities per annum. The IGasE(1989) does, however, allow that severity of consequence does affect the perception of risk, and mentions risk of loss of gas supply as being of lower consequence than personal injury.

For the purposes of determining required levels of safety integrity a balance must be struck between the, possibly irrational or uninformed, views of the general public and the real risks and benefits as far as they are known.

Work has been done by DIN (1988) on the use of risk-parameter graphs as a means of combining the diverse factors influencing the level of risk. I believe that an approach based on that suggested above is superior to the risk-parameter graph for combining Frequency and Hazard Severity. The risk-parameter graph approach may well, however, be extremely well adapted to the combination of the diverse social factors influencing Benefit and Hazard Severity.

Various factors associated with the social system are identified in Fig. 1. Based on these I suggest a break down into the factors and categories shown in Table 1.

These factors can then be combined using a risk-parameter graph such as that shown for Benefit in Fig. 2. to modify the transformation from Risk Level to integrity level shown in Fig. 1. The risk-parameter graph shown completes all legs for all parameter except perception of benefit. The values of the modifiers are not identified.

In order for this approach to be successful further work is required on;

i)   the choice of appropriate parameters

ii)   the relative importance of the parameters (this may affect the order in which they are applied)

iii) whether any parameters may be excluded from consideration in certain circumstances

iv) the values of the modifiers to be applied and whether they can be applied directly to the Risk Level or whether they should be applied as modifiers to the integrity level

v) the ranges for perception of harm and risk.

CONCLUSION

Concern about safety and particularly the safety of systems which incorporate software is becoming more widespread. Safety is a essentially subjective characteristic which it is not possible to measure directly. The approach often taken is to lay down standards which must be complied with for specific applications in particular circumstances. This approach has served well in the past but is not necessarily suitable for the wide variety of applications and circumstances in which software may be used.

The concept of integrity levels has been developed to enable some form of objectivity to be applied to the safety of software. These integrity levels provide a possible means of taking account of a variety of social factors when deciding on the level of safety required in any particular case. As yet no means of objectively determining the integrity level required in a particular case has been developed.

I suggest a first attempt to define a method of determining the integrity level which I believe shows promise. The method is based on the estimation of Frequency and Hazard Severity in the range [0,1] and the combination of these by multiplication to give a Risk level. The Risk level is then transformed to an Integrity Level using a risk-parameter approach which incorporates provision for socially determined risk parameters of hazard and benefit.

APPENDIX A

WORST CATASTROPHES

Provided that we accept that there is a continuum of severity of hazard then there is a range on that continuum which is meaningful. What that range is depends on one's viewpoint. As individuals we can appreciate each others viewpoint to a greater or lesser extent, but the viewpoints are nevertheless different, in fact a single person's viewpoint may be different at different times.

Driver's Viewpoint

When considering a car driver on the motorway the meaningful range of severity is likely to encompass the range from

car radio breaks down (benign)

to multiple pile-up occurs around vehicle (worst catastrophe).

These days of course one might also consider Jumbo jet crash lands on my bit of motorway or earthquake opens up road in front as alternative candidates for "worst catastrophe".

Environmentalist's Viewpoint

Consider another viewpoint, that of a person participating in a discussion about the environment, the range might be;

reduction in fish stocks (benign)

to no unpolluted fish (worst catastrophe)

or reduced tree life from acid rain (benign)

to Chernobyl style accident makes Europe uninhabitable (worst catastrophe).

Engineer's Viewpoint

Given this variety what viewpoint should be taken by engineers considering the application of Programmable Electronics in safety related systems? If it is accepted that safety includes safety from environmental damage and financial hardship there is a practical range which needs to be considered.

no discernible harm to persons, the environment or any person's standard of living (benign)

to Catastrophic accident in the most damaging circumstances to the most dangerous system currently controlled using programmable electronics or which any society would in the future countenance controlling with programmable electronics (worst catastrophe)

REFERENCES

IEC (1989). *Software for Computers in the Application of Industrial Safety-Related Systems, Proposal for a Standard.* International Electrotechnical Commission, Sub Committee 65A, Working Group 9.

HSE (1988). *The Tolerability of Risk in Nuclear Power Stations.* HMSO, London.

Bennett, P.A. (1984). *The Safety of Industrially Based Controllers Incorporating Software.* PhD Thesis, Open University, Milton Keynes.

HSE (1987). *Programmable Electronic Systems in Safety Related Applications.* HMSO, London.

DIN (1988). *Prestandard DIN V 19 251.* Deutsche Electrotechnische Kommission, Frankfurt.

Table 1 Social factors influencing integrity level

	Category	Range	
Benefit	Number of persons	All ⋮ One individual	N1 N2 N3
	Type of Benefit	Essential Desirable Luxury	T1 T2 T3
	Extent of Benefit	Great ⋮ Little	E1 E2 E3
	Perception of Benefit	???	P?
Hazard	Number of persons	All ⋮ One individual	N1 N2 N3
	Degree of harm	Death Injury Environment Economic	H1 H2 H3 H4
	Perception of harm	???	P?
	Degree of choice	Involuntary Difficult to avoid Entirely voluntary	C1 C2 C3

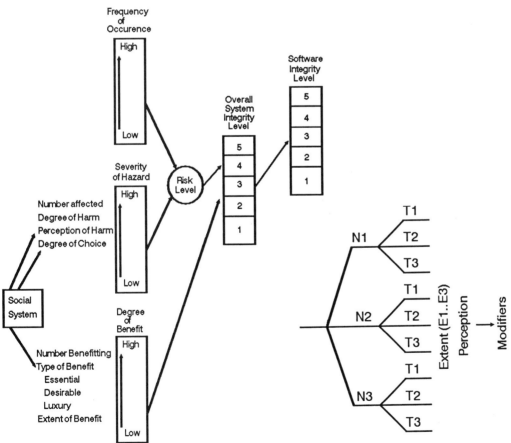

Fig. 1. Risk and integrity level mappings   Fig. 2. Risk-parameter graph deriving modifiers

KEYNOTE ADDRESS

SKILLS AND TECHNOLOGIES FOR THE DEVELOPMENT AND EVALUATION OF SAFETY CRITICAL SYSTEMS

J. A. McDermid

Department of Computer Science, University of York, UK

Abstract. This paper briefly outlines some of the fundamental problems of producing computer based safety critical systems. It then gives the author's views on likely key developments in safety critical systems technologies over the next decade. Finally it addresses the skills needed by developers and evaluators of safety critical systems, relating education and training issues to the predicted advances in technology.

Keywords. Safety, software development, software engineering, specification languages, program translators, formal languages, redundancy.

1. Introduction

The development and evaluation of safety critical computer based systems is difficult! More significantly, the development and evaluation of a safety critical system is considerably harder than the development and evaluation of another system of similar functional complexity which does not have to meet safety requirements. A brief discussion of the reasons behind this difference will serve to set the context for a discussion of the relevant skills and technologies for developing safety critical systems.

1.1. The Problems of Producing Safety Critical Systems

The primary distinction between critical and non-critical applications seems to be that in critical applications we need to consider what happens when things go wrong, e.g. system components fail, as well as what happens when all parts of the system are functioning as intended. The need to take into account failures applies not only to safety critical systems but also to other dependable applications, e.g. where military security is a requirement or where loss of valuable data could cause severe commercial damage to an enterprise. In all cases we need to take into account failures that arise from events in the environment of the system, e.g. accidents causing physical damage to the computer system, and malicious behaviour by users of the system or by other individuals who can gain access to some component of the system. The need to take failures into account makes the specification of the requirements for the system more difficult but it also affects the rest of the development process.

In all developments the requirements for the system represent "the interface" between the system and its environment. The production of requirements specifications is never straightforward. However, for many non-critical systems, once a requirements specification has been produced, it is relatively straightforward to carry out the development without considering further the environment or ways in which the system can go wrong. For some non-critical applications it may be necessary to detect and report various sorts of errors to users but it is not necessary to provide continued operation, or controlled termination, in the presence of faults.

When developing safety critical applications the requirements specification must take into account failures of the system itself and of its environment, including malicious behaviour. Also safety critical systems need to ensure safe operation even in the presence of failures of internal components. The set of internal components will not, in general, be completely known when the requirements specification is produced so additional failure modes need to be taken into account as we proceed through the design and development process.

Further, we not only need to take into account failure modes of system components due to internal mechanisms, but we also need to consider the effect of external mechanisms on the design components. For example, consider the use of a programmable peripheral chip as an output device. In a non-critical application we would probably be willing to set the device into output mode at system initialisation and then continue to write data to the output device through the rest of the system execution. In a safety critical application, before each write operation, we would need to check that the programmable peripheral device was still in output mode and that it had not been switched into some other mode by noise on power lines or electro-magnetic radiation. Thus, in developing safety critical systems, we have to take into account new possible failure modes at each level of abstraction that we reach in the development process. Put simply this means that there are vastly more eventualities that we have to take into account when developing critical systems than when developing non-critical systems. Worse, it is often very difficult to identify what are possible, or potential, failure modes.

Whilst there are many differences between the development of critical and non-critical systems the author believes that it is the need to ensure safety in the presence of a potentially unbounded set of failure modes that is the underlying cause of difficulty in developing such classes of system. It is also interesting to note how the need to treat failures constrains our ability to deal with complexity. The two primary weapons we have in dealing with system complexity are structuring and abstraction. Failures can "break" abstractions that are used in system development, e.g. we can no longer assume the availability of reliable virtual machines, and failures can also destroy system structure. Thus the need to deal with failures increases the complexity of the design and evaluation process as well as of the system itself and reduces the effectiveness of our two main weapons for addressing complexity.

The set of potential failure modes and combination of failure modes for a system is essentially unbounded. Consequently we can never produce complete guarantees that systems will be safe under any possible circumstances so we need to assess the extent to which a system is safe. The aim of development and evaluation techniques is to increase *confidence* or *assurance* that a system is safe and perhaps to quantify our degree of confidence. In order to put the discussion of skills and technologies more firmly in context, we first amplify on the notion of achieving assurance in software development and evaluation.

1.2. Assurance in Software Development and Evaluation

Software cannot directly cause loss of life but it may control some equipment that can cause loss of life, hence software can contribute to the safety (or otherwise) of a system. Software is quite different from hardware in that its only "failure modes" are through design and specification faults, rather than any form of physical mechanism such as ageing. Some classes of software faults, e.g. timing faults, may only be exposed as hardware ages, so software can *appear* to age, but this is simply belated exposure of a latent design flaw, not genuine ageing. In reality, therefore, software failures reflect human fallibility.

Software development is concerned with avoiding failures which might arise due to human fallibility and with providing mechanisms for tolerating failures which might arise due to failures of other system components (software and hardware) whether arising from human fallibility or otherwise. Evaluation is concerned with giving *confidence* or *assurance* that critical failures of the system have been avoided, by avoiding component failures or through mechanisms for tolerating any possible residual component failures.

Note that the above points cannot be established for software in isolation, but we will deal with software as independently of its operational environment as possible. In practice high levels of confidence cannot be achieved in evaluation unless the software is designed with evaluation in mind so we drop the distinction between the two activities in the following discussion.

To gain confidence we would like to achieve, and to demonstrate, for the software in a system, that:

- none of its specifications (at any level of abstraction) admit (allow) executions which would lead to catastrophic failure in its operational context;
- it is free from implementation flaws which could lead to catastrophic failure in its operational context, i.e. that it satisfies its specifications (or, at least, the safety relevant portion of them);
- it can protect itself against the failures of other components of the system (which are not trapped by other means, e.g. hardware memory protection), and from external threats or attacks which could cause catastrophic failure.

Due to limitations on space we focus on the first two objectives and discuss the degree to which the objectives are attainable.

1.3. The Limits to Assurance

Assurance is based on a number of issues including the level of trust we have in the individuals carrying out the development and evaluation, etc. However one of the main contributing factors to assurance is the *evidence* of safety produced during software development — and this in turn derives from the verification and validation activities which we carry out throughout the software development process.

It is common to equate validation with answering the question "are we building the right thing?" and verification with answering the question "are we building the thing right?". Whilst these definitions are intuitively appealing they are not very precise and it is difficult, for example, to say whether demonstration of safety is an aspect of verification or validation. There appears to be a much more appropriate (helpful) distinction, based on the long-established (philosophical) distinction between *analytic* and *synthetic* reasoning (which dates back to Leibniz and Kant).

Analytic reasoning is something that can be carried out entirely within a logical framework, e.g. predicate calculus or differential calculus. Synthetic reasoning requires one to look at the "real world" — e.g. the statement "this computerised control system will be safe when used in this factory" can only be validated by inspecting the system, the factory, its environment and the relationships between them: no amount of logical/analytical reasoning will check the validity of this statement. There is a strong link between validation and synthetic reasoning, and similarly between verification and analytic reasoning, but the terms are not identical. It is worth pointing out that, within a particular analytical framework, it is usually possible to judge "how well we have done" against the theoretical capability of the framework when using a given analytical technique. However with synthetic reasoning there is no equivalent "yardstick" against which to judge efficacy and the best we can do is to rely on experience.

Safety is a property of the "real world" — thus it is (ultimately) the province of synthetic, not analytic, reason. Design, and design verification, techniques work within a logical framework. Most obviously, formal verification/program proving are analytical techniques and cannot *of themselves* ensure safety of a software controlled system. (This seems to be essentially the point that Fetzer (1988) was making, albeit at excessive length.)

Synthetic reasoning is inherently uncertain — we can never be certain that we have "apprehended the real world" correctly (Wittgenstein 1969). Thus the production of specifications for safety critical systems is inherently uncertain, and so, therefore, is the production and assessment of the systems themselves. Our aim as responsible software engineers is to reduce the uncertainty, or increase the assurance, in the dependability of some critical item of software as much as possible. Pragmatically this may mean using a combination of formal verification, testing, and so on — and one of the requisite skills is the determination of an appropriate mix of techniques for a project. Also we must recognise that we deploy safety critical systems when we have "enough assurance", not when we know the system "is safe" so there is a requirement for human judgement as well as analysis in evaluating safety critical systems.

1.4. Future Skills and Technologies

The aim of this paper is to identify trends in skills and technologies which indicate how safety critical systems will be developed in the 1990s and beyond. We can now give a short answer to the implied question — improved analytical techniques and better skills in selecting appropriate techniques for the system being developed or evaluated. In providing a fuller and more detailed answer to the question we must consider the capabilities of current techniques, their limitations and the ways in which they can be improved. As many of the skills will be concerned with applying technologies we consider the technological dimension first.

2. Technologies

It is difficult to assess trends in the technology for developing and evaluating safety critical systems over the next decade and beyond, both due to the timescales involved and due to the number of factors that can influence technical and commercial progress. It would be interesting to address the problem by producing a fault tree for the development process and trying to assess what technological improvements are required to counter or to reduce the potential sources of failure present in the process, given current technology. However it is not only difficult to produce such a fault tree but it is also an unwarranted assumption that our progress will address the most serious problems! Consequently the following topics represent the author's bias about key issues rather than the results of any systematic analysis of the problems facing current developers and evaluators of safety critical systems.

2.1. Requirements Analysis

Requirements analysis is one of the first technical activities in system or software development and is concerned with deriving a requirements specification for the system to be produced. Requirements analysis consists of three processes:

- *Elicitation* - extraction, usually from potential users of the system, of information about what the system is required to do;
- *Verification* - checking internal consistency and, as far as possible, completeness of the specification;
- *Validation* - checking that the specification accurately reflects the requirements of the potential system users, or some other authority.

In practice elicitation and validation are essentially synthetic activities, whereas verification, if the specification is written in a suitable notation, is analytic. In addition analytic activities, e.g. deriving consequences of specifications, may be used in support of validation. More concretely, analytical techniques may be used to show that no sequence of activities carried out by the system can bring it into an unsafe state.

Requirements specifications for safety critical systems are typically couched in fairly low level, detailed, technical and computational terms. However the true safety requirements, e.g. "the reactor core shall not melt down", are properties of the "real world". One of the difficulties of developing requirements specifications for safety critical systems is that the "intellectual gap" between the real world concepts and the computational level specification is so great that it is very easy to capture incorrectly the real world notions of safety. A major advance in requirements specification will come from an ability to express requirements in terms much closer to the "real world concepts" whilst still presenting them in an analytical framework so that we can evaluate the consequences of our specifications.

Extrapolating from today's research programmes it would seem that most advances are likely to come from the use of non-standard logics, e.g. those that have a more sophisticated range of basic notions than first order predicate calculus. One interesting approach is that taken by the FOREST project (Maibaum 1986, Potts 1986). Here a deontic logic is used where it is possible to represent notions of permission and obligation. Using the FOREST specification language it is possible to specify that some set of actions is not permitted or to say that some given action, e.g. tripping a reactor, is obliged to be carried out by the system. This seems to be much closer to the real world notions of safety than the functional properties that are expressible within a first order logical framework.

It is also useful to be able to represent causality at the requirements level. A number of formalisms, e.g. Petri nets, are used to represent data flow oriented causal relations in systems. Interestingly certain non-standard logics, e.g. those incorporating praxiological operators, are capable of representing notions of causality and authority. Thus, for example, one can formalise notions such as "sees to it that" or "brings it about that" which seem to capture fairly directly the requisite properties of safety monitoring devices. Similarly temporal logics can be used to represent causal models.

Development of sophisticated logical frameworks for presenting requirements specifications will be of little value unless we have adequate ways of eliciting and validating such specifications. Consequently it is to be expected that these logics will be related to more traditional techniques such as fault tree analysis and failure modes and effects analysis as ways both of deriving the specification and evaluating their fault coverage. Thus, pragmatically, we might expect to see enhancements to existing techniques for safety analysis which have much stronger logical underpinnings. Further, it seems possible to establish links with Markov models in order to gain quantitative specifications of reliability, or at any rate the probability of violating the safety specification.

Also it seems plausible that we could develop new techniques for mapping between requirements specifications in non-standard logics and the formalisms conventionally used for software specification and verification. This will extend our ability to carry out specification and design in an analytic manner to a larger proportion of the system and software lifecycle.

Finally, as we indicated earlier, these "real world issues" really need to be taken into account at each level in the design process. Thus a long term aim would be to model the environment at each level in the design process and to represent causal relationships between the system and its environment at each stage as a basis for analysing the safety properties of the system at that level of abstraction. This remains a long term aim, but one that represents our closest approach to eliminating the "synthetic gap" between our real world concerns and the available analytical frameworks for developing and verifying the software.

2.2. Transformation and Verification

Currently our most powerful techniques for analysing the behaviour of software come from the so-called program verification environments such as Gypsy (Good 1984). Within these frameworks a specification of program functionality is produced, typically using a language based on set theory and first order predicate logic, and the behaviour of the program is shown, by reasoning within the logic, to conform to its specification. The terms program verification or program proving are often used for such techniques as the programs are proven, or verified, to be consistent with their specifications. Of course these techniques do not cater for errors in specifications, but due to their analytical power they are sometimes mandated for use on safety critical systems, e.g. in the Interim Draft Defence Standard 00-55 produced by the UK MoD.

Although these techniques have some attractions they also have their problems and limitations. They are extremely difficult and expensive to apply and the biggest system that can be handled with today's technology is about 10,000 lines of code. Perhaps worse, the proofs which are generated are often very difficult to understand thus, in this case, comprehensibility seems to be at variance with analytical power. At present most of the verification or proof is carried out by using theorem proving tools which are guided at a strategic level by the human user, but which carry out most of the details of the proof automatically. In principle, improvement to the current situation could be brought about by a vast, e.g. three orders of magnitude, improvement in theorem proving technology, but there are technical reasons which suggest that such improvements are likely to be difficult to achieve and it is interesting to look at alternative approaches.

Using program verification approaches, both a specification and a program are produced then an attempt is made to find proofs that show that specification and program correspond. An alternative approach, usually referred to as *transformational*, is to try to generate or to construct a program from its specification using rules of known (proven) properties. The potential substantive advantage of this approach is that the rules only have to be proven once, then they can be used in many applications in the secure knowledge that the programs that they generate from specifications will faithfully implement the required functionality. Currently there are systems, e.g. Refine (Abraido-Fandino 1987), which embody such principles. However they are limited in a number of ways, e.g. by an inability to deal with timing properties of programs or failure behaviour. (Incidentally this is also a limitation of current program verification systems so one should not take too negative an attitude of the current transformational systems.) However this does mean that it is difficult to take into account issues of interest in developing safety critical systems, e.g. recall the need to deal with failures that could affect the status of a programmable peripheral chip as indicated in the introduction.

In the author's view future generations of formal program development systems will use transformational techniques, rather than relying on *post hoc* verification. Whilst there are technical difficulties in developing transformational systems, the inherent attraction of freeing the software engineer from the burden of carrying out extensive formal proofs indicates that work to overcome these problems will be handsomely repaid.

At present it is relatively difficult to see how these transformational techniques can be extended to deal with timing and failure behaviour. However there are some possible avenues to explore. For example it seems possible to represent failure behaviour simply as another specification – in other words one specifies normal, or expected, behaviour of a software component and also its behaviour under failure conditions. The transformational tool then has to select, from a library of standard techniques such as TMR approaches, a strategy for overcoming the predicted failure modes. Whilst such an approach is not without its difficulties, the attraction of encapsulating the complexity and difficulty inside the tool rather than presenting it to the system developer is a major potential benefit.

It is possible that production quality transformational tools will be available within the decade but there is a further problem with regard to trustworthiness of the tools themselves. Program verification and transformational tools tend to make use of a large number of heuristics and thus it is unclear how trustworthy the tools are themselves. To put it bluntly, to what extent should one trust a 12 megabyte LISP program developed over many decades by a large variety of researchers of varying abilities and attitudes? An approach to the problem of the trustworthiness of tools has been found in the program verification community. For example work with m-EVES (Craigen 1987) has shown that it is possible to build a verified proof checker for a system where the theorem prover, i.e. the program which endeavours to find the proofs in the first place, is a large, heuristic LISP program. Whilst it is difficult to see what the exact analogy would be for a transformational system, it may be possible to prove the transformational rules used and to bootstrap such a tool using a trustworthy program verification system such as m-EVES.

Finally, the nearer we can "push" the language supported by transformational systems towards the non-standard logics needed for adequate requirements specification, the nearer we get to a "full-blown" computer aided design system for the software for safety critical systems.

2.3. Application Oriented Technology

It has long been recognised in software development that application oriented technology and tools can be very effective. Perhaps the most obvious examples of this general rule are the use of parser generators in developing compilers or other language processors and the use of fourth generation languages (4GLs) in database and transaction processing applications. These tools facilitate fast and reliable program generation through re-use of tried and trusted software components and through re-use of existing design decisions. Re-use is made possible by working within a well-defined application domain; it is interesting to speculate to what extent such gains in power and dependability could be achieved in the safety critical systems context.

Certain classes of safety critical application are also well defined problem domains, e.g. simple control problems. Indeed it could be argued that the bulk of the safety critical systems market-place is comprised of simple control systems implemented using programmable logic controllers (PLCs). Thus we might imagine developing a "4GL for safety critical control applications" with the following characteristics:

- *Direct specification of control laws* – the requirements would be expressed directly in terms used by control engineers (this is another example of moving requirements closer to "real world concepts");
- *Pre-defined design rules* – a set of rules would be defined for generating programs to implement the control laws, probably by making use of state machine techniques which are a well understood formal system;
- *Implementation* – a high integrity kernel for executing the control/state machine programs on a PLC, perhaps developed and verified using classical program verification techniques.

In short, such a tool would employ the basic principles set out above but would work in a narrow application domain and present to its user an interface that would be easy to understand, i.e. in terms of control laws in this case, to improve confidence that the specification as understood by the system represented the required real world properties.

Whilst the most obvious application of this idea seems to be in the domain of simple control problems, e.g. three term controllers, there seems to be no reason why this could not be generalised for use in other well understood problem domains. Other possible applications include more sophisticated control regimes, e.g. adaptive and robust control, and fault diagnosis and monitoring systems. Looked at from a different point of view such systems may be thought of as ways of bringing transformational technology into practical use within a relatively short time scale.

2.4. Theory of Composition

Most safety critical systems are built out of components which are designed, constructed and evaluated independently. Whilst our development and evaluation techniques may tell us what components do when developed independently, they do not tell us what they will do when they are composed, or integrated, into a complete system – this is the domain of the theory of composition. More specifically such a theory would enable us to answer questions such as "given two components developed and evaluated against the same safety specification do they jointly satisfy that specification when operated in conjunction?" Surprisingly, in general, the answer to such questions is "no"! We can see this by means of a simple example.

Imagine a system for which the safety requirement is that a valve is opened, by turning it a quarter turn anti-clockwise, when a certain input is received (which might be a temperature reading from some system under control). If we develop a simple control system which monitors the temperature and sends appropriate drive signals to the motor controller we may be able to produce a safe component. If one such component is attached to one end (the appropriate end!) of the valve drive shaft then we would have a safe system. However if another identical system is attached to the other end of the drive shaft then the motors will oppose one another and the system will not be safe. In other words we have composed two systems which satisfy their individual safety specifications, but have produced a system which is not safe – even though we might have thought we improved overall system reliability, and hence safety, by use of redundancy. Whilst the above example may seem straightforward, or even trivial, the relationships between components of systems is often far from clear.

It seems to the author that problems of composition are treated on an essentially *ad hoc* basis for development of safety critical systems and that there is no well understood theory on which we can draw, especially for treating composition on an analytical basis. However similar problems have been identified in the field of computer security where they are often referred to as "hook up problems" because they relate to the problems of achieving security when you "hook up" two systems which are secure when treated in isolation. The notion of trust domains (Neely 1985) developed in the security community may give a basis for a composition theory. In addition, work based on Cliff Jones' ideas of rely and guarantee conditions, which indicates properties guaranteed by one component and relied on by others, might give an appropriate basis for such a theory. Clearly, however, we need to be able to relate such a theory to issues of failure modes effect analysis and fault trees. Thus, for example, we would need to be able to determine whether or not the properties relied on by some component are guaranteed even in the presence of all anticipated failures.

Work towards a theory of composition is essentially speculative. However it is also essential as a way of factoring the problem of building large scale, trusted computer systems whether used in safety critical or other dependable applications. In the author's view this is one of the greatest intellectual challenges facing us in the development of large scale safety critical systems.

2.5. Complete Verified Systems

The primary benefit of formal methods and formal verification is that we are able to show that some artefact has the expected properties and, to a limited extent, that it has no other properties. Historically, formal verification has been applied to programs but this leaves a number of other doubts about the behaviour of a computer system: for example does the compiler preserve program semantics, does the computer correctly implement its instruction set, and so on? The intention in aiming to develop complete verified systems is to encompass as much as possible of the system development within a formal, analytic framework in order to reduce to an unavoidable minimum uncertainty about correspondence between specification and realisation.

In practice the two main sources of problems arise from transformational tools, e.g. compilers, and from interpreters, e.g. central processing units in computers. The reason for selecting these as critical stages in the development or execution process is that, typically, there is no independent check on the correctness or appropriateness of the transformation or of the processor execution. Clearly the behaviour of other tools in the software development process can affect the soundness of the resultant product, however, typically, there is independent assessment of the correctness or adequacy of the processing carried out by these tools. Thus for a complete verified system we require a set of verified specification-to-specification transformation tools, a verified compiler and a verified hardware implementation. (Note that a viable alternative might be to carry out periodic validity checks on program execution; however space does not permit us to investigate this option.)

With specification transformation tools the property that we wish to verify is that the semantics of the specifications are preserved, or extended in acceptable ways, by the transformation processes. For a compiler we wish to show that the semantics of the object program correspond to that of the source program (or specification) submitted to it. Neither of these are trivial exercises, however they do appear to be tractable for simple specification and programming languages. For the hardware we require verification down to the lowest level at which the computer can be represented as a state machine. Trying to go to any lower levels, e.g. the levels of analogue waveforms, would take us outside the realms of phenomena that we are able to handle within the logical formalisms available to us. Again, in principle, it is possible to carry out such exercises and arguably this is rather easier than dealing with software of similar complexity as we do not have awkward issues such as recursion and aliasing to trouble us. The link between these two activities is that the semantics of the processor instruction set must be exactly the same as those that are used in verifying the correctness of the compilation step.

There are currently two projects aimed at the production of complete verified systems. A group at Computational Logic Inc. in Austin Texas are building a so-called "tower" of verified tools and components. This involves the specification and verification of a CPU and provision of support for software development through a number of stages of compilation and transformation as outlined above. The spirit behind this work derives from earlier experiences in using the Gypsy verification system and with the Boyer-Moore theorem proving system. In Europe the SafeMOS and ProCoS projects have a similar scope and objective, but deriving from a different basis. The work here is strongly influenced by previous experiences with CSP, occam and the transputer.

Work in this area is inevitably long term and somewhat speculative. It must also be recognised that the production of "complete verified systems" does not guarantee that they will behave as we intended. We still have the "synthetic problems" of deriving and validating appropriate initial specifications and taking into account relevant failure modes at each stage of development. Similarly it is possible that we have misunderstood the semiconductor and device physics, and that the gates we produce to implement the low level state machines defined for our processors do not behave as expected. Similarly, so far as we are aware, neither of the projects alluded to above takes cognisance of the effect of hardware failures on system execution (although ProCoS is considering timing problems). Nonetheless, in the next century, such approaches may represent the most effective tools we have for building systems where we need to have a high degree of confidence that they do what they were specified to do.

2.6. Maintenance and Evolution

It has long been recognised that a large proportion of the effort and cost associated with ownership of software systems arises from the need to maintain or evolve the system once it has been delivered. This observation is true for safety critical systems as well as for other forms of computer based system. Whilst there are cost issues associated with the problems of maintenance in the context of safety critical systems, the primary concern is that it is all too easy to introduce flaws which deleteriously affect the safety of the system whilst carrying out maintenance or evolving the system to satisfy new requirements. Indeed representatives of Merlin-Gerin, who developed the trip systems for the French nuclear reactors used in commercial power generation, have indicated that a relatively large proportion of the "anomalies" discovered with the software in their trip systems have arisen due to changes made in maintenance†.

From a philosophical point of view the difficulty is how to handle changes to systems in such a way that we don't compromise the safety integrity of the system. From a pragmatic point of view the difficulty is how to deal with change in an incremental way, i.e. how to carry out re-verification and validation with a level of effort commensurate with the scale of the change, not the scale of the overall system. It is perhaps worth noting that this is probably one of the greatest limitations of current formal techniques, especially formal verification, where the cost of re-verification after a small change is often nearly the same as that for the initial development.

From a theoretical point of view an attractive way of addressing this problem would be to carry out proofs that the changes don't affect the proofs concerning the safety of the system. These proofs about proofs, or meta-proofs, unfortunately tend to be very difficult to carry out and are not supported by most verification systems. Pragmatically therefore, at least for the foreseeable future, we will need to carry out such re-verification or re-analysis within the framework used for the initial development.

Recognising these constraints, it seems that the ideas of the theory of composition may give an appropriate "weapon" for attacking this problem. Clearly here we would endeavour to make the components that we compose of a sufficiently small size that we can easily handle re-analysis and verification of a particular component, then show that the composition of this new component with the existing ones has the appropriate properties. This should make the re-analysis and re-verification effort much more nearly commensurate with the scale of change. Alternatively we can try to solve these problems via a transformational approach. Specifically we can carry out transformations from the old program, or specification, to the new one and endeavour to show that these transformations preserve the requisite properties of the programs. Since such approaches would be largely constructive they enable us to avoid some of the problems of meta-proofs alluded to above. Clearly this sort of approach would be appropriate within the context of a transformational approach to system and software development. Such techniques are under investigation at the Centre for Software Maintenance at the University of Durham (Ward 1989).

There is, however, a further problem – that of installing the change in the system whilst it is still running. Whilst this is not a problem for all applications, there are a number of situations, e.g. on autonomous vehicles, where it may be necessary to change the software while the system is operational. Clearly this is potentially a problem for any system which is monitoring or controlling some continuously running process which would be very difficult or impossible to stop whilst the software or system is changed. A number of techniques have been developed for this sort of change management. Probably the most relevant and appropriate approach is that embodied by the Conic project (Kramer 1988). Pragmatically the issues that have to be addressed are the ability to remove a modular program from a running system and to replace it with a new one whilst preserving the integrity of the data structures and without disturbing the communication pattern between the different parts of the system. There are both low level, detailed issues such as the representation of data structures and high level issues to do with scheduling and avoiding deadlocks in communication.

In the long run, probably in the next century, it ought to be possible to make changes to safety critical systems with the same confidence that we can develop the system in the first place and with a cost of change commensurate with the scale of change. It then ought to be possible to install these changes in a continuously running system without compromising its integrity and safety. However this is one of the more significant technical challenges that we have to address.

2.7. ASICs and Standard VLSI Components

Historically it has been assumed that the predominant failure modes of hardware components of safety critical systems arise through some forms of physical fault, e.g. ageing or the effect of ionising radiation. However VLSI components, especially central processing units, are now becoming extremely complex and there are many recorded instances of hardware "bugs". Examples of such faults include differences in the floating point arithmetic semantics between different variants of the same processor. It seems clear that the increase in VLSI component complexity is the primary reason for the incidence of failures related to design errors. For example the Motorola 68040 processor is reputed to contain 1.2M transistors and has a vast number of possible internal states, so it is clear that it is very difficult to design such devices correctly and impractical to test them exhaustively.

For some time it has been possible to develop application specific integrated circuits (ASICs) which are VLSI components aimed at carrying out a specific functional task. For example it would be possible to implement a simple control algorithm entirely in hardware using an ASIC as opposed to using software on a conventional VLSI processor. The potential advantage is that by making an application specific, as opposed to general purpose, component the complexity of the circuit is reduced. On the other hand there are increased risks, or at any rate uncertainties, introduced by the custom design process as opposed to being able to rely on well tried and tested commercial components.

Historically ASICs have not enjoyed great commercial success and even where custom integrated circuits have been produced they have tended to be developed by modifications to standard CPU components rather than by developing application specific components. One of the possible technical reasons for this might be the difficulty of providing arithmetic capabilities "from scratch" for the ASIC. However probably the driving force has been a commercial one – it is simply perceived as being cheaper and less risky to modify existing chips than to develop ASICs. However it seems possible that the use of ASICs will become more prevalent in safety critical systems as part of an overall drive to reduce system complexity. In some senses the technology is available now, e.g. through silicon compilers, however it seems likely that in order to develop ASICs to the degree of

---

† This fact was reported at a workshop of the ESPRIT Basic Research Project PDCS – Predictably Dependable Computing Systems.

integrity required for safety critical applications it might be necessary to extend hardware verification techniques so that they can be used to develop this sort of system component.

There is an additional technical factor which may lead to wider use of ASICs. In many application domains it is common to use diverse designs in order to try to reduce the possibility of common mode failures in systems. This is most commonly applied to software however experimental evaluations of such approaches (Knight 1986) suggest that the degree of diversity achieved in practice is not very great and that common mode failures may still arise in supposedly diverse software. The use of conventional software development on a standard CPU in conjunction with an ASIC may give much more genuine diversity and greatly reduce the likelihood of common mode failures. Thus it seems quite possible that ASICs will become one of the armoury of techniques that we use in producing safety critical systems, especially as the speed and reliability of the (computer aided) design process for such circuits improves.

2.8. System Safety Arguments

It has long been recognised that development and evaluation of safety critical systems is a multi disciplinary exercise. In principle, in designing or evaluating a safety critical system, we need to draw together arguments from a range of disciplines that provide a convincing set of reasons in favour of deployment of a system. In practice not all the arguments are explicitly recorded and, all too often, they are not properly linked or correlated. However it is these chains of arguments between disciplines and between sub-systems which are critical to evaluating safety. For example, arguments about the design of a particular piece of hardware might lead to the conclusion that a given processor will "failstop" and this may enable us to use a particular software algorithm for failure detection which, through the medium of formal proof and timing analysis, would then allow us to determine the latency of error detection ... and eventually lead to a Markov model of system reliability. Thus chains of arguments can be constructed enabling us to derive important system properties.

Where arguments about safety, or safety related aspects of systems, are produced they tend to be developed in a fragmentary way and are represented on different media, e.g. as documents or as databases of test results. Major benefit would accrue from the ability to record all the arguments in one place and to automate analysis of argument consistency and, perhaps further, to indicate the ramifications of change. Thus we would like a technology for a system safety argument which is capable of ranging over a wide number of technologies and system components. Technically the development of such a support tool involves the establishment of an "inter-lingua" or canonical representation for arguments and a set of rules for argument consistency checking and change propagation.

The development of automated support for safety argument management is very challenging primarily because of the range of argument types to be supported and the difficulty of finding an appropriate unifying basis for argument representation. However there are at least two projects investigating this important topic. The Eureka supported safety engineer's workbench project, which involves a number of organisations including Adelard in the UK, is addressing the problem from the relatively pragmatic point of view of looking at commonalities in existing safety analysis techniques. On the other hand the safety argument manager project undertaken by Logica, the Civil Aviation Authority and the University of York is looking at the problem from a more philosophical point of view. Specifically this project is using the general model of arguments developed by Toulmin (1957) to try to provide an underlying basis for representing broad classes of safety relevant arguments. Whilst this work is highly speculative it may have major benefit in providing designers and evaluators with powerful tools for analysing consistency and, to some extent, completeness of safety arguments.

2.9. Summary and Conclusions

The author strongly believes that advances in the above technologies would greatly improve and enhance the development and evaluation of safety critical systems. However he is far less confident that the requisite advances will necessarily be made! In addition it is clear that the above points, while important, do not cover all aspects of development and evaluation of safety critical systems. For example most safety systems are also time critical and advances in timing analysis of programs and scheduling theory are likely to be pertinent also. Further, user interfaces are a key facet of safety critical systems. Operational errors can arise due to inappropriate operator action, perhaps caused by the difficulty of interpreting the data displays. Design to reduce cognitive load and to reduce the potential for operator errors is therefore very important. Thus the above should really be interpreted as an identification of some interesting technologies, not an exhaustive analysis of priorities for long term research programmes.

There are, however, a number of themes running through the above topics which we would also expect to apply to other major technologies. Four key points are:

- improvements to the analytical power of available methods plus a broadening of their scope in terms of the facets of critical systems that can be handled within an analytic framework;
- expansion of concepts that can be represented formally in order to reduce the "synthetic gap" between real world concepts and the ideas that we can handle within our analytical frameworks;
- improvements in usability of the analytical techniques especially through embodying the mathematics within "computer aided design tools";
- reduction in complexity both of systems and of the development approaches.

These principles should perhaps be taken as key criteria in assessing new technologies and proposed research programmes.

3. Skills

A large proportion of the skills required by software engineers engaged in development and evaluation of safety critical systems are determined by the technologies. Thus, for example, they require (or will require) knowledge of the underlying principles of discrete mathematics and skills in using the appropriate transformation and verification tools. It is relatively easy to determine the set of required skills by consideration of the technologies discussed above. However a number of other types of skill are required and these cannot simply be elaborated by considering the technologies above. We focus on these issues, especially those that are "over and above" the skills we would expect of a typical professional software engineer.

3.1. Choice of Technologies

In principle, when developing computer based systems, we need to design a process for each project we undertake. This means determining the stages of the process, the methods and tools to be used at each stage, the quality assurance procedures, and so on. In practice, design involves selecting from available alternatives constrained by issues such as staff skills and costs. For many projects the driving factors in the design of the process are likely to be productivity and cost effectiveness. In the case of safety critical systems the primary driving factor in process design will be the need to ensure safety and to do so in a way that is inspectable and independently verifiable by an evaluation team.

Pragmatically this means that we need to select methods and techniques for each stage of the life-cycle and each facet of the system, e.g. functionality, timing, failure behaviour, etc. The methods need to cover not only direct technical development issues but also technical management, such as configuration management, and management issues including cost estimation and planning. Whilst we recognise that all these issues are important we focus on technical issues in order to clarify the concept of process design.

A principle in designing a process and selecting appropriate methods and tools is to avoid having "weak links in the chain".

For example there is no point carrying out a formal development and verifying source code programs if we cannot place trust in the accuracy of the compilation. Similarly we should endeavour to achieve similar levels of confidence in the veracity of our specifications as we have that the implementation satisfies those specifications. This is clearly an area where skill and judgement is required as there is no mechanistic way of checking such properties.

A second principle is to ensure that all actions are, in some sense, independently verified. This might mean that some activity is carried out automatically and checked manually, or vice versa. It might also mean, in the context of a verified compiler, that the independent verification is carried out indirectly by the compiler development and evaluation team. In a way this is simply a variation on the idea of having "no weak links in the chain" by endeavouring to ensure that there is no single point of failure in the development process.

In the context of configuration management the primary concerns are to ensure that we can unequivocally identify what items of software have gone into a particular released version of a system and that we have a full derivation history for the items of software through source code, specifications, etc. In addition, by suitable choice of technical activities and configuration management mechanisms, we should know unequivocally how the programs that end up in the system as deployed have been derived and have confidence that all relevant aspects of the software and systems have been independently verified and validated.

Whilst the above may seem to be common sense, a key point to note is that there is no simple or mechanistic way of making the choice of process. The set of techniques that are relevant will depend on the scale of the system, the nature of the threats, the sort of functionality it is intended to achieve, whether the system can fail safe or has to fail active, and so on. There is no straightforward codification of rules for making such selections and, as a consequence, we rely on the skill and judgement of experienced staff to design an effective process.

3.2. Application Domain Knowledge

Whilst we have focused largely in our discussion of technology on analytical issues, clearly it is extremely important to ensure that the specifications derived reflect correctly the needs of the application. This can only be achieved through suitable knowledge of the application domain. Put more bluntly knowledge of all the interesting and relevant analytical techniques for developing safety critical systems does not, of itself, qualify someone for developing safety critical systems. The judgement to know whether or not the specification is appropriate, has covered likely failure modes, has adequately addressed risks and threats, and so on, can only come from knowledge of the particular application domain. In other words to make judgements about the "real world" one needs to have knowledge and experience of the relevant aspects of the application.

The conclusion from this is that a software engineer should not be entitled to work on the development of a safety critical system just because he has knowledge of the relevant software development techniques. Preferably he ought to build up expertise and knowledge in the application domain, perhaps through working on non-critical projects over a number of years. Failing this he ought at least to have a clear understanding of the limitations of his knowledge and to know when he needs to seek advice from problem domain experts. This naturally leads us on to the issues of judgement and ethics.

3.3. Judgement and Ethics

There are two specific aspects of judgement and ethics which relate to development and evaluation of safety critical systems. These both relate to deciding when something ought *not* to be done.

There are circumstances where computers have been used in safety critical applications where, quite simply, the value of using the computer system was not outweighed by the attendant risks. Indeed there appears to be evidence that fatal accidents have occurred because of inappropriate use of computer systems in safety critical devices. Thus one form of ethical judgement that a software engineer should be willing to make is to determine when it is *not* appropriate to use computers in a safety critical application. Pragmatically this means that software engineers need to be willing to make judgements that some computerised safety critical application should not be built because the technology to develop it to the requisite level of assurance is not available. Given commercial and other pressures that may be placed on software engineers this requires a certain amount of moral courage as well as skill and judgement.

A related, and similarly sensitive, issue relates to assessments of one's own skills and limitations. Again a software engineer ought to be willing to say that working on some particular system is beyond his skill and competence. Psychologically this is perhaps rather difficult to do but nonetheless it is part of the meaning of being a professional engineer. Perhaps the simplest way to look at this, and the least emotionally loaded, is to evaluate whether or not further training or education would be appropriate before undertaking some particular project. If an engineer decides that he does require training then he should not work on the project, particularly if it is safety critical, until or unless that training is provided and satisfactorily completed.

3.4. Management and Team Skills

Although we have focused largely on individual skills it is improbable that any single software engineer will have all the necessary skills for a given project. Thus an important managerial skill is the ability to select a team with an appropriate and balanced set of skills. However it is also important that the team is harmonious and has the right *attitude* to safety. This latter point is very important.

Managers need to foster a "safety culture" where all members of the team see safety as the key objective in their work so that concern for safety permeates all their actions and decisions. There is no easy way of establishing such a culture – but it is crucial. Arguably such a culture is the single most significant factor contributing to the production of safe systems.

3.5. Education and Training

An extensive programme of education and training is required both to transfer the technology specific skills and to stimulate people to think about the wider issues of judgement and ethics. The volume of material to be taught is quite large and it is hard to see how to fit this in to a conventional working environment. However the BCS has studied education and training requirements for safety critical systems and has proposed a syllabus for such a course, based on the structure of a series of week-long intensive residential modules over a period of perhaps 18 months. The BCS is now actively seeking to promote this curriculum and to establish a suitable set of courses. It is to be hoped that this initiative represents the beginning of a wider awareness of the need for education and training associated with the development of safety critical systems.

4. Conclusions

There are many possible directions in which the technologies for developing and evaluating safety critical systems can evolve over the next decade. A primary motivation behind any such developments should be to improve professional and public confidence in the safety of the systems that are being deployed. A major factor in gaining such confidence is the ability to analyse the software in systems that have been produced, including dealing with functional, timing and failure behaviour. Another important facet of such developments is to bring the analytical frameworks closer to "real world concerns" so that there is a much greater likelihood that analysable specifications accurately reflect the requirements of the applications.

The requisite skills cover both the ability to use the technologies and to judge their limitations, i.e. to determine situations whereby the technologies available do not or cannot give adequate confidence that the system can safely be deployed in a critical application. This involves not only understanding the technologies themselves, but their limitations and broader issues such as sources of error in the development process. It also requires an appropriate ethical and moral attitude to the work being undertaken. In summary, software engineers working on safety critical developments should continually be asking themselves "do we have the capability and knowledge to build a system of this nature to the standards required?". By doing this, and by having the moral courage to act if the answer appears to be no, we will see both improvements in the development standards of safety critical systems and, hopefully, less of a propensity to deploy computers in safety critical applications where the risks of using such complex technology are not outweighed by the attendant benefits.

5. References

Abraido-Fandino, L.M. (1987). "An Overview of Refine", in *Proc. 2nd International Symposium on Knowledge Engineering-Software Engineering*, Madrid, Spain.

Craigen, D., S. Kromodimoeljo, I. Meisels, A. Neilson, W. Pase and M. Saaltink (1987). "m-EVES: A Tool for Verifying Software", CP-87-5402-26, IP Sharp Associates Ltd.

Fetzer, J. H. (1988). "Program Verification: The Very Idea", *CACM* 31(9).

Good, D. (1984). "Mechanical Proofs about Computer Programs", Technical Report 41, Institute for Computing Science, The University of Texas at Austin.

Knight, J.C. and N.G. Leveson (1986). "An Empirical Study of Failure Probabilities in Multiversion Software", *Proc. 16th IEEE International Symposium on Fault-Tolerant Computing*, IEEE Press.

Kramer, J. and J. Magee (1988). "A Model for Change Management", *Proceedings of the IEEE Distributed Computing Systems in the '90s Conference*.

Maibaum, T.S.E., S. Khosla and P. Jeremaes (1986). "A Modal [Action] Logic for Requirements Specification", in *Software Engineering 86*, ed. P J Brown and D J Barnes, Peter Peregrinus.

Neely, R.B. and J. W. Freeman (1985). "Structuring Systems for Formal Verification", *Proceedings of the 1985 Symposium of Security and Privacy*, IEEE.

Potts, C.J. and A. Finkelstein (1986). "Structured Common Sense", in *Software Engineering 86*, ed. P.J. Brown and D.J. Banres, Peter Peregrinus.

Toulmin, S.E. (1957). *The Uses of Argument*, Cambridge University Press.

Ward, M. (1989). "Transforming a program into a specification", *submitted to IEEE (available as a technical report from the University of Durham)*.

Wittgentstein, L. (1969). *On Certainty*, Blackwell.

AUTHOR INDEX

Anders, U. 113
Anderson, T. 1

Coen-Porisini, A. 105

Dale, C. 77
De Paoli, F. 105
Dittrich, E. 59
Drager, K. H. 125

Erb, A. 59

Fergus, E. 119
Ficheux, F. 151
Fichot, F. 71
Finnie, B. W. 95
Fritz, P. 59

Gaj, K. 43
Georges, C. 71
Gil, P. J. 89
Goring, C. J. 145
Górski, J. 7
Gorski, K. 43
Grams, T. 37
Grasegger, S. 59
Gulliksen, R. 125

Halang, W. A. 25
Hedley, D. 55
Hennell, M. A. 119

Johnston, I. H. A. 95, 157
Jung, Soon-Key 25

Karger, B. von 19
Kopp, H. 59
Kossowski, R. 43

Koutny, M. 1
Kropfitsch, D. 59

List, G. 99

Mainka, E.-U. 113
Mayadoux, Y. 151
McDermid, J. A. 163

Obreja, I. 131
Ors, R. 89

Padfield, G. D. 135
Pain, C. 151

Rabe, G. 113
Ravn, A. P. 13
Rischel, H. 13
Romain, M. 71

Saeed, A. 1
Säflund, M. 31
Santonja, V. 89
Schoitsch, E. 59
Serrano, J. J. 89
Smith, P. R. 135
Sobczyk, J. 43
Soma, H. 125
Stålhane, T. 83
Stålmarck, G. 31
Stavridou, V. 13

Talbot, R. J. 49
Tomlinson, B. N. 135

Veevers, A. 67

Wright, J. B. 71

KEYWORD INDEX

Active control technology, 135
Ada, 105
Artificial intelligence, 131
Assembly language, 19

Benefit, 157

Coding errors, 83, 95
Compiler verification, 13
Compilers, 19
Computer architecture, 25
Computer control, 1, 25, 99, 113
Computer evaluation, 95, 113
Computer selection and evaluation, 157
Computer simulation, 125
Computer software, 19, 37, 83, 99, 157
Computer viruses, 43
Computer-aided instruction, 125
Confidence building, 77
Consequence, 157
Control system design, 135
Coverage metrics, 67
Covert channels, 49

Dependability evaluation, 89
Development methods, 49
Diagnosis, 131
Dynamic analysis, 55
Dynamic analysis and conformance, 119

Educational aids, 125
Expert systems, 125, 131

Failure detection, 95, 131
Faults, 7
Flight control, 135
Formal languages, 163
Formal specification, 7, 13

Hard tautologies, 31
Helicopter control, 135
High integrity software, 25
High level languages, 25
Human factors, 157

MOD 00-55, 13
Maintenance engineering, 151
Markov modeling, 89

Nuclear plants, 131

Operating systems, 43

Prediction, 83
Probabilistic logic, 37
Process control, 1
Program testing, 55, 67, 83, 105
Program transformation, 13

Program translators, 163
Program validation, 13, 55
Program verification, 105
Programmable controllers, 25, 113
Programming environments, 25
Proof procedures, 31
Propositional logic, 31

Qualitative modeling, 131
Quality control, 71, 113, 151

Railway interlocking, 59
Railways, 31
Real time computer systems, 25
Real-time embedded, 55
Redundancy, 59, 163
Regression testing, 59
Reliability, 67, 71, 77, 83, 89
Reliability theory, 37
Requirements analysis, 49
Requirements analysis and formal methods, 1
Requirements analysts, 49
Requirements capture, 13
Risk, 157

Safety, 1, 7, 43, 77, 89, 95, 99, 125, 157, 163
Safety critical automation, 25
Safety licensing, 25
Security, 49
Social and behavioural sciences, 157
Software correctness, 37
Software development, 71, 151, 163
Software diversity, 59
Software engineering, 25, 67, 71, 77, 83, 95, 163
Software metrics, 77
Software reliability, 37
Software safety validation, 99
Software tools, 19, 71, 113
Specification languages, 163
Standards, 49, 95
Static analysis, 119
Symbolic execution, 105
System architecture, 59
System failure and recovery, 43
System integrity, 43, 95
System maintenance, 99
System verification, 31

Test tools, 59

Validation, 7
Verification, 7, 19
Verification and validation, 59

Watchdog processors, 89

IFAC Publications, Published and Forthcoming Symposia/Workshop volumes

SYMPOSIA VOLUMES

ADALI & TUNALI: Microcomputer Application in Process Control
AKASHI: Control Science and Technology for the Progress of Society, 7 Volumes
ALBERTOS & DE LA PUENTE: Components, Instruments and Techniques for Low Cost Automation and Applications
ALONSO-CONCHEIRO: Real Time Digital Control Applications
AMOUROUX & EL JAI: Control of Distributed Parameter Systems (1989)
ATHERTON: Multivariable Technological Systems
BABARY & LE LETTY: Control of Distributed Parameter Systems (1982)
BALCHEN: Automation and Data Processing in Aquaculture
BANKS & PRITCHARD: Control of Distributed Parameter Systems (1977)
BAOSHENG HU: Analysis, Design and Evaluation of Man–Machine Systems (1989)
BARKER & YOUNG: Identification and System Parameter Estimation (1985)
BASANEZ, FERRATE & SARIDIS: Robot Control "SYROCO '85"
BASAR & PAU: Dynamic Modelling and Control of National Economies (1983)
BAYLIS: Safety of Computer Control Systems (1983)
BEKEY & SARIDIS: Identification and System Parameter Estimation (1982)
BINDER & PERRET: Components and Instruments for Distributed Computer Control Systems
CALVAER: Power Systems, Modelling and Control Applications
Van CAUWENBERGHE: Instrumentation and Automation in the Paper, Rubber, Plastics and Polymerisation Industries (1980) (1983)
CHEN HAN-FU: Identification and System Parameter Estimation (1988)
CHEN ZHEN-YU: Computer Aided Design in Control Systems (1988)
CHRETIEN: Automatic Control in Space (1985)
CHRISTODULAKIS: Dynamic Modelling and Control of National Economies (1989)
COBELLI & MARIANI: Modelling and Control in Biomedical Systems
CUENOD: Computer Aided Design of Control Systems†
DA CUNHA: Planning and Operation of Electric Energy Systems
DE CARLI: Low Cost Automation
De GIORGIO & ROVEDA: Criteria for Selecting Appropriate Technologies under Different Cultural, Technical and Social Conditions
DUBUISSON: Information and Systems
EHRENBERGER: Safety of Computer Control Systems (SAFECOMP '88)
ELLIS: Control Problems and Devices in Manufacturing Technology (1980)
FERRATE & PUENTE: Software for Computer Control (1982)
FLEISSNER: Systems Approach to Appropriate Technology Transfer
FLORIAN & HAASE: Software for Computer Control (1986)
GEERING & MANSOUR: Large Scale Systems: Theory and Applications (1986)
GENSER, ETSCHMAIER, HASEGAWA & STROBEL: Control in Transportation Systems (1986)
GERTLER & KEVICZKY: A Bridge Between Control Science and Technology, 6 Volumes
GHONAIMY: Systems Approach for Development (1977)
HAIMES & KINDLER: Water and Related Land Resource Systems
HARDT: Information Control Problems in Manufacturing Technology (1982)
HERBST: Automatic Control in Power Generation Distribution and Protection
HRUZ & CICEL: Automatic Measurement and Control in Woodworking Industry — Lignoautomatica '86
HUSSON: Advanced Information Processing in Automatic Control
ISERMANN: Automatic Control, 10 Volumes
ISERMANN: Identification and System Parameter Estimation (1979)
ISERMANN & KALTENECKER: Digital Computer Applications to Process Control
ISIDORI: Nonlinear Control Systems Design
JANSSEN, PAU & STRASZAK: Dynamic Modelling and Control of National Economies (1980)
JELLALI: Systems Analysis Applied to Management of Water Resources
JOHANNSEN & RIJNSDORP: Analysis, Design, and Evaluation of Man–Machine Systems
JOHNSON: Adaptive Systems in Control and Signal Processing
JOHNSON: Modelling and Control of Biotechnological Processes
KAYA & WILLIAMS: Instrumentation and Automation in the Paper, Rubber, Plastics and Polymerization Industries (1986)

KLAMT & LAUBER: Control in Transportation Systems (1984)
KOPACEK et al.: Skill Based Automated Production
KOPACEK, TROCH & DESOYER: Theory of Robots
KOPPEL: Automation in Mining, Mineral and Metal Processing (1989)
KUMMEL: Adaptive Control of Chemical Processes (ADCHEM '88)
LARSEN & HANSEN: Computer Aided Design in Control and Engineering Systems
LEININGER: Computer Aided Design of Multivariable Technological Systems
LEONHARD: Control in Power Electronics and Electrical Drives (1977)
LESKIEWICZ & ZAREMBA: Pneumatic and Hydraulic Components and Instruments in Automatic Control†
LINKENS & ATHERTON: Trends in Control and Measurement Education
MACLEOD & HEHER: Software for Computer Control (SOCOCO '88)
MAHALANABIS: Theory and Application of Digital Control
MANCINI, JOHANNSEN & MARTENSSON: Analysis, Design and Evaluation of Man–Machine Systems (1985)
MARTOS, PAU, ZIERMANN: Dynamic Modelling and Control of National Economies (1986)
McGREAVY: Dynamics and Control of Chemical Reactors and Distillation Columns
MLADENOV: Distributed Intelligence Systems: Methods and Applications
MUNDAY: Automatic Control in Space (1979)
NAJIM & ABDEL-FATTAH: System Approach for Development (1980)
NIEMI: A Link Between Science and Applications of Automatic Control, 4 Volumes
NISHIKAWA & KAYA: Energy Systems, Management and Economics
NISHIMURA: Automatic Control in Aerospace
NORRIE & TURNER: Automation for Mineral Resource Development
NOVAK: Software for Computer Control (1979)
O'SHEA & POLIS: Automation in Mining, Mineral and Metal Processing (1980)
OSHIMA: Information Control Problems in Manufacturing Technology (1977)
PAUL: Digital Computer Applications to Process Control (1985)
PERRIN: Control, Computers, Communications in Transportation
PONOMARYOV: Artificial Intelligence
PUENTE & NEMES: Information Control Problems in Manufacturing Technology (1989)
RAMAMOORTY: Automation and Instrumentation for Power Plants
RANTA: Analysis, Design and Evaluation of Man–Machine Systems (1988)
RAUCH: Applications of Nonlinear Programming to Optimization and Control†
RAUCH: Control of Distributed Parameter Systems (1986)
REINISCH & THOMA: Large Scale Systems: Theory and Applications (1989)
REMBOLD: Robot Control (SYROCO '88)
RIJNSDORP: Case Studies in Automation Related to Humanization of Work
RIJNSDORP et al.: Dynamics and Control of Chemical Reactors (DYCORD '89)
RIJNSDORP, PLOMP & MÖLLER: Training for Tomorrow— Educational Aspects of Computerized Automation
ROOS: Economics and Artificial Intelligence
SANCHEZ: Fuzzy Information, Knowledge Representation and Decision Analysis
SAWARAGI & AKASHI: Environmental Systems Planning, Design and Control
SINHA & TELKSNYS: Stochastic Control
SMEDEMA: Real Time Programming (1977)†
STRASZAK: Large Scale Systems: Theory and Applications (1983)
SUBRAMANYAM: Computer Applications in Large Scale Power Systems
TAL': Information Control Problems in Manufacturing Technology (1986)
TITLI & SINGH: Large Scale Systems: Theory and Applications (1980)
TROCH, KOPACEK & BREITENECKER: Simulation of Control Systems
UHI AHN: Power Systems and Power Plant Control (1989)
VALADARES TAVARES & EVARISTO DA SILVA: Systems Analysis Applied to Water and Related Land Resources
van WOERKOM: Automatic Control in Space (1982)
WANG PINGYANG: Power Systems and Power Plant Control

WESTERLUND: Automation in Mining, Mineral and Metal Processing (1983)
YANG JIACHI: Control Science and Technology for Development

YOSHITANI: Automation in Mining, Mineral and Metal Processing (1986)
ZWICKY: Control in Power Electronics and Electrical Drives (1983)

WORKSHOP VOLUMES

ASTROM & WITTENMARK: Adaptive Systems in Control and Signal Processing
BOULLART et al.: Industrial Process Control Systems
BRODNER: Skill Based Automated Manufacturing
BULL: Real Time Programming (1983)
BULL & WILLIAMS: Real Time Programming (1985)
CAMPBELL: Control Aspects of Prosthetics and Orthotics
CHESTNUT: Contributions of Technology to International Conflict Resolution (SWIIS)
CHESTNUT et al.: International Conflict Resolution using Systems Engineering (SWIIS)
CHESTNUT, GENSER, KOPACEK & WIERZBICKI: Supplemental Ways for Improving International Stability
CICHOCKI & STRASZAK: Systems Analysis Applications to Complex Programs
CRESPO & DE LA PUENTE: Real Time Programming (1988)
CRONHJORT: Real Time Programming (1978)
DI PILLO: Control Applications of Nonlinear Programming and Optimization
ELZER: Experience with the Management of Software Projects
GELLIE & TAVAST: Distributed Computer Control Systems (1982)
GENSER et al.: Safety of Computer Control Systems (SAFECOMP '89)
GOODWIN: Robust Adaptive Control
HAASE: Real Time Programming (1980)
HALME: Modelling and Control of Biotechnical Processes
HARRISON: Distributed Computer Control Systems (1979)
HASEGAWA: Real Time Programming (1981)†
HASEGAWA & INOUE: Urban, Regional and National Planning—Environmental Aspects
JANSEN & BOULLART: Reliability of Instrumentation Systems for Safeguarding and Control
KOTOB: Automatic Control in Petroleum, Petrochemical and Desalination Industries
LANDAU, TOMIZUKA & AUSLANDER: Adaptive Systems in Control and Signal Processing
LAUBER: Safety of Computer Control Systems (1979)

LOTOTSKY: Evaluation of Adaptive Control Strategies in Industrial Applications
MAFFEZZONI: Modelling and Control of Electric Power Plants (1984)
MARTIN: Design of Work in Automated Manufacturing Systems
McAVOY: Model Based Process Control
MEYER: Real Time Programming (1989)
MILLER: Distributed Computer Control Systems (1981)
MILOVANOVIC & ELZER: Experience with the Management of Software Projects (1988)
MOWLE: Experience with the Management of Software Projects
NARITA & MOTUS: Distributed Computer Control Systems (1989)
OLLUS: Digital Image Processing in Industrial Applications—Vision Control
QUIRK: Safety of Computer Control Systems (1985) (1986)
RAUCH: Control Applications of Nonlinear Programming
REMBOLD: Information Control Problems in Manufacturing Technology (1979)
RODD: Artificial Intelligence in Real Time Control (1989)
RODD: Distributed Computer Control Systems (1983)
RODD: Distributed Databases in Real Time Control
RODD & LALIVE D'EPINAY: Distributed Computer Control Systems (1988)
RODD & MULLER: Distributed Computer Control Systems (1986)
RODD & SUSKI: Artificial Intelligence in Real Time Control
SIGUERDIDJANE & BERNHARD: Control Applications of Nonlinear Programming and Optimization
SINGH & TITLI: Control and Management of Integrated Industrial Complexes
SKELTON & OWENS: Model Error Concepts and Compensation
SOMMER: Applied Measurements in Mineral and Metallurgical Processing
SUSKI: Distributed Computer Control Systems (1985)
SZLANKO: Real Time Programming (1986)
TAKAMATSU & O'SHIMA: Production Control in Process Industry
UNBEHAUEN: Adaptive Control of Chemical Processes
VILLA & MURARI: Decisional Structures in Automated Manufacturing

†*Out of stock—microfiche copies available. Details of prices sent on request from the IFAC Publisher.*

IFAC Related Titles
BROADBENT & MASUBUCHI: Multilingual Glossary of Automatic Control Technology
EYKHOFF: Trends and Progress in System Identification
NALECZ: Control Aspects of Biomedical Engineering